Proceedings in Life Sciences

The Chemistry and Biology of Isoquinoline Alkaloids

Edited by
J. D. Phillipson, M. F. Roberts,
and M. H. Zenk

With 178 Figures

Springer-Verlag
Berlin Heidelberg NewYork Tokyo

Professor Dr. J. David Phillipson
Dr. Margaret F. Roberts
Department of Pharmacognosy
The School of Pharmacy
Brunswick Square
London, WC1N 1AX, United Kingdom

Professor Dr. M. H. Zenk
Pharmazeutische Biologie
Universität München
Karlstraße 29
8000 München 2, FRG

ISBN-13: 978-3-642-70130-6 e-ISBN-13: 978-3-642-70128-3
DOI: 10.1007/978-3-642-70128-3

Library of Congress Cataloging in Publication Data. Main entry under title: The chemistry and biology of isoquinoline alkaloids (Proceedings in life sciences). Based on lectures presented at a symposium arranged by the Phytochemical Society of Europe in Apr. 1984. Includes bibliographies and index. 1. Isoquinoline–Congresses. 2. Alkaloids–Congresses. 3. Botanical chemistry–Congresses. 4. Plants–Composition–Congresses. I. Phillipson, J.D. (John David) II. Roberts, M.F. (Margaret F.) III. Zenk, M.H., 1933–. IV. Phytochemical Society of Europe. V. Series. QK898.I78C48 1985 581.19'242 84-23622

2131/3130-543210

Preface

Isoquinolines form one of the largest groups of plant alkaloids and they include a number of valuable clinical agents such as codeine, morphine, emetine and tubocurarine. Research into different aspects of isoquinolines continues in profusion, attracting the talents of botanists, chemists, biochemists, analysts, pharmacists and pharmacologists. Many of these aspects are of an interdisciplinary nature, and in April 1984, The Phytochemical Society of Europe arranged a 3-day symposium on *The Chemistry and Biology of Isoquinoline Alkaloids* in order to provide a forum for scientists of differing disciplines who are united by a common interest in this one class of natural product. Each chapter in this volume is based on a lecture given at this symposium. Attempts have been made to make the aims and objectives, experimental findings and conclusions reached, intelligible to scientists of differing backgrounds.

The introductory chapter, which is mainly based on a historical discussion, stresses that plants containing isoquinolines have proved to be both a boon and a curse to mankind. The Opium Poppy, *Papaver somniferum*, produces the medicinally used alkaloids morphine, codeine, noscapine and papaverine whilst it also continues to provide drugs of abuse, particularly morphine and its readily prepared O,O-diacetyl derivative, heroin. Numerous other alkaloids have been isolated from other members of the Papaveracea, and a knowledge of their presence and distribution within the various species has proved a useful adjunct to systematic botanical studies.

There is considerable variation in structure between the groups of isoquinoline alkaloids, some being simple isoquinolines while others are complex dimers. About 50 simple isoquinolines occur in plants and they are characteristic of some families such as the Cactaceae. Interest in these compounds has stemmed from the well-known hallucinogenic phenethylamine, mescaline, and in one chapter, the occurrence of these simple isoquinolines is reviewed.

The family Annonaceae comprises some 2000 species and only about 150 have been investigated for their chemical constituents. Studies during the past decade have indicated that this family can perform biosynthetic sequences which yield an array of isoquinolines. As examples of the more unique types of isoquinolines, *Erythrina* alkaloids and the cularines form the subjects of two separate chapters. Relatively little information exists on the pharmacology of the cularines. There are now about 225 bisbenzylisoquinolines in Nature and they occur mainly in species of the Berberidaceae, Menispermaceae, Monimiaceae and Ranunculaceae. The distribution, structure, structure determination and new alkaloids isolated in the period 1981–1983 form the basis of the chapters on bisbenzylisoquinolines.

The aporphine alkaloids form one of the largest groups of isoquinolines and their biosynthesis has been extensively studied. However, less is known about the catabolism of these alkaloids and this interesting topic is briefly covered. Differently substituted aporphines have been used as rigid analogues of dopamine in order to investigate the nature of dopamine receptors. Natural and synthetic aporphines have been used to ascertain the structural requirements of dopamine agonists and antagonists. Such studies are useful in designing new drugs for the treatment of vascular diseases, convulsions and ulcers.

The chemistry and pharmacology of the morphinans continues to be a subject of considerable scientific endeavour, and two chapters from scientists at the National Institutes of Health reflect this interest. The whole background to the synthetic approaches to the morphinans is reviewed. In one chapter, the mode of action of different morphinans is discussed, while the other deals with recent synthetic studies. Methods are now available for the large scale production of the natural and the unnatural isomers of all of the medically valuable opium derived morphinan derivatives.

Bridging the gap between pure synthetic chemistry and biosynthetic studies is the area of biomimetic syntheses. The Ipecacuanha alkaloids can be obtained from substituted dopamines and secologanin. Similar studies have resulted in the isoquinoline analogues of the heteroyohimbine alkaloids which have not yet been isolated as natural products.

The biosynthesis of the various isoquinolines is reviewed in two chapters, one devoted solely to the morphinans and the other to the other skeletal types. Attempts to produce morphinan alkaloids by plant cell culture techniques has produced a flurry of activity in several laboratories throughout the world. Four chapters are concerned with plant cell culture techniques for the production of isoquinolines. A series of enzymes which catalyse specific steps in isoquinoline alkaloid biosynthesis have been isolated in recent years and the stage is now set for the isolation and characterization of the enzymes responsible for morphinan alkaloid biosynthesis. The utilization of these enzymes for biotechnological conversions is now envisaged.

We express our sincere thanks to Hoechst (Frankfurt), Nattermann (Cologne), Schering (Berlin) and Wacker (Munich) for their generous financial support of the Symposium and to all speakers and chairmen of the sessions. We are indebted to Dr. F. Fish, Dean of The School of Pharmacy, for his permission to hold the Symposium at The School and we are most grateful to Mrs Linda Lisgarten, librarian, and her staff for invaluable help in the editing of the manuscripts.

Spring 1985 J.D. Phillipson M.F. Roberts M.H. Zenk

Contents

List of Contributors

You will find the address at the beginning of the respective contribution

Plants as a Source of Isoquinoline Alkaloids

N. G. BISSET [1]

1 Introduction

Mankind has long made use of plants containing isoquinoline alkaloids as active principles for medicinal and other purposes, and they have proved to be both a boon and a curse. In the sections that follow these two aspects will be illustrated in a largely historical discussion using *Papaver somniferum, Argemone mexicana,* and Curare as examples; at the same time, some of the reasons for the continuing interest in this field will become clear. An additional motive for such a discussion is that it is with isoquinoline compounds that the whole field of alkaloid research may be said to have begun. The final section is a more general account of the distribution of isoquinoline alkaloids in the plant kingdom, together with an indication of their supposed significance.

2 Papaver

2.1 Early History

2.1.1 Archaeological Evidence

Poppy remains – seeds and capsules – have been recovered from neolithic sites chiefly in western and western central Europe. Two of the finds include capsules of the dehiscent type. The carbonized specimen from the lakeside dwelling at Robenhausen, near Zürich in Switzerland, is of an unripe fruit, but its incomplete state precludes a precise identification. The capsules from the Cueva de los Murciélagos (Cave of Bats), near Albuñol in the southern province of Granada, Spain, date from the Late Neolithic, ca. 2500 BC, and have been identified as being of *P.somniferum* L. The poppy finds from the Rhineland region, from lakeside dwellings in France, Italy and Switzerland, and from Poland, some of which have been reported as *P.setigerum* (but cf. Sect. 2.1.3), are far from the Mediterranean distribution range of that species and may therefore have come from cultivated plants or from weeds among other culti-

1 Pharmacognosy Research Laboratories, Department of Pharmacy, Chelsea College, University of London, Manresa Road, London SW3 6LX, United Kingdom

The Chemistry and Biology of Isoquinoline Alkaloids, ed. by Phillipson et al.

vated plants (Schultze-Motel 1979). The massive finds of seeds in Germany and Switzerland suggest use as a food or a source of oil rather than as a medicine.

No poppy material has been detected at any neolithic site in the eastern Mediterranean or Near East (Krikorian 1975), and it is not until the Bronze Age that knowledge of the poppy and its products appears in the eastern Mediterranean. Possibly the earliest indication recognized so far is a portrayal of the plant, including capsule, at the palace of Knossos in Crete; it may belong to the latter part of the Middle Minoan III period, ca. 1600 BC (Merrillees 1979). Better known is the remarkable terra-cotta figure of a goddess adorned with three scarified poppy capsules, clearly recognizable as those of *P. somniferum*, which has been recovered from a site at Gazi, west of Heraklion, also in Crete. This figure is of the Late Minoan III period, ca. 1400–1350 BC. The circumstances of the find and the associated artefacts, as well as finds elsewhere on the island, lead to the conclusion that during this period the use of opium for religious, and probably also medicinal, purposes was known on the island (Kritikos and Papadaki 1967).

Contact between Crete and Cyprus goes back to at least the 16th century BC and recent work in north-western Cyprus suggests that Cretan traders may have brought the opium poppy to the island at about that time or a little earlier (Merrillees 1979).

From a study of the opium trade in the eastern Mediterranean Merrillees (1962, 1968, 1979) concludes that small long-necked juglets manufactured in Cyprus, which in shape closely resemble a poppy capsule on its stem and which have been discovered in Egypt and elsewhere, were used for the transport of opium. Moreover, the decoration on later types of juglets suggests the scarification and flow of latex on the surface of the capsule. These imported juglets appear in Egypt as early as the first half of the XVIIIth dynasty, ca. 1567–1320 BC. Other artefacts – faience beads, earrings, temple decorations – show that it was during this period that the Egyptians became acquainted with the plant itself (Gabra 1956, Merrillees 1968). The reported discovery in a tomb dating from the time of Tuthmosis III (ca. 1504–1450 BC) of an unscarified capsule is further evidence (Merrillees 1968); however, its identification appears not to be entirely free of doubt (Germer 1979). In the tomb of the royal architect Cha, who died during the reign of Amenophis (Amenhotep) III (ca. 1417–1379 BC), seven small alabaster vases were found, including one remarkably similar in shape to the Cypriote juglets. One of these vases[2] contained a product which according to chemical tests comprised vegetable fat together with iron and morphine – hence opium (alkaloid extraction, and alkaloid and colour tests); pharmacological tests also indicated the presence of the alkaloid (Muzio 1925, Schiaparelli 1927). Later chemical studies on the contents of Cypriote juglets have implied the presence of opium in some cases but not in others (Merrillees 1968).

---

2 Neither the excavation report (Schiaparelli) nor the report on the chemical and pharmacological work (Muzio) indicates which vase contained the material that was examined. Merrillees (1962 pp. 290–291) remarks rather wistfully how tempting it is to identify the container concerned as the one vase which appears to be an imitation of the Cypriote juglets. It is perhaps suggestive that it is with this particular vase that Muzio illustrated her original report. (Merrillees presumably only saw the report as given by Schiaparelli which omitted the illustration since he himself had a picture of it as part of his report (Schiaparelli 1927 Fig. 138, middle juglet of the bottom row).

In Mycenae, brooch pins have been recovered dating from the 16th century BC and having the characteristic feature of a grooved poppy capsule and its stem (Kritikos and Papadaki 1967). It is suggested that the scarcity of Cypriote juglets in the Aegean, as compared with the Levant, may have been because the opium poppy was already being grown, and its latex tapped, in Crete (cf. above) and Greece (Merrillees 1979). It is also suggested that certain containers known as "aryballoi", although usually believed to have been used for perfume, might also have held preparations of opium. In Greece, there is extensive archaeological and literary (8th century BC and onwards) evidence regarding the poppy and its derived product opium, although the latter was probably not used much before the 4th century BC (Kritikos and Papadaki 1967).

Evidence for the existence of the poppy in Anatolia during the time of the Hittite Empire, ca. 1400–1200 BC, is provided by two identical gold pins. These faithfully reproduce details of the poppy capsule and its stalk down to a loose object inside (presumably representing seeds) so that the pins can rattle like the real thing (Merrillees 1979).

2.1.2 Linguistic Evidence

Supposed recognition of Sumerian and Akkadian words meaning "poppy" or "opium poppy" or similar terms is now discounted, and the same is true of supposed representations of the poppy on Assyrian reliefs (Krikorian 1975).

Ingredients included in certain prescriptions found in the great Smith and Ebers ancient Egyptian medical papyri (ca. 1600 BC) have been translated as "poppy flower", "capsule", or "seed" (Gabra 1956), but here also there appears to be some doubt about the validity of the translations (Germer 1979). It also has to be borne in mind that, if Merrillees's views on the Levant opium trade are correct (Sect. 2.1.1), then the Smith and Ebers papyri may have been written just before the opium poppy and opium reached Egypt. Further investigation of the question is warranted.

2.1.3 Botanical Evidence

It has long been realized that the plant most closely related to *Papaver somniferum* is *P.setigerum*. According to Danert (1958) and Hammer and Fritsch (1977), the morphological differences between the two appear to be very slight. Vesselovskaya (1975), however, regards them as being more distinct. Diploid (2n = 22) and tetraploid (2n = 44) forms of both taxa occur, so that it is not possible to separate them on the basis of the chromosome number; nevertheless, most *P.setigerum* material is tetraploid. Differences in the chromosome morphology of the *P.somniferum* and *P.setigerum* diploids are likewise very slight (Hammer and Fritsch 1977; Malik et al. 1979), but again Vesselovskaya's opinion (1975) is somewhat at variance.

Malik et al. take the view that *P.somniferum* or an allied form was involved in the origin of *P.setigerum*, rather than the other way round, or that the two taxa had a common parent; while Vesselovskaya concludes that, although genetically related, the two species are isolated and her subsp. *somniferum* rather than *P.setigerum* is the primary ancestral form.

Fritsch (1979) has compared the seed morphology (length, width, seed coat) of diploid and tetraploid forms of *P.somniferum* and *P.setigerum* and he emphasizes that there is considerable and overlapping variation. What this means in practical terms is that there are great difficulties in the way of identifying archaeological material and Fritsch suggests that only if the seeds are ca. 0.95 mm long with a length/width index of ca. 1.50 is it reasonably certain that they belong to a diploid form of *P.somniferum*. This reinforces the conclusion drawn by Schultze-Motel (1979) that samples from the lakeside dwellings and elsewhere (Sect. 2.1.1) identified as being *P.setigerum* might not be that and that it is better to include them in a broadly conceived *P.somniferum*.

In the taxonomic classification proposed by Vesselovskaya (1975) there are six subspecies and *P.setigerum* is excluded. The more recent one of Hammer (1981) divides *P.somniferum* into three subspecies; subsp. *somniferum* and subsp. *songaricum* Basil., two cultivated races which differ in geographical distribution and in features of the stigma rays, and subsp. *setigerum* (DC.) Corb., the putative ancestral wild race. The two cultivated subspecies are each further divided into two convarieties, one with indehiscent and one with dehiscent capsules. Each convariety is in turn subdivided into 13 varieties on the basis of seed and flower colours, as originally proposed by Danert (1958).

According to Hammer and Fritsch (1977) the present-day distribution of *P.setigerum (P.somniferum* subsp. *setigerum),* now relatively rare, comprises the northern part of the Mediterranean with the centre of gravity in the west – ranging from the Canary Islands as far as Greece and Cyprus. Although they decline to draw any conclusion about the origin of the opium poppy (and the cytological evidence discussed above does not make it any easier), both Vesselovskaya (1975) and Schultze-Motel (1979) indicate that the western Mediterranean must be recognized as the region where it originated.

In spite of the rather inconclusive nature of the foregoing discussion, it may nevertheless be accepted that members of the isoquinoline alkaloid-bearing genus *Papaver* were among the earliest medicinal plants used by man.

2.2 The Beginnings of Alkaloid Chemistry

A jump of almost 3500 years to the beginning of the 19th century AD introduces another facet of the history of the poppy, for it is then that efforts to isolate the active principle of opium were finally crowned with a measure of success. Derosne (1803) described how from an aqueous opium extract he obtained a precipitate which in solution turned violet juice green, thus indicating alkaline properties. However, he thought that the effect was due to adhering potash and he called his product "*sel* d'opium... pour éviter des circonlocutions et faute de savoir le nom propre à lui assigner". Experiments on dogs showed that the material had properties similar to opium itself.

A few years later, Sertürner (1806) reported on a series of 57 experiments which were evidently also carried out in 1803 (Coenen 1954), when he was only 20 years old. By the use of ammonia he obtained a precipitate, again from an aqueous opium extract, which he could crystallize from ethanol. Experiments on several dogs showed that it was the *principium somniferum*. In contrast with Derosne, however, he drew the conclusion that the basic nature was an integral property of the substance – through the facts that the material could form salts and that it did not redden litmus. At the same time, Sertürner recognized the differing solubilities of the base and its salts and he made use of this in purifying the substance via the alternating precipitation with base and dissolution with acid – a method that has been widely applied ever since by alkaloid chemists. Nevertheless, at that stage he was willing to allow Derosne the priority for having isolated the narcotic principle of opium. This 1806 paper unfortunately attracted little attention, and it was not until 11 years later that, as a result of more detailed work, Sertürner (1817) withdrew his recognition of Derosne's priority and, at the same time, showed even more clearly the essential properties of " das neue Pflanzen-Alkali": turning red litmus paper blue, formation of neutral salts with various organic and inorganic acids. He placed the new base next to ammonia and thought that, although he had not had time to carry out adequate experiments, in addition to carbon, hydrogen, and oxygen, it contained nitrogen. Experiments, this time on himself and three young volunteers who were not yet 17 years old, again demonstrated the activity of the substance.

It was as a result of this 1817 paper that Sertürner's fame became assured, not only as the discoverer of morphine (1) but also as the founder of the whole field of plant alkaloids and their chemistry; for brief accounts of the discovery of morphine, see Lockemann (1924, 1951) and Coenen (1954). Sertürner clearly understood that his new substance "morphium" was a plant alkali rather than an acid, as active principles in plants were up until then thought to be. The term "alkaloid" for such substances was introduced by Meissner (1819), because he considered that they differed greatly in many of their properties from true alkalis. Derosne's product was probably a mixture of morphine (1) and noscapine (narcotine) (2).

Since its discovery, morphine, together with its diacetyl derivative diamorphine or heroin (1a), has been a major weapon in the alleviation of pain, particularly in terminal illness, the euphoric effects playing an important part in addition to the analgesia that it brings about. Nevertheless, the other side of the coin has to be recognized and the psychological (and physical) dependence on, i.e. addiction to, these substances has become an international social, and criminal, problem that is increasingly difficult to combat. On the other hand, continuing research into the constituents of opium led to the isolation of further alkaloids which have proved beneficial in medicine. Thus, codeine (1b), besides being a moderate analgesic, has antitussive and antidiarrhoeal properties. From thebaine (3), in an attempt to separate the analgesic from the addictive properties (the boon and the curse!), Diels-Alder adducts have been prepared – etorphine (4) and related compounds – which are many hundreds of times more potent than morphine but which unfortunately are still addictive. Noscapine (2) and papaverine (5) are devoid of analgesic properties, but they both have some vasodilatory activity and noscapine finds use as a cough suppressant (Bowman and Rand 1980; J.E.F. Reynolds 1982).

Thus, *Papaver,* in addition to having been one of the first medicinal plants, was also the starting point for what has become the vast field of alkaloid chemistry.

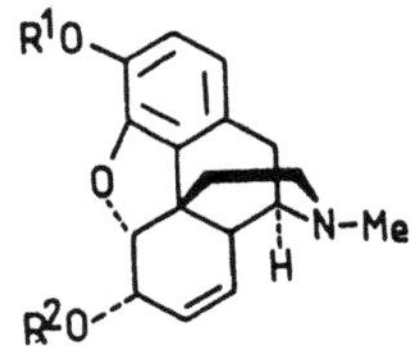

1 $R^1 = R^2 = H$
1a $R^1 = R^2 = CO.Me$
1b $R^1 = Me, R^2 = H$

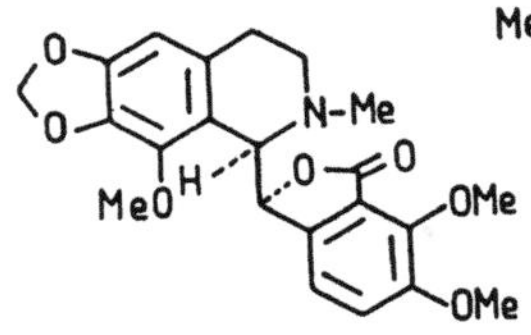

2.

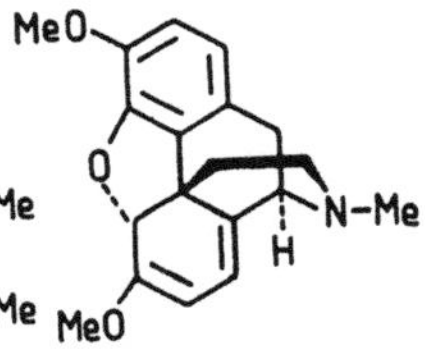

3.

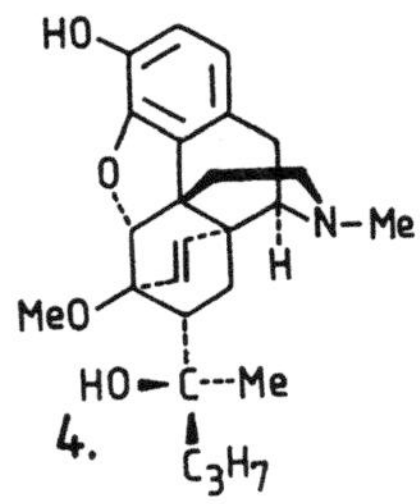

4.

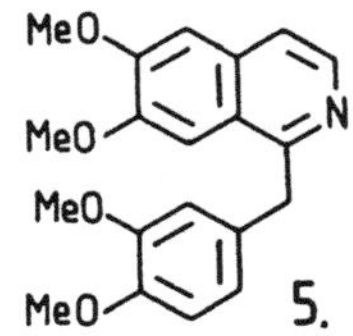

5.

Fig. 1

3 Argemone

3.1 Historical

A number of plants with yellow to red-coloured stems, roots, or latex, e.g. *Chelidonium,* have been applied as remedies for diseased eyes; most of them have been identified as members of the Papaveraceae or Fumariaceae. Dioscorides, a Greek army physician of the 1st century AD whose work influenced European medicine for many centuries to come, recommended a poultice of the leaves of a Mediterranean plant which he called *argemone* and which "being applyed . . . doe take away the Argema & the Nubeculas, & doe assuage inflammations" (Gunther 1933).[3] Argema is a white corneal ulcer and nubecula is cloudiness of the cornea. Many attempts have been made to identify the plant concerned and among the suggestions put forward have been species of *Thalictrum, Papaver, Adonis,* and *Argemone. Papaver argemone* L. appears to be a likely candidate (cf. Basmadjian 1938, Dubler 1955).

When at the end of the 16th century the prickly poppy came to Europe from the Caribbean, it was recognized as being related to the poppies and given the name *Papaver spinosum.* In herbals and botanical works of the same period, such as those by Dodoens, Gerarde, Parkinson, L'Obel, Bauhin, etc., certain plants which are true poppies had descriptive names starting with *Argemone.* Tournefort (1694) sought to remedy the situation – he transferred these plants to his genus *Papaver,* but most

3 This quotation comes from the translation made from the Greek by John Goodyer in 1652/55. According to Riddle (1971), however, no accurate translation of Dioscorides has yet appeared in any modern European language and he stigmatizes that by Goodyer as being "woefully inadequate". In the Spanish translation by Andrés de Laguna, dating from 1555, the above passage reads more convincingly: "Su rayz es redonda, y el liquor que de si derrama, de color de açafran, y agudo: el qual quita los fluecos, & las nuues que offuscan la vista. Sus hojas aplicadas en forma de emplastro, son vtiles a las inflammationes" (Dubler 1955).

unfortunately chose the name *Argemone* thus freed for a new genus comprising the single species *Argemone mexicana,* the prickly poppy. Linnaeus (1753) accepted Tournefort's genus and species when he came to write the *Species Plantarum.* Through association of its name with the *argemone* that was used to treat eye diseases, the plant was introduced by the Portuguese into East Africa and later, in the 18th century, into Goa from where it has spread to become a common weed in the drier parts of the Indian subcontinent.

In India, the supposed value of *A. mexicana* in dealing with eye conditions, resulted in a second, and equally unfortunate, misinterpretation, this time by the practitioners of indigenous medicine. They mis-identified the plant as one of the *svarṇakṣīrī,* golden milk or latex, drugs recommended in the early Sanskrit medical works by Caraka, Suśruta, and Vāgbhaṭa for skin and eye diseases; up to that time none of them had been identified. Nowadays, *svarṇakṣīrī* is taken as a Sanskrit synonym of the vernacular names usually applied to the prickly poppy; and all parts of the plant are used in the Ayurvedic and Siddha, as well as Unani, systems of medicine for treating skin and eye diseases and a wide range of other conditions (see, e.g. Nadkarni 1954). More recently, Singh and Chunekar (1972) have concluded that the *svarṇakṣīrī* of the ancients were in fact various species of *Euphorbia* and *Garcinia.*

3.2 Epidemic Dropsy (and Glaucoma)

Since 1866 outbreaks of "epidemic dropsy" have been recorded in India; in 1935, for example, as many as 7000 people were affected – some 1500 died and of those who survived 700 became blind. The disease is characterized by oedema in the lower extremities, gastro-intestinal disturbances, and changes in the skin; and frequently associated with it is a primary high-tension glaucoma which often leads to permanent damage or blindness. In 1926 it was established that the culprit was cooking oil, especially mustard seed oil, contaminated with the oil from the seeds of *A.mexicana* (Manson-Bahr 1946, Hakim 1954, Dyke 1978). According to Kaul and Ahuja (1980), it is oil seeds reaching the markets in April and May that are most likely to be involved, since it is at that time of year that the prickly poppy matures.

The seeds of *A.mexicana* may yield up to 60% oil (Murav'eva and Bui 1974), which may contain up to ca. 0.9% alkaloid. Sarkar (1948) isolated the benzophenanthridine alkaloids sanguinarine (6), which comprised 5%, and dihydrosanguinarine (7), which accounted for over 85%, of the total alkaloid present. Seeds from Vietnamese plants when grown in Pyatigorsk, in southern Russia between the Black Sea and the Caspian Sea, yielded oil containing 0.35% sanguinarine (6) and 0.31% of the protopine base allocryptopine (8a) (Bui 1974).

Hakim (1954) found that young rats given orally either 1 ml of seed oil or 0.35 mg sanguinarine daily over 225 days exhibited some increase in ocular tension; and subcutaneous and intravenous injections of the alkaloid also produced temporary rises in tension. Shenolikar et al. (1974) have calculated that in human beings a daily intake of as little as 0.02 ml argemone oil or 0.11 mg sanguinarine could lead to the development of clinical symptoms of epidemic dropsy. Kaul and Ahuja (1980) recommend that oils containing more than 0.01% argemone oil should be processed in order to remove any sanguinarine present.

Analyses of other parts of the plant have revealed the presence in addition to sanguinarine of protopine (8), cryptopine (8b), berberine (9), coptisine (9a), and a host of related minor bases (Haisová and Slavík 1975; and references cited therein).

From the data given by Hakim (1954) it is evident that argemone oil is several times more toxic than can be accounted for by the amount of sanguinarine present. The co-occurrence of a "potentiating factor" in the oil has been postulated (Rukmini 1971, Shenolikar et al. 1974), and one compound obtained from the oil which appears to have such properties has been shown to be (+)-6-hydroxy-6-methyl-9-oxo-octacosanoic acid (10) (Rukmini 1975).

Seed oils and capsule extracts of several other papaveraceous plants, from the genera *Chelidonium, Eschscholzia, Papaver,* and *Platystemon,* are also active. While dihydrosanguinarine is inactive, chelerythrine (6a) and berberine (9) are active (Hakim 1954). Intravenous injections of other types of isoquinoline alkaloids have also been shown to be active (but in much higher doses than sanguinarine) – they include codeine (1b), noscapine (narcotine) (2), papaverine (5), bulbocapnine (11), corydine (11a), and chelidonine (12) (Lieb and Scherf 1956).

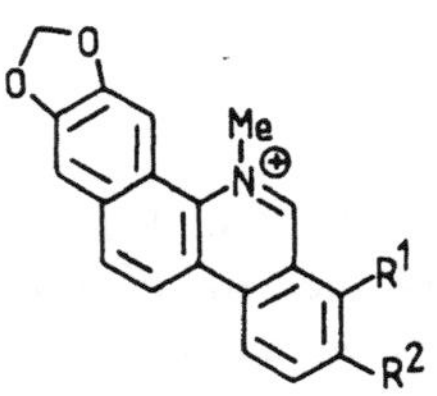

6 $R^1, R^2 = OCH_2O$
6a $R^1 = R^2 = OMe$

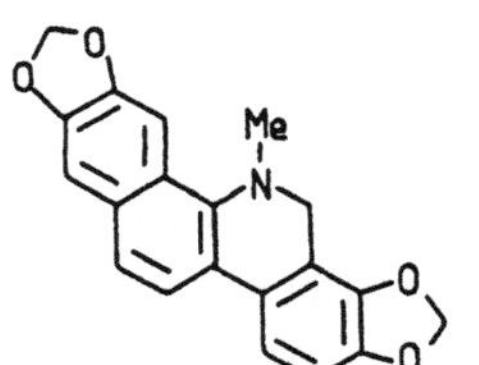

7

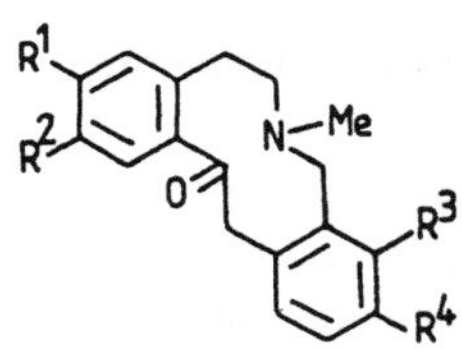

8 $R^1, R^2 = R^3, R^4 = OCH_2O$
8a $R^1, R^2 = OCH_2O, R^3 = R^4 = OMe$
8b $R^1 = R^2 = OMe, R^3, R^4 = OCH_2O$

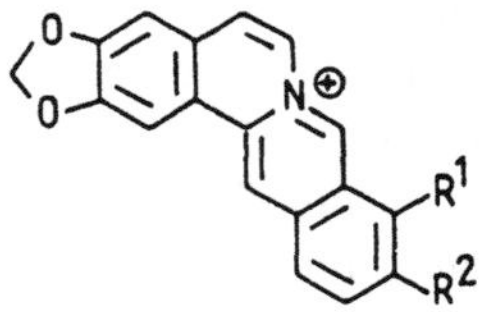

9 $R^1 = R^2 = OMe$
9a $R^1, R^2 = OCH_2O$

$Me(CH_2)_8CO(CH_2)_2\text{-}C(OH)(Me)\text{-}(CH_2)_4COOH$

10

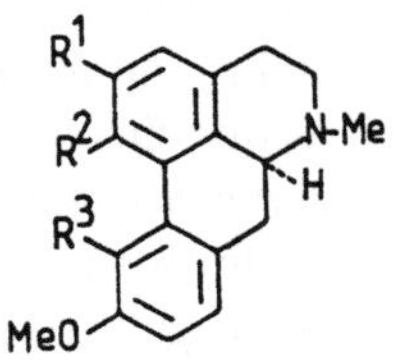

11 $R^1, R^2 = OCH_2O, R^3 = H$
11a $R^1 = R^3 = OMe, R^2 = OH$

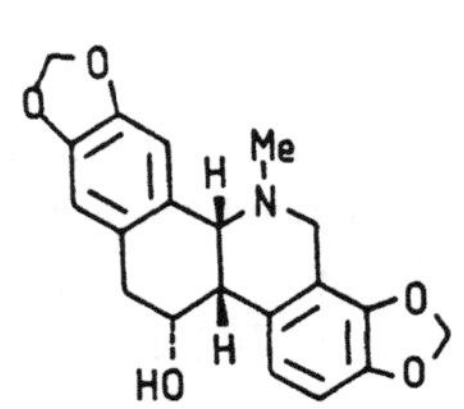

12

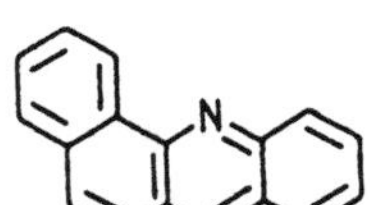

13

Fig. 2

3.3 Occurrence of Sanguinarine and Endemic Glaucoma

The fact that sanguinarine produces glaucoma experimentally raised the possibility that it might be behind the endemic occurrence of the disease elsewhere in the world. Hakim et al. (1961b) screened 53 different species belonging to the Papaveraceae and Fumariaceae and obtained evidence for the presence of the alkaloid in 49 of them, especially in the stems and leaves, including members of the following genera:

Papaveraceae – *Argemone, Chelidonium, Dicranostigma, Eschscholzia, Glaucium, Macleaya, Meconopsis, Papaver, Stylophorum;*
Fumariacaeae – *Corydalis, Dicentra, Fumaria, Hypecoum.*

Arising from the implied more widespread occurrence of substances able to bring about increased ocular tension, as well as the fact that sanguinarine can be transmitted in milk (Hakim et al. 1961a), is the question to what extent dietary habits and other variables may be concerned in endemic glaucoma not involving *A.mexicana.*

Epidemiological studies show that in Europe the inhabitants of the Faroe Islands have the highest incidence of glaucoma, but this is hardly surprising as the main fodder for their milk-producing animals is *Papaver nudicaule* whose leaves and other tissues contain sanguinarine and related compounds (Hakim et al. 1961b). Other ethnic groups with a higher than average incidence of glaucoma are the Jains of India, who are vegetarians and who consume large quantitites of milk; opium addicts – here the cause is likely to be one or other of the opium alkaloids, since sanguinarine appears to be largely absent from the capsules, and hence the latex, of *Papaver somniferum;* and poorly fed communities on a low protein diet – it seems that sanguinarine in milk has less effect than sanguinarine on its own and this is ascribed to the albumin present, an indication which is supported by the fact that to some extent the amino-acid cysteine is able to reduce or inhibit the effect of the alkaloid. Argemone seed oil is also used in Africa and India as a purgative and the stems and leaves of the plant are eaten as a vegetable. It is clear that there is a variety of sources through which sanguinarine and other glaucoma-producing alkaloids can be ingested (Hakim 1954, Moody 1965, Dyke 1978).

In spite of all the work that has been done so far, the mechanism that brings about the glaucoma is still not properly understood. There is some evidence that it may be due to disturbance of a hormone secreted by the hypothalamus, possibly one of the prostaglandins, which regulates the intra-ocular pressure (Dyke 1978).

Yet another twist to this sorry tale is that milk containing sanguinarine also contains a green-fluorescing metabolite formed in the liver. This substance has been shown to be benz[c]acridine (13) (Hakim et al. 1961a) – a known carcinogen – and tests with its precursor sanguinarine have demonstrated that on topical application it also leads to the development of tumours in experimental animals.

4 Curare

4.1 Plant Sources

Curare! Called the "flying death" and any number of other fanciful and evocative designations, the arrow and dart poisons of the South American Indians caught the imagination of the early explorers and of scientists and laymen alike. And it is in no small measure due to the study of this group of remarkable poisons[4] that wide fields of botany, chemistry, physiology, and pharmacology have been opened up.

The aura of mystery and obscurity surrounding the poisons grouped together under the name curare has now been largely dispelled and the main sources of the active constituents are known to be the stem bark and root bark of plants belonging to the two families Loganiaceae and Menispermaceae. In the case of the Loganiaceae, the plants are all members of the genus *Strychnos* L. and among the alkaloids they contain are highly active bis-quaternary dimeric indole alkaloids with muscle-relaxant properties; these will not be considered further here. On the other hand, several genera belonging to the Menispermaceae are involved:

Abuta Barrère ex Aublet, *Chondrodendron* Ruiz et Pavón, *Cissampelos* L., *Curarea* Barneby et Krukoff, *Sciadotenia* Miers, and *Telitoxicum* Mold.

Curarea is a fairly recent segregate from *Chondrodendron* based on differences in perianth and fruit characters (Barneby and Krukoff 1971).

In the early days of curare research, the poisons were classified according to the kind of receptacle into which they were put, the three main types being tube, calabash, and pot curare. Of these, tube curare gradually came to be equated with the product derived from *Chondrodendron*, etc., while calabash curare became more or less synonymous with that prepared using *Strychnos*. Nowadays, the type of container is no longer diagnostic and the curare maker is just as likely to use a handy bottle.

Curares based on Menispermaceae predominate in the Peruvian Amazon region (in the area bounded by the rivers Napo, Marañon, and Ucayali); mixed Loganiaceae/Menispermaceae curares are found chiefly in the area covered by the middle reaches of the Brazilian Amazon and its tributaries; and *Strychnos* curares come principally from the region between the Orinoco in the north and the Negro and lower reaches of the Amazon in the south (Vellard 1965, 1973). There is, however, evidence that mixed curares were also made in Guyana (Moody 1965, Snedden et al. 1970).

4 The arrow poisons first encountered by the Spanish Conquistadores were not curares in the modern sense, as the victims usually died in the course of 24 h or more in great pain – effects which are not characteristic of an overdose of muscle-relaxant. These poisons were used in warfare and their active principles were the fruit and latex of certain Euphorbiaceae, e.g. *Hippomane, Hura*. Their effects were probably due to the combined action of highly toxic lectins and irritant diterpene esters. True curares were not prepared by the coastal Indians but rather by the forest Indians for hunting; they were rarely used against fellow human beings (see Vellard 1965, 1973).

4.2 Muscle-Relaxants

Research during the 1930s and 1940s, largely by King (1935, 1948, and related papers), Wintersteiner and Dutcher (1943), Dutcher (1946) and others, on samples of tube curare and material of various *Chondrodendron* (and *Curarea*) species finally led to the recognition that their muscle-relaxant principles were bis-benzylisoquinoline alkaloids. Several such compounds were isolated and structures proposed for them. Among these compounds was (+)-tubocurarine, which was thought to be a bis-quaternary compound and was therefore assigned structure (14b) [now known to belong to chondrocurarine (Reynolds et al. 1975)]. Attention remained focused on the (+)-isomer, since biological testing on the isolated rat-diaphragm preparation showed that it was 30–60 times more powerful than the (–)-isomer (14a) (King 1947).

When the neuromuscular-blocking activity of the supposed bis-quaternary (+)-tubocurarine and related compounds, in which the two nitrogens are situated at a distance of ca. 10 Å from each other, came to be determined, it appeared that members of the bis-quaternary series were more active than the mono-quaternary derivatives and that O-methylation also increased activity. Among the earlier synthetic compounds produced was gallamine triethiodide or flaxedil (15), which, although rather less active than tubocurarine, was subsequently introduced into

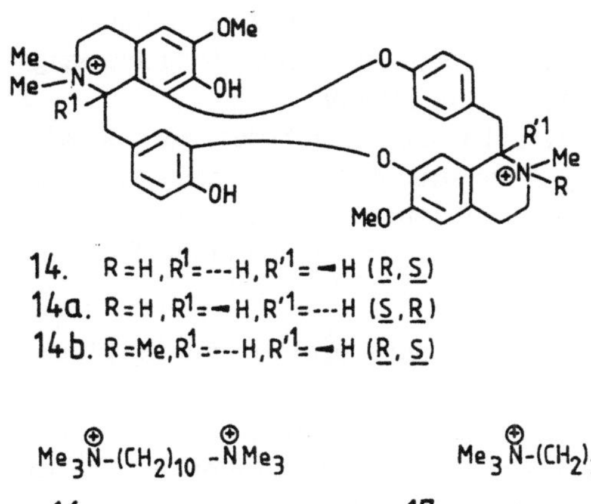

14. R = H, R¹ = ---H, R′¹ = —H (R, S)
14a. R = H, R¹ = —H, R′¹ = ---H (S, R)
14b. R = Me, R¹ = ---H, R′¹ = —H (R, S)

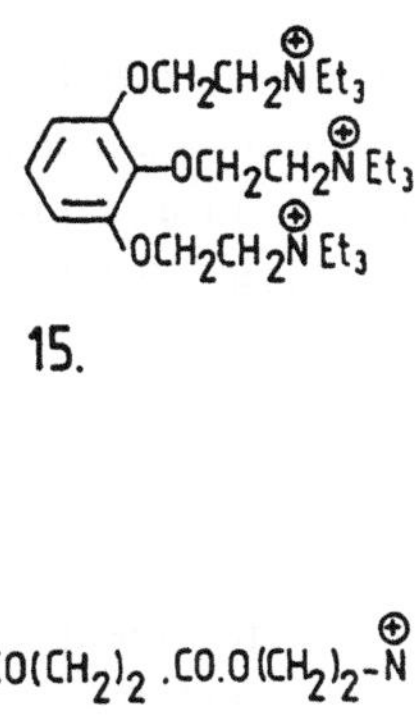

15.

$Me_3\overset{\oplus}{N}\text{-}(CH_2)_{10}\text{-}\overset{\oplus}{N}Me_3$

16.

$Me_3\overset{\oplus}{N}\text{-}(CH_2)_2\,O.CO(CH_2)_2\,.CO.O(CH_2)_2\text{-}\overset{\oplus}{N}\,Me_3$

17.

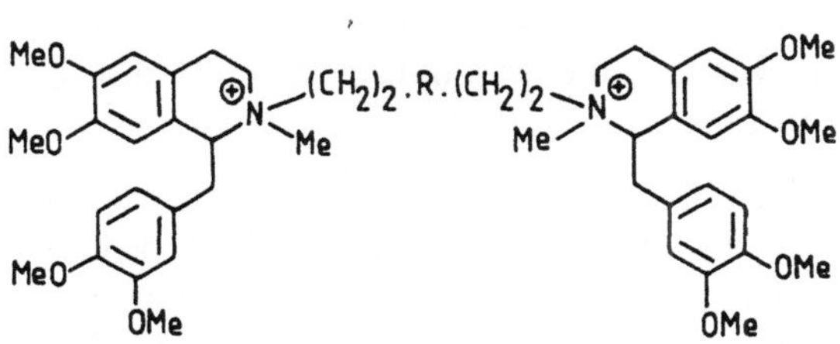

18. R = $(CH_2)_6$
18a. R = $CO\text{-}O\text{-}(CH_2)_5\text{-}O\text{-}CO$

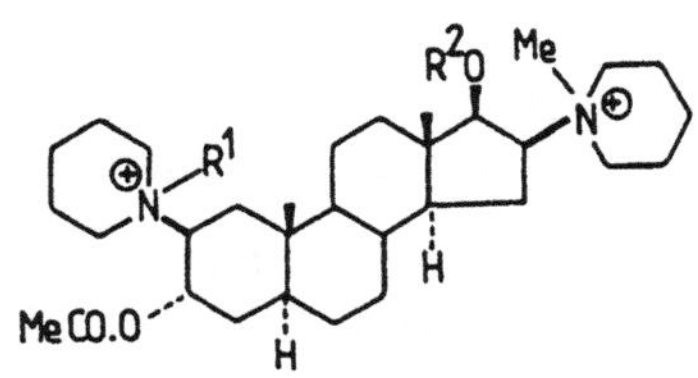

19. R^1 = Me, R^2 = CO.Me
19a. R^1 = Me, R^2 = H
19b. R^1 = H, R^2 = CO.Me

Fig. 3

clinical practice and is still in use (Barlow 1964). (+)-Tubocurarine is now known to be a mono-quaternary base (14) but in the body the tertiary nitrogen will normally be protonated (C.D. Reynolds et al. 1975, and references cited therein). The mechanism of its action and that of gallamine is of the competitive or non-depolarizing type, and in addition to binding to and blocking the recognition sites of the neurotransmitter acetylcholine on the motor end-plates at the neuromuscular junction, they can also block the acetylcholine-activated ionic channels (Lambert et al. 1983).

Further work soon revealed complications. On the one hand, it was found that relatively simple bis-quaternary molecules like decamethonium (16) and esters of the suxamethonium (or succinyldicholine) type (17) and their analogues were also active, but for rather short periods of time. They act by a depolarizing mechanism, i.e. by mimicking the effect of acetylcholine itself, which when present for long enough in high concentration causes blockade by preventing action potentials from being propagated away from the zone surrounding the motor end-plate. In the case of acetylcholine, hydrolysis by cholinesterases soon puts an end to its effects, but with decamethonium and suxamethonium the effects are more prolonged (Bowman and Rand 1980). Both decamethonium and suxamethonium were also used in surgery; and while decamethonium has been largely superseded, suxamethonium because of the brief duration of its action – a few minutes only – is still used in minor surgical procedures. In homologues of these two compounds, particularly when N-methyl functions are successively replaced by N-ethyl groups, the depolarizing blocking activity may gradually be replaced by non-depolarizing activity.

On the other hand, more complex structures based on the tubocurarine molecule led to laudexium (18); and as knowledge of the structural requirements for non-depolarizing neuromuscular-blocking drugs became clearer, it laid the basis for a less empirical and more positive approach to the design of appropriate drugs. The considerations taken into account include the following:

a) quaternary nitrogen atoms should be present, rather than other onium functions, as they afford greater blocking power;

b) bis-quaternary structures tend to be more potent than mono-quaternary ones;

c) the two nitrogen functions should be separated by a distance of ca. 11 Å to give maximum neuromuscular- rather than ganglion-blocking potency;

d) bulky molecules and bulky substituents on the nitrogens tend to favour a non-depolarizing rather than depolarizing mode of action;

e) inclusion of acetylcholine-like moieties allows hydrolysis in a way similar to that suffered by acetylcholine itself and would at the same time increase the affinity of the substance for the appropriate receptors.

These considerations led to the synthesis and development of pancuronium (19), the required dose of which is about one-fifth of that of tubocurarine. In it the bulk is supplied by the steroid skeleton, which also situates the nitrogen functions at about the correct distance from each other, and bulky substituents are provided by N-piperidinium groups (Bowman and Rand 1980). Among the newer drugs of this type are the bis-quaternary dacuronium (19a) and the mono-quaternary vecuronium (19b) (Bowman 1980, Symposium 1980).

In contrast, following on from laudexium there is the more recent atracurium (18a), which appears to be a considerable advance in that, among other things, it can be used for patients with serious liver and kidney dysfunction. The substance combines features of a suxamethonium homologue with the bulk of two quaternary benzylisoquinoline moieties. Its period of activity is self-limiting, since in the body tissues at physiological pHs, in addition to enzymic ester hydrolysis, most of it undergoes spontaneous Hofmann elimination which in human blood is complete after ca. 35 mins. (Hughes and Chapple 1980, Hunt et al. 1980, Stenlake 1982).

4.3 Other Curare Ingredients Containing Isoquinoline-Related Alkaloids

As indicated above, plants from other genera of the Menispermaceae are incorporated into curares and they likewise contain bis-benzylisoquinoline alkaloids. Those so far isolated from *Abuta* species have their two halves joined head-to-head and tail-to-tail or just tail-to-tail, but only one of them, macoline (20), from the stem wood of *A.grisebachii* Triana et Planchon, is a monoquaternary base (Galeffi et al. 1977); all the others are bis-tertiary alkaloids. *Cissampelos ovalifolia* DC. is reported to have been a constituent of the curare which used to be prepared by the Macusi Indians of Guyana and an ethanolic extract has been shown to have neuromuscular-blocking activity. Only bis-tertiary bis-benzylisoquinolines have been extracted from the plant (Moody 1965, Snedden et al. 1970). *Sciadotenia toxifera* Krukoff et

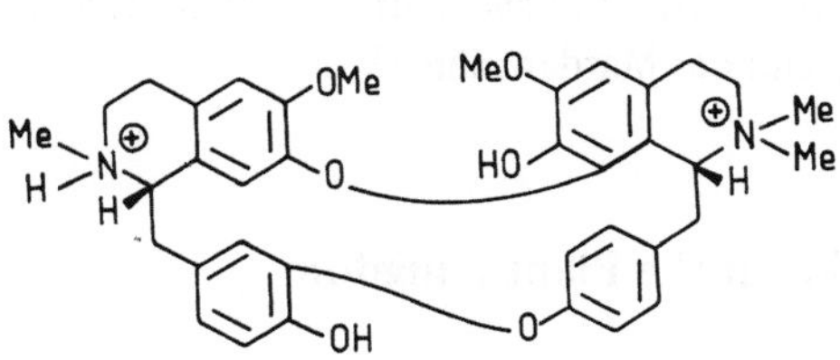

20.

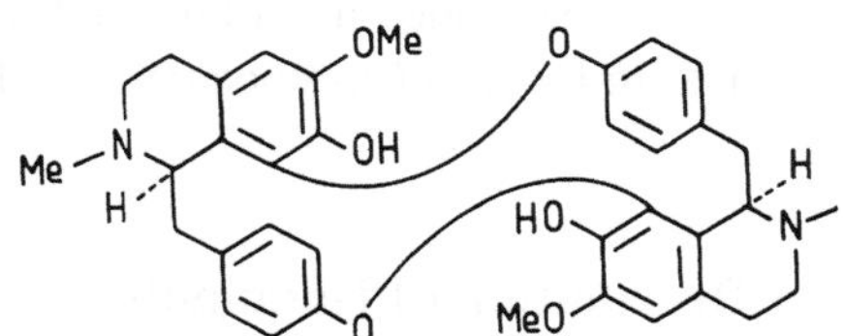

21.

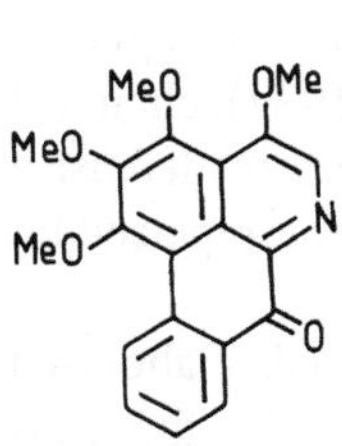

22.

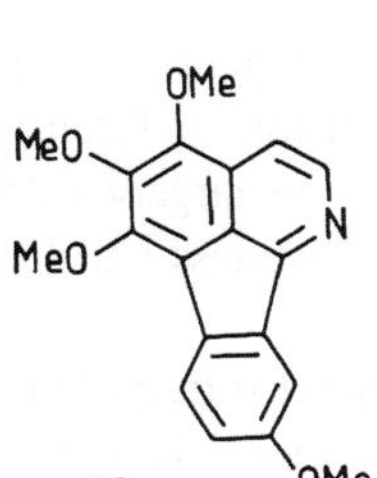

23.

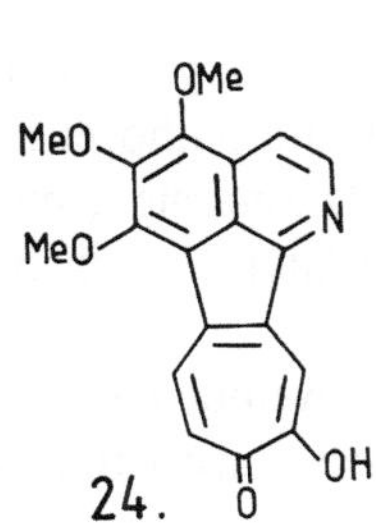

24.

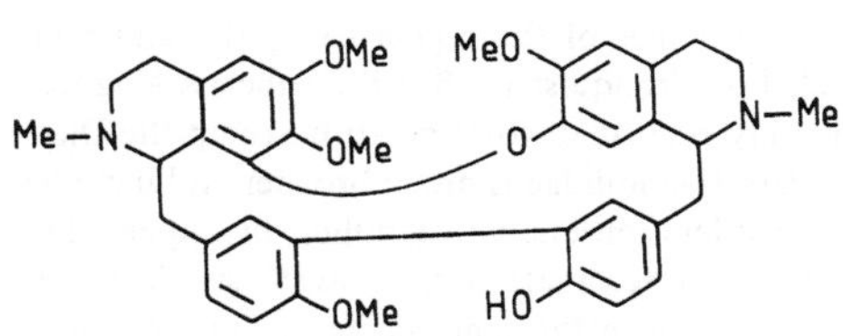

25.

Fig. 4

A.C. Smith has furnished a number of imino-bis-benzylisoquinoline derivatives, with the two parts arranged in a head-to-tail fashion; in addition, isochondrodendrine (21) is present in the plant (Takahashi and Cava 1976, Takahashi et al. 1976, Galeffi et al. 1978).

Bis-benzylisoquinoline alkaloids are present in many other genera of the Menispermaceae besides those discussed above, mostly accompanied by other types of isoquinoline alkaloid. Thus, species of *Abuta,* in addition to bis-benzylisoquinolines, biosynthesize a variety of oxoaporphines such as imenine (22) and azafluoranthenes like rufescine (23) (Glick et al. 1969, Cava et al. 1972, 1975, Skiles et al. 1979), and more recently a tropolono-isoquinoline, grandirubrine (24), has been obtained from one species (Menachery and Cava 1980). The only species of *Telitoxicum* that has been investigated, *T.peruvianum* Mold., also contains oxoaporphines and azafluoranthenes and they are the same as or similar to those occurring in *Abuta* species (Menachery and Cava 1981).

Other plants containing isoquinoline alkaloids enter into the composition of South American dart poisons. Thus, species from three genera of the Annonaceae, *Duguetia* A.St.Hil., *Guatteria* Ruiz et Pavón, and *Unonopsis* R.E.Fries also find use, and isochondrodendrine (21) and related compounds have been isolated from the stem bark of *Guatteria megalophylla* Diels (Galeffi et al. 1975). *Ocotea* Aubl., a genus belonging to the Lauraceae, has provided yet another plant, *O.venenosa* Kosterm. et Pinkley (Kostermans et al. 1969); in this case, the main alkaloid present is rodiasine (25) and, although it is a bis-tertiary base and unlike the other dimeric alkaloids mentioned has a biphenyl bridge, it is stated to have neuromuscular effects similar to those of (+)-tubocurarine (Murthy and der Marderosian 1973).

5 Distribution of Isoquinoline Alkaloids in the Plant Kingdom

5.1 Plant Chemical Defenses

It has been argued that secondary metabolites in plants are important in repelling and discouraging predators and pathogens. According to Cronquist (1977), chemical differences among major taxonomic groupings are largely differences in defensive weapons, which, sooner or later, lose their effectiveness and have to be replaced by new ones.

Thus, when resistance developed to the benzylisoquinoline alkaloids elaborated by the subclass Magnoliidae sensu Cronquist,[5] generally conceded to be the most

5 In the classifications of Cronquist and Dahlgren the names of the higher categories and their coverage of orders, families, etc. is not identical. For Cronquist (1981) the subclass Magnoliidae, further divided into orders and families, is only one of six that go to make up the Dicotyledons. For Dahlgren (1980), however, the subclass Magnoliidae is much broader and includes all the Dicotyledons and his subdivisions are superorders, orders, and families; his superorders Magnoliiflorae, Nymphaeiflorae, and Ranunculiflorae approximate to Cronquist's subclass. The orders in the two classifications do not necessarily comprise the same range of families. However, the overall tendencies shown by the two systems are similar.

primitive group of Angiosperms, tannins (including ellagic acid and proanthocyanin derivatives) arose in the other subclasses of the Angiosperms (Hamamelidae, Caryophyllidae, Rosidae, etc., sensu Cronquist) as they became differentiated from the Magnoliidae. Predators which in turn developed resistance to tannins may perhaps simultaneously have lost some of their resistance to benzylisoquinoline alkaloids, so that these substances have been tried out with some success by certain more advanced groups of plants (Cronquist 1977). The sporadic appearance of these alkaloids in more advanced orders may also be interpreted as an indication of the antiquity of the character (Gomes and Gottlieb 1980).

From the biochemical point of view, one of the postulates relating to the evolution of such allelochemics (chemical substances produced by plants and animals for interaction with other living organisms) in higher plants is that they tend to arise by oxidation processes (Gottlieb 1982). Thus, as shown in Fig. 5, the shikimate pathway leads via chorismic acid to phenylalanine, which can produce cinnamic acids and cinnamyl alcohols, the precursors of lignins and condensed tannins. Alternatively, blocking the final steps in this pathway will lead to the accumulation of the phenylalanine, which would then be available for the production of isoquinoline alkaloids; an additional oxidation resulting in the condensation of two benzylisoquinoline units would yield bis-benzylisoquinoline bases. As the woody gymnosperms gave way first to woody and then herbaceous angiosperms, there was evidently a decrease in the production of lignins in favour of phenylalanine-based metabolites,

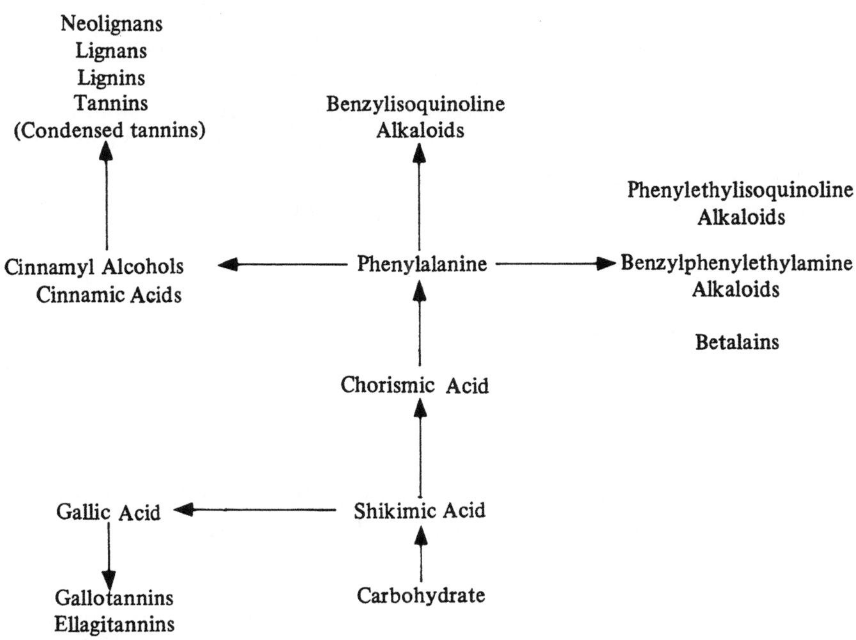

Fig. 5. Biosynthesis of some phenylalanine-derived secondary metabolites (allelochemics) (After Gottlieb 1982)

and this was followed by shortening of the pathway, through blockage at the shikimic acid stage, giving rise to gallic acid and ultimately to hydrolysable tannins (gallotannins, ellagitannins, etc.), which, since they do not require nitrogen, are more economical defense weapons.

5.2 Distribution of Isoquinolines and Associated Compounds

The interest and appreciation of taxonomists and others in chemistry as an aid in botanical classification, the field of chemotaxonomy or biochemical systematics, has been growing steadily over the last 20 years and has concentrated mainly on the secondary plant constituents (Cronquist 1977, 1980, 1981, Dahlgren 1980, Dahlgren and Clifford 1982, Takhtajan 1973, but cf. Heywood 1973). The distribution of these constituents correlates in varying degrees with taxonomic groupings recognized on classical morphological and other grounds. Chemical features as taxonomic characters are most readily accepted when they are used in conjunction with other characters. In their general classifications of higher plants, Cronquist, Takhtajan, and Dahlgren often include chemical information in describing various groupings. The discussion which follows is based on the system of Dahlgren, simply as a matter of convenience.

Phenylethylamines and simple isoquinolines are found scattered among ca. 40 families and in over half of these one or other or both types of compound are not accompanied by any more complex alkaloids. The families in which they occur are distributed throughout the plant kingdom. They are especially common in the Cactaceae and Chenopodiaceae (Caryophyllales), Fabaceae (Fabales), and Rutaceae (Rutales), families that belong to orders fairly close to those in which the more complex isoquinoline bases are mainly concentrated (Dahlgren 1980).

These more complex phenylalanine-derived alkaloids, benzylisoquinolines and others, are also found in ca. 40 families; the exact number depends on how the various families are delimited. In about 12 of them the occurrence of the alkaloids is rare. The main distribution in the subclass Magnoliidae, Dictoyledons (Dahlgren 1980), is as follows:

Magnoliiflorae

Annonales:	**Annonaceae, Eupomatiaceae,** Canellaceae
Aristolochiales:	**Aristolochiaceae**
[Rafflesiales:	–]
Magnoliales:	**Magnoliaceae**
[Illiciales:	–]
Laurales:	**Monimiaceae** (incl. Atherospermataceae and Siparunaceae), **Lauraceae, Hernandiaceae** (incl. Gyrocarpaceae)
Nelumbonales:	**Nelumbonaceae**

Nymphaeiflorae

Piperales:	Piperaceae
[Nymphaeales:	–]

Ranunculiflorae

Ranunculales:	**Menispermaceae, Ranunculaceae** (incl. Hydrastidaceae), **Berberidaceae** (incl. Leonticaceae), **Nandinaceae**
Papaverales:	**Papaveraceae, Fumariaceae** (incl. Hypecoaceae)

Note: Families not known to contain benzylisoquinoline alkaloids are omitted – they are mostly small ones. Those families whose names are printed in boldface are the chief sources of the alkaloids.

These three superorders correspond approximately with the subclass Magnoliidae of Cronquist (1981) which is divided into only eight orders – the Annonales is combined with the Magnoliales, the Nelumbonales with the Nymphaeales, and the Rafflesiales are removed elsewhere.

As already mentioned, there are other, more advanced, families scattered throughout the plant kingdom in which benzylisoquinoline alkaloids occur in a rather more sporadic fashion. The more distant of these families are otherwise quite unrelated to those in which the benzylisoquinoline alkaloids are concentrated. Some examples, in order of increasing distance, are:

Rutiflorae:	Rutales – Rutaceae *(Euodia, Fagara, Phellodendron, Toddalia, Zanthoxylum)*
Malviflorae:	Rhamnales – Rhamnaceae *(Colletia, Colubrina, Discaria, Phylica, Retanilla, Talguenea, Zizyphus)*
	Euphorbiales – Euphorbiaceae *(Croton)*
Fabiflorae:	Fabales – Papilionaceae *(Andrachne, Erythrina)*
Araliiflorae:	Araliales – Araliaceae *(Hedera)* – Apiaceae *(Heracleum)*
Santaliflorae:	Celastrales – Celastraceae *(Euonymus)*
Corniflorae:	Cornales – Symplocaceae *(Symplocos)* – Phellinaceae *(Phelline)*
	Dipsacales – Caprifoliaceae *(Symphoricarpos)*
Gentianiflorae:	Gentianales – Rubiaceae *(Cephaelis)*

In the subclass Liliidae, Monocotyledons (Dahlgren and Clifford 1982), phenylalanine-derived alkaloids (Fig. 5) have so far been found in two superorders:

Ariflorae:	Arales – Araceae (*Lysichiton:* aporphines)
Liliiflorae:	Asparagales – Amaryllidaceae (e.g. *Crinum, Galanthus, Haemanthus, Hymenocallis, Narcissus, Nerine, Zephyranthes,* etc.: benzylphenylethylamines)
	Liliales – Colchicaceae (e.g. *Androcymbium, Colchicum, Gloriosa, Iphigenia, Merendera, Schelhammera,* etc.: phenylethylisoquinolines)

5.3 Chemotaxonomy of (Benzyl)isoquinoline Alkaloids

It is the widespread occurrence and great variation in skeletal structure that makes alkaloids such an attractive area for chemotaxonomic discussion. It has repeatedly been emphasized that it is not only the types of alkaloids present in a plant that are important, but also the biosynthetic pathways involved in their production.

Many writers have used (benzyl)isoquinoline alkaloids to illustrate the potentialities and limitations of the chemotaxonomic approach utilizing micromolecules, and the most detailed analysis of their occurrence has been made by Gottlieb and his coworkers (Rezende et al. 1975, Gomes and Gottlieb 1980, Ferreira et al. 1980, Gottlieb 1982). In their analysis, Rezende et al. arrange the various groups of alkaloids according to the presumed biogenetic relationships of their skeletons in three major groups: (1) aporphines, (2) berberines, and (3) morphinans, erythrinans, etc., and each of these groups is in turn broken down into appropriate biogenetically derived subgroups, those further along the pathway being more complex and specialized. The latest treatment (Gottlieb 1982) deals with the literature up to 1978 and includes 57 different skeletal types; but additional ones have, of course, been discovered since then, among them several with a rearranged isoquinoline moiety, but this does not affect the main features of the analysis.

Bis-benzyltetrahydroisoquinolines and aporphines are of very general occurrence, so that these alkaloid types are to be considered primitive features of the benzylisoquinoline biogenetic group; they are found in the Magnoliiflorae and Ranunculiflorae and in most families elsewhere that contain benzylisoquinoline bases. Tetrahydroprotoberberines and their oxidation products are present in some families (Annonaceae, Aristolochiaceae, Magnoliaceae) of the Magnoliiflorae, but a more varied range of structures is found in the Ranunculiflorae. Within the latter superorder, evolution seems to have followed increased skeletal specialization (benzophenanthridines, protopines, spirobenzylisoquinolines, etc.) in the case of the Papaverales, and increased skeletal specialization and oxidation levels mark the Ranunculales (Gottlieb 1982). The increasing alkaloid complexity in the direction Ranunculales and Papaverales supposedly also leads to greater toxicity (Gomes and Gottlieb 1980, Rezende et al. 1975).

Another line of development from phenylalanine has been followed in the monocotyledons with the formation of phenylethylisoquinolines in the Liliales and benzylphenylethylamines in the Asparagales (Gomes and Gottlieb 1980).

Superorders outside the Magnoliiflorae and Ranunculiflorae have so few families containing benzylisoquinolines that their occurrence has little significance, in spite of apparent similarities, e.g. between Malviflorae and Magnoliiflorae and between Rutiflorae and Ranunculiflorae. Erythrinans are present in both the Fabiflorae (Fabaceae) and Magnoliiflorae (Menispermaceae), but their common occurrence in the two superorders does not imply a relationship (Gottlieb 1982).

Certain parts of the main benzylisoquinoline-containing families appear to be entirely or almost entirely devoid of alkaloids, e.g. the subtribe Anibinae of the Lauraceae, the tribe Sarumeae of the Aristolochiaceae, the subfamily Podophylloidae of the Berberidaceae. Instead, there are metabolites arising through the deamination of phenylalanine: arylpyrones, cinnamamides, and neolignans in *Aniba* and *Piper*, and lignans in *Asarum* and *Podophyllum* (Rezende et al. 1975). More recent research has shown that in addition to the arylpyrones and neolignans, whose presence indicates an early divergence from the other subtribes of the Lauraceae, *Aniba* species also contain simple benzyltetrahydroisoquinolines. Thus, within the Magnoliiflorae there appears to be a diversification in one direction towards benzylisoquinoline alkaloids (Ranunculales and Papaverales) and in another towards cinnamic acid-derived metabolites, phenylpropanoids, this being the more advanced character (Ferreira et al. 1980).

6 Conclusions and Summary

The theme running through the present paper is that isoquinoline alkaloids have been both a boon and a curse to mankind. The cultivated opium poppy, *Papaver somniferum,* which appears to have originated several thousand years ago in the western Mediterranean region, has given rise to serious and large-scale social problems, but at the same time its dried juice, opium, has provided no less than five alkaloids – morphine, codeine, thebaine, papaverine, and noscapine (narcotine) – that are currently used in medicine.

The case of the prickly poppy, *Argemone mexicana,* with its involvement in epidemic dropsy and glaucoma, has few redeeming features, but has nevertheless brought awareness and understanding of related problems in other parts of the world.

Curare continues to serve as a model for the development of improved muscle-relaxants and is still recognizable in the newer generation of products such as atracurium and vecuronium.

The final brief survey of the evolution and occurrence of benzylisoquinoline alkaloids completes the circle and provides a basis for confronting the problem of where in the plant kingdom to search for new and potentially useful derivatives in the endeavour to separate the boon from the curse.

Acknowledgment. I am indebted to Prof. J.D. Phillipson, Department of Pharmacognosy, The School of Pharmacy, University of London, for helpful discussion during the preparation of this paper.

References

Barlow RB (1964) Introduction to chemical pharmacology, 2nd edn. Methuen, London; John Wiley, New York, chap 5, pp 87–139

Barneby RC, Krukoff BA (1971) Supplementary notes on American Menispermaceae. VIII. A generic survey of American Triclisieae and Anomospermeae. Mem NY Bot Gard 22(2): 1–89, 7–15

Basmadjian KJ (1938) L'identification des noms des plantes du *Codex Constantinopolitanus* de Dioscoride. J Asiat 230: 577–621, 587 no. 46

Bowman WC (1980) A new non-depolarizing neuromuscular blocking drug. Trends Pharmacol Sci 1: 263–266

Bowman WC, Rand MJ (1980) Textbook of pharmacology, 2nd edn. Blackwell, Oxford, pp 16 1–3, 7–8, 17 33–49

Bui TY(1974) (Chemical study of the seeds of *Argemone mexicana* L. cultivated in the USSR and indigenous to Vietnam). Farmatsiya (Moscow) (2): 36–38

Cava MP, Buck KT, daRocha AI (1972) Azafluoranthene alkaloids. A new structural type. J Am Chem Soc 94: 5931

Cava MP, Buck KT, Noguchi I, Srinivasan M, Rao MG (1975) The alkaloids of *Abuta imene* and *Abuta rufescens*. Tetrahedron 31: 1667–1669

Coenen H (1954) Über das Jahr der Morphiumentdeckung Sertürners in Paderborn. Arch Pharm 287: 165–180

Cronquist A (1977) On the taxonomic significance of secondary metabolites in angiosperms. In: Kubitzki K (ed) Flowering plants. Evolution and classification of higher categories. Springer, Berlin Heidelberg New York, pp 179–189

Cronquist A (1980) Chemistry in plant taxonomy: an assessment of where we stand. In: Bisby FA, Vaughan JG, Wright CA (eds) Chemosystematics. Principles and practice. Academic Press, London New York, pp 1–27

Cronquist A (1981) An integrated system of classification of flowering plants. Columbia Univ Press, New York, pp 20–150

Dahlgren RMT (1980) A revised system of classification of the angiosperms. Bot J Linn Soc 80; 91–124

Dahlgren RMT, Clifford HT (1982) The monocotyledons. A comparative study. Botanical systematics, vol II. Academic Press, London New York

Danert S (1958) Zur Systematik von *Papaver somniferum* L. Kulturpflanze 6: 61–88

Derosne (JF) (1803) Mémoire sur l'opium. Ann Chim 45: 257–285

Dubler CE (1955) La 'Materia Médica' de Dioscórides. Transmisión medieval y renacentista, vol III. La 'Materia Médica' de Dioscórides traducida y comentada por D.Andrés de Laguna (Texto Critíco). Barcelona, p 253, lib II, cap 168

Dutcher JD (1946) Curare alkaloids from *Chondodendron tomentosum* Ruiz and Pavón. J Am Chem Soc 68: 419–424

Dyke S (1978) Poppies and glaucoma. New Sci 79: 679–680

Ferreira ZS, Gottlieb OR, Roque NF (1980) Chemosystematic implications of benzyltetrahydroisoquinolines in *Aniba*. Biochem System Ecol 8: 51–54

Fritsch R (1979) Zur Samenmorphologie des Kulturmohns (*Papaver somniferum* L.). Kulturpflanze 27: 217–227

Gabra S (1956) Papaver species and opium through the ages. Bull Inst Egypte 37 (1): 39–56

Galeffi C, Marini-Bettòlo GB, Vecchi D (1975) (*R,R*)(–,–)-12′-*O*-Methylcurine and (*R,R*) (–,–)-*O,O*-dimethylcurine, two new natural alkaloids from *Guatteria megalophylla* Diels. Gazz Chim Ital 105: 1207–1213

Galeffi C, Scarpetti P, Marini-Bettòlo GB (1977) New curare alkaloids. II. New bisbenzylisoquinoline alkaloids from *Abuta grisebachii* (Menispermaceae). Farmaco Ed Sci 32: 853–865

Galeffi C, La Bua R, Messana I, Alcazar RZ, Marini-Bettolo GB (1978) The alkaloids of *Sciadotenia toxifera* Krukoff and A.C. Smith. Gazz Chim Ital 108: 97–100

Germer R (1979) Untersuchung über Arzneimittelpflanzen im Alten Ägypten. Dissertation, Univ Hamburg, pp 326–328

Glick MD, Cook RE, Cava MP, Srinivasan M, Kunitomo J, DaRocha AI (1969) Imenine, a ring-B substituted aporphine alkaloid. J Chem Soc Chem Commun 1217–1218

Gomes CMR, Gottlieb OR (1980) Alkaloid evolution and Angiosperm systematics. Biochem Syst Ecol 8: 81–87

Gottlieb OR (1982) Micromolecular evolution, systematics and ecology, chap 5. Springer, Berlin Heidelberg New York, pp 35–54

Gunther RT (1933) The Greek herbal of Dioscorides. Englished by John Goodyer, AD 1655. Edited and first presented. Harper, New York, p 221, book II, no 208

Haisová K, Slavík J (1975) On the minor alkaloids from *Argemone mexicana* L. Coll Czech Chem Commun 40: 1576–1578

Hakim SAE (1954) Argemone oil, sanguinarine, and epidemic-dropsy glaucoma. Br J Ophthalmol 38: 193–216

Hakim SAE, Mijović V, Walker J (1961a) Experimental transmission of sanguinarine in milk: detection of a metabolic product. Nature (London) 189: 201–204

Hakim SAE, Mijović V, Walker J (1961b) Distribution of certain poppy-fumaria alkaloids and a possible link with the incidence of glaucoma. Nature (London) 189: 198–201

Hammer K (1981) Problems of *Papaver somniferum* classification and some remarks on recently collected European poppy land-races. Kulturpflanze 29: 287–296

Hammer K, Fritsch R (1977) Zur Frage nach der Ursprungsart des Kulturmohns *(Papaver somniferum* L.). Kulturpflanze 25: 113–124

Heywood VH (1973) Chemosystematics – an artificial discipline. In: Bendz G, Santesson J (eds) Chemistry in botanical classification. Nobel Symp (Med Nat Sci) 25: 17–29

Hughes R, Chapple DJ (1980) Experimental studies with atracurium, a new neuromuscular blocking agent. Br J Anaesth 52: 238P

Hunt TM, Hughes P, Payne JP (1980) Preliminary studies with atracurium in anaesthetised man. Br J Anaesth 52: 238P-239P

Kaul A, Ahuja MM (1980) An extensive survey for argemone contamination in rape and mustard oil seed and oil. Indian Chem J 15(1): 36–38. Cited from: Chem Abstr 93: 166344 (1980)

King H (1935) Curare alkaloids. Part I. Tubocurarine. J Chem Soc 1381–1389

King H (1947) Curare alkaloids. Part VI. Alkaloids from *Chondrodendron tomentosum* R. and P. J Chem Soc 936–937

King H (1948) Curare alkaloids. Part VII. Constitution of *dextro*-tubocurarine chloride. J Chem Soc 265–266

Kostermans AJ, Pinkley HV, Stern WL (1969) A new Amazonian arrow poison: *Ocotea venenosa.* Bot Mus Leafl Harv Univ 22: 241–252

Krikorian AD (1975) Were the opium poppy and opium known in the ancient Near East? J Hist Biol 8: 95–114

Kritikos PG, Papadaki SP (1967) The history of the poppy and of opium and their expansion in antiquity in the eastern Mediterranean area. Bull Narcot 19(3): 17–38, (4): 5–10

Lambert JJ, Durant NN, Henderson EG (1983) Drug-induced modification of ionic conductance at the neuromuscular junction. Annu Rev Pharmacol Physiol 23: 505–539

Lieb WA, Scherf HJ (1956) Papaveraceae alkaloids and eye pressure. Klin Monatsbl Augenheilkd 128: 686–705. Cited from: Chem Abstr 54: 11271 (1960)

Linnaeus C (1753) Species Plantarum. Impensis Laurentii Salvii, Holmiae, vol I, pp 508–509

Lockemann G (1924) Friedrich Wilhelm Sertürner. Ein Beitrag zu seiner wissenschaftlichen Würdigung. Z Angew Chem 37: 525–532

Lockemann G (1951) Friedrich Wilhelm Sertürner, the discoverer of morphine (translated by RE Oesper). J Chem Educ 28: 277–279

Malik CP, Mary TN, Grover IS (1979) Cytogenetic studies in *Papaver.* V. Cytogenetic studies on *P.somniferum* x *P.setigerum* hybrids and amphiploids. Cytologia 44: 59–69

Manson-Bahr PH (ed) (1946) Manson's tropical diseases, 12th edn. Cassell, London, pp 787–789

Meissner W (1819) Ueber ein neues Pflanzenalkali *(Alkaloid).* J Chem u Phys 25: 379*–381 (* incorrectly numbered 377 in the original)

Menachery MD, Cava MP (1980) Grandirubrine, a new tropoloisoquinoline alkaloid. Heterocycles 14: 943–945

Menachery MD, Cava MP (1981) The alkaloids of *Telitoxicum peruvianum.* J Nat Prod 44: 320–323

Merrillees RS (1962) Opium trade in the Bronze Age Levant. Antiquity 36: 287–292

Merrillees RS (1968) The Cypriote Bronze Age pottery found in Egypt. Stud Mediter Archaeol 18: 154–157, 176, 179, 196, pl XXXV. 1 and 2, XXXVI. 1

Merrillees RS (1979) Opium again in antiquity. Levant 11: 167–171

Moody DP (1965) Chemotherapeutic consequences of culture collisions. Proc R Anthrop Inst 33–45

Murav'eva DA, Bui TY (1974) Study of alkaloid composition of *Argemone mexicana.* Aktual Vopr Farm 2: 24–26. Cited from: Chem Abstr 84: 147614 (1976)

Murthy SSN, der Marderosian A (1973) The isolation and identification of rodiasine from *Ocotea venenosa.* Lloydia 36: 440

Muzio I (1925) Su di un olio medicato della tomba di Cha. Atti Soc Ligustica Sci Lett [n.s.] 4: 249–253

Nadkarni AK (1954) KM Nadkarni's Indian materia medica, 3rd edn, vol I. Popular Book Depot, Bombay, pp 133–136

Reynolds CD, Palmer RA, Gorinsky BA, Gorinsky C (1975) X-ray structure of the curare alkaloid (+)-tubocurarine dibromide. Biochem Biophys Acta 404: 341–344

Reynolds JEF (ed) (1982) Martindale. The Extra Pharmacopoeia, 28th edn, Pharmaceutical Press, London, pp 1004, 1018, 1059, 1263

Rezende CMA da M, Gottlieb OR, Marx MC (1975) Benzyltretrahydroisoquinoline alkaloids as systematic markers. Biochem Syst Ecol 3: 63–70

Riddle JM (1971) Dioscorides. In: Gillispie CC (ed) Dictionary of scientific biography, vol IV. Scribner's Sons, New York, pp 119–123

Rukmini C (1971) Sanguinarine potentiating factor in Argemone oil. Indian J Med Res 59: 1676–1680

Rukmini C (1975) New, unusual long chain fatty acid (argemonic acid) from *Argemone mexicana*. J Am Oil Chem Soc 52: 171–173

Sarkar SN (1948) Isolation from argemone oil of dihydrosanguinarine and sanguinarine: toxicity of sanguinarine. Nature (London) 162: 265–266

Schiaparelli E (1927) La tomba intatta dell'architetto *Cha* nella necropoli di Tebe. In: Relazione sui lavori della missione archeologica italiana in Egitto (anni 1903–1920), vol II. R Mus Antichità, Torino, pp 154–158

Schultze-Motel J (1979) Die urgeschichtlichen Reste des Schlafmohns (*Papaver somniferum* L.) und die Entstehung der Art. Kulturpflanze 27: 207–215

Sertürner (FWAF) (1806) Darstellung der reinen Mohnsäure (Opiumsäure) nebst einer chemischen Untersuchung des Opiums mit vorzüglicher Hinsicht auf einen darin neu entdeckten Stoff und die dahin gehörigen Bemerkungen. J Pharm Aerzte Apotheker Chem 14/I: 47–93

Sertürner (FWAF) (1817) Ueber das Morphium, eine neue salzfähige Grundlage, und die Mekonsäure, als Hauptbestandtheile des Opiums. Ann Phys [n.s.] 25: 56–90

Shenolikar IS, Rukmini C, Krishnamachari KAVR, Satyanarayana K (1974) Sanguinarine in the blood and urine of cases of epidemic dropsy. Food Cosmet Toxicol 12: 699–702

Singh TB, Chunekar KC (1972) Glossary of vegetable drugs in *Bṛhattrayī*. Chowkhamba Sanskrit Series Office, Varanasi, pp 461–462, cf. 64, 71, 90, 127, 441

Skiles JW, Saá JM, Cava MP (1979) Splendidine, a new oxoaporphine alkaloid from *Abuta rufescens* Aublet. Can J Chem 57: 1642–1646

Snedden W, Parker RB, Gorinsky C (1970) Electron-impact studies in medicine and biochemistry – II: The mass spectra of the alkaloids from *Cissampelos ovalifolia* D.C. Org Mass Spectr 4: Suppl 607–614

Stenlake JB (1982) Atracurium: A contribution to anaesthetic practice. Pharm J 229: 116–120

Symposium [on Org NC 45] (1980) Br J Anaesth 52: Supplement 1, 1S–72S

Takahashi K, Cava MP (1976) Sciadoline, a new type of bis-benzylisoquinoline alkaloid. Heterocycles 5: 367–371

Takahashi K, Mitchell MJ, Cava MP (1976) Sciadenine, a new bis-benzylisoquinoline alkaloid from *Sciadotenia toxifera*. Heterocycles 4: 471–474

Takhtajan A (1973) The chemical approach to plant classification with special reference to the higher taxa of Magnoliophyta. In: Bendz G, Santesson J (eds) Chemistry in botanical classification. Nobel Symp (Med Nat Sci) 25: 17–29

Tournefort JP de (1694) Elemens de botanique. Imprimerie Royale, Paris, vol I, p 204; vol II, pl 121

Vellard J (1965) Histoire du curare. Gallimard, Paris, pp 7, 17–34

Vellard (J) (1973) Les curares indiens. Leur préparation, leurs variations, et leur mode d'action. Anesth Analg Réanim 30: 237–245

Vesselovskaya MA (1975) (Poppy (variability, classification, evolution)). Tr Prikl Bot Genet Sel 55(1): 175–223

Wintersteiner O, Dutcher JD (1943) Curare alkaloids from *Chondodendron tomentosum*. Science 97: 467–470

Chemotaxonomy of the Papaveraceae Alkaloids

V. PREININGER[1]

1 Introduction

The classification of plant taxa has mostly been carried out on the basis of morphological characteristics, and in practice such characteristics are of great importance for the determination of individual species. In recent years, chemotaxonomy has become significant in systematic botanical studies and both botanists and chemists have found many correlations between chemical characteristics and botanical features. A summarising survey which brings together the classical divisions of plant taxa and a knowledge of secondary metabolites has been produced in a series of volumes (Hegnauer 1962–1973).

Phytochemical studies of plants of the family Papaveraceae have been directed primarily towards the alkaloids which have been investigated intensively during the past 25 years (Saxton 1971–1975, Grundon 1976–1983, Šantavý 1970 and 1979, Shamma 1972, Shamma and Moniot 1978, Preininger 1975). The numerous alkaloids obtained represent a rich source of potential medicinal compounds on the one hand, while on the other hand, they may be used for furthering our knowledge of plant systematics within the Papaveraceae.

The taxonomic delineation of the family Papaveraceae has not been without its problems. Fedde (1909) classified the fumariaceous plants within the Papaveraceae, as the subfamily Fumarioideae, whereas Hutchinson (1921) considered that they were an independent family, the Fumariaceae. The occurrence of isoquinoline alkaloids in many of the species indicates a close relationship between the Papaveraceae and the Fumariaceae and points to relationships with the Polycarpicae and with the orders Ranales and Magnoliales (Hegnauer 1961). Traditionally, the order Rhoeadales included the Papaveraceae and the Fumariaceae, together with the Brassicaceae, Capparaceae, Moringaceae, Resedaceae and Tovariaceae. More recently, the order Rhoeadales has been replaced by the Papaverales and the Capparales; the Papaverales includes Papaveraceae and Fumariaceae, while the remaining five families are grouped into the Capparales. Species from families within the Capparales contain glucosinolates rather than alkaloids whereas species from the Papaverales contain alkaloids and not glucosinolates. These clear chemical distinctions closely follow the systematics of Hutchinson (1959) and of Takhtajan (1959), which were based on botanical characteristics.

1 Institute of Chemistry, Medical Faculty, Palacký University, Olomouc, Czechoslovakia

The Chemistry and Biology of Isoquinoline
Alkaloids, ed. by Phillipson et al.

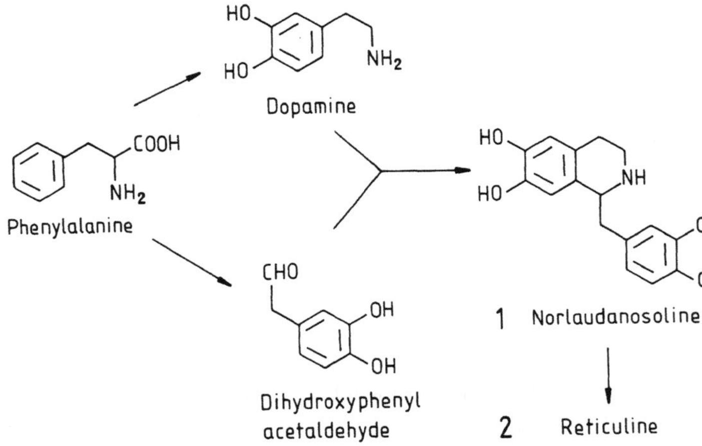

Fig. 1. First steps in the biosynthesis of 1-benzyltetrahydroisoquinoline alkaloids

The family Papaveraceae Juss., which is divided into subfamilies Hypecoideae Prantl et Kündig and Papaveroideae A. Br., embraces some 23 genera and 430 species, while the family Fumariaceae DC. includes 7 genera and 350 species. Literature exists on the alkaloids of about 165 species of the Papaveraceae and 77 species of the Fumariaceae (Preininger 1984). A characteristic feature of these two families is the presence of tetrahydroisoquinoline alkaloids and their derivatives. The general pathways to these alkaloids arises from phenylalanine by condensation of dopamine with 3,4-dihydroxyphenylacetaldehyde and subsequent Mannich condensation to the product (Fig. 1). Other intermediates formed during the biosynthesis of more complicated constitutional types of isoquinoline alkaloids are R-(–) or S-(+)-reticuline (2) and orientaline. In a few isolated cases, other types of alkaloids, including lupinane and indoles, have been reported from species of the Papaveraceae. The alkaloids of the Papaveraceae can be divided into a series of constitutional types according to their basic skeleta (Fig. 2) i.e. simple isoquinolines, benzylisoquinolines, pavines and isopavines, cularines, proaporphines, aporphines, promorphinans, morphinans, protoberberines, and retroprotoberberines, secoberbines, benzophenanthridines, protopines, phthalideisoquinolines, secophthalideisoquinolines, indenobenzazepines, spirobenzylisoquinolines and rhoeadines.

For chemotaxonomic considerations, it is necessary to know about the combinations of certain alkaloidal types and about their biosynthetic inter-relationships. The types of alkaloid present may be used as characters for assessing relationships of taxa at the levels of subfamily, tribe, genus, species, subspecies or chemical race. It is not always possible to make any significant chemotaxonomic conclusions where data is missing from the literature, e.g. the yields of individual alkaloids, the nature of the minor alkaloids present and the plant part analysed. It is also important to know about the locality of growth and the season of collection of the plant material.

The most widely spread alkaloids of the families Papaveraceae and Fumariaceae are protopine (3), magnoflorine (4), quaternary protoberberines (5) and quaternary

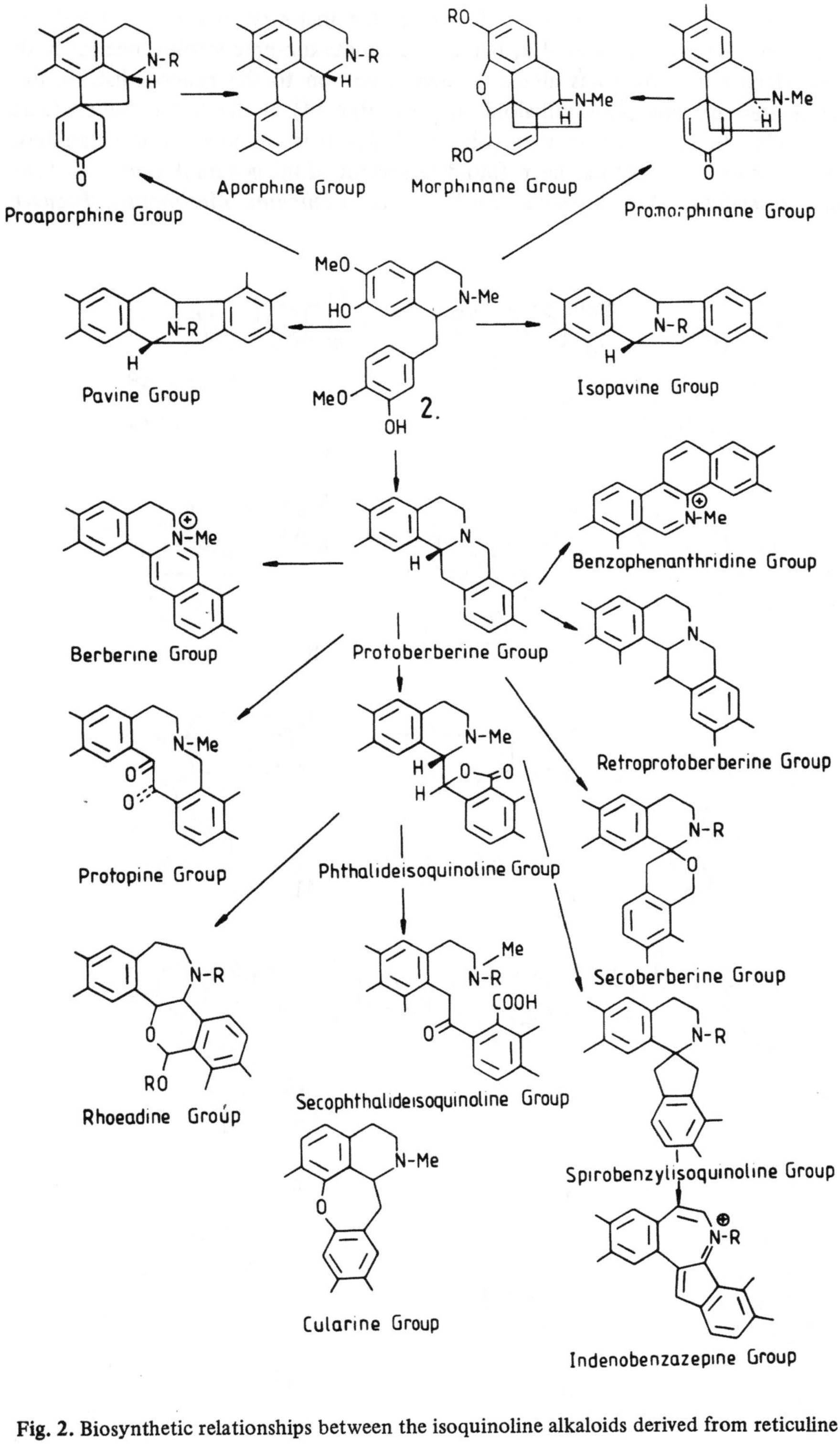

Fig. 2. Biosynthetic relationships between the isoquinoline alkaloids derived from reticuline (2)

enzophenanthridines (6) which are found in the majority of genera. Therefore, ıese alkaloids can be considered as being of chemotaxonomic significance for both amilies. Hegnauer (1961) has already drawn attention to the general distribution f the quaternary aporphine alkaloid magnoflorine (4) in the order Polycarpicae nd according to more recent work (Slavík 1980a), it is common to both families. 'o date, magnoflorine (4) has been found in species of the genera *Argemone, Cheli- 'onium, Corydalis, Dicranostigma, Eschscholzia, Glaucium, Meconopsis, Papaver*

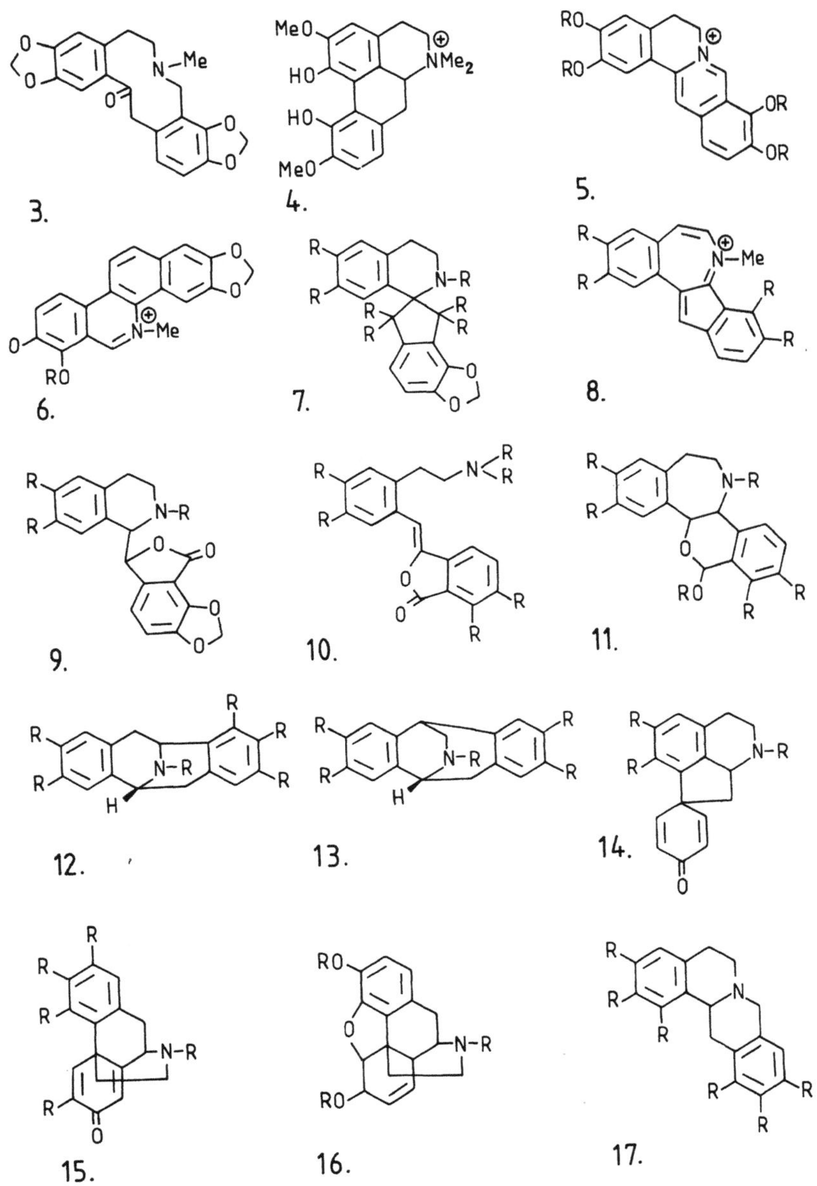

Fig. 3

and *Pteridophyllum* and in some species it occurs as the dominant alkaloid. In contrast, some alkaloids are limited in their distribution to only one of the two families. The spirobenzylisoquinoline (7) and indenobenzazepine (8) alkaloids are found only in the Fumariaceae while the phthalideisoquinoline (9) and secophthalideisoquinoline (10) alkaloids are mainly, but not exclusively, found in the same family. The rhoeadane (11) alkaloids and their N-demethyl-derivatives, the papaverrubines, the pavine (12), isopavine (13), proaporphine (14), promorphinan (15), morphinan (16) and retroprotoberberine (17) alkaloids are found only in the Papaveraceae.

2 Chemotaxonomic Evaluation of the Family Papaveraceae Juss.

Most of the chemotaxonomic findings have been made regarding the genera *Papaver* (Tourn. ex L.), L., *Eschscholzia* Cham., *Bocconia* (Plum. ex L.) L., *Glaucium* (Tourn. ex Haller) Adans., *Meconopsis* Viguier, *Argemone* (Tourn. ex L.) L. and *Hypecoum* (Tourn. ex L.) L.

2.1 The Genus Papaver

Fedde (1909) divided the genus *Papaver* into nine sections: Orthorhoeades Fedde, Argemonorhoeades Fedde, Carinatae Fedde, Mecones Bernh., Miltantha Bernh., Pilosa Prantl., Macrantha Elk., Scapiflora Reichb. and Horrida Elk. This classification is still in use today and forms the basis for this genus; however, in recent years, chemotaxonomic studies have given sufficient impetus to warrant a newer classification (Günther 1975, Preininger et al. 1981) (Table 1).

Table 1. Survey of the sections of the genus *Papaver* L.

Fedde (1909)	Present state (Günther 1975, Preininger et al. 1981)
Orthorhoeades Fedde	*Rhoeadium* Spach.
Argemonorhoeades Fedde	*Argemonidium* Spach.
Carinatae Fedde	*Carinatae* Fedde
Mecones Bernh.	*Papaver*
	Glauca J. Nov. et V. Prein.
Miltantha Bernh.	*Miltantha* Bernh.
Pilosa Prantl.	*Pilosa* Prantl.
	Pseudo-pilosa M. Pop.
Macrantha Elk.	*Macrantha* Elk.
Scapiflora Reich.	*Meconella* Spach.
Horrida Elk	*Horrida* Elk.
	Roemeria (Medic.) Günther

2.1.1. Section Orthorhoeades

The major alkaloids of the section Orthorhoeades are rhoeadine (18) and protopine (3) while isorhoadine (19) and the papaverrubines are minor alkaloids. These alkaloids can be considered as chemotaxonomic characters for the whole section. Some differences have been noted in the alkaloids present in *P. dubium* L. and its varieties and in *P. commutatum* Fisch. et Mey. Some authors have found that *P. dubium* contains either rhoeadine alkaloids or proaporphine with aporphine alkaloids, or all of these alkaloid types occurring together. *P. commutatum* has been shown to contain either alkaloids of the rhoeadine type or aporphines with papaverine or all of these alkaloids together. If we exclude the possibility of erroneous botanical determination, then these different findings can be explained by the fact that during development, these plants are able to elaborate (or to accumulate) several types of alkaloid. Thus, some plants may have been collected at a time when the alkaloids of one type prevailed, while others were collected at a time when alkaloids of another type predominated. There is also the distinct possibility that these species exist in different chemical races. The finding of the alkaloid-glycoside latericine (20) in *P. californicum* and of the secophthalideisoquinoline alkaloid adlumidiceine (21) in *P. rhoeas* is of interest because latericine is characteristic of plants in the section Pilosa and adlumidiceine of the genera *Fumaria* and *Corydalis*.

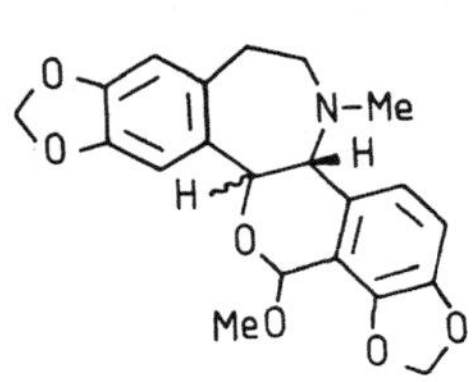

18. C-1H beta
19. C-1H alpha

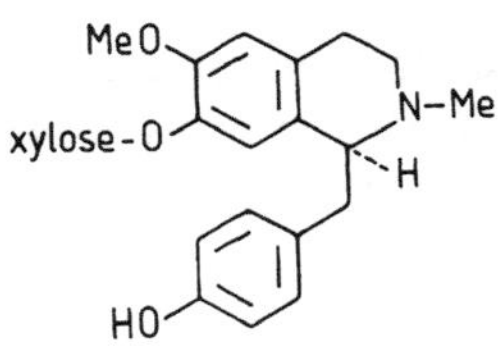

20.

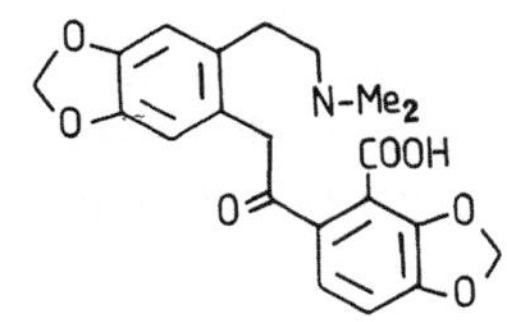

21.

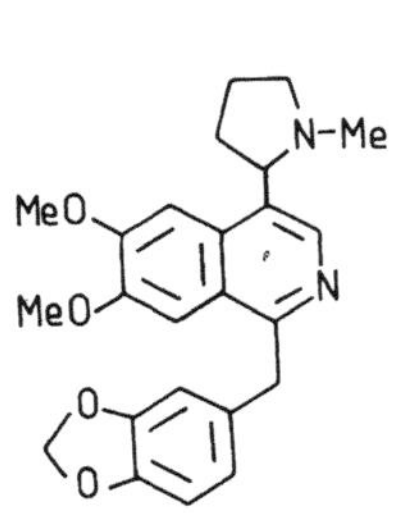

22.

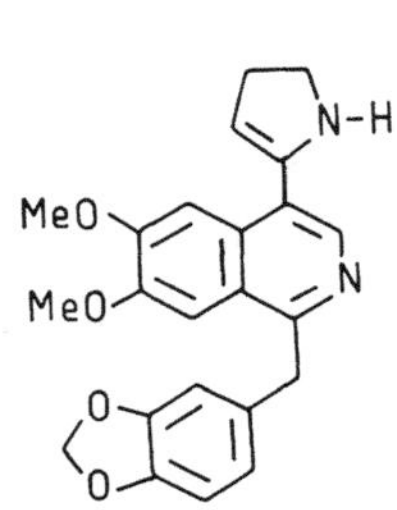

23.

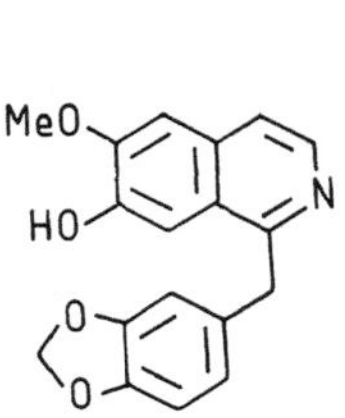

24.

Fig. 4

2.1.2 Section Argemonorhoeades

Studies of plants from the section Argemonorhoeades show that these plants contain only a very few alkaloids including those of the protoberberine (5), rhoeadine (11), papaverrubine and protopine (3) types. These alkaloids are similar to those found in the section Orthorhoeades but they occur in markedly lower yields.

2.1.3 Section Carinatae

Of the seven species classified by Fedde (1909) to this section, only *P. macrostomum* Boiss. et Huet. has been studied for its alkaloid content. Plants cultivated in Central Europe (ČSSR, GDR), were found to contain rhoeadine (18), papaverrubine and protopine (3), alkaloids, similar to the species from the sections Orthorhoeades and Argemonorhoeades. Plants collected in the vicinity of Lake Sevan (Armenia) contained the l-benzylisoquinoline alkaloid macrostomine (22) as the major alkaloid, together with dehydronormacrostomine (23) and sevanine (24). We did not dectect rhoeadine- or papaverrubine-type alkaloids in this plant material. These differences in alkaloid content may be explained by the fact that *P. macrostomum* from Armenia was growing in its natural habitat whereas the plants obtained from C. Europe were cultivated artificially. Therefore, it is proposed that alkaloids of the macrostomine type are chemotaxonomic characters for the section Carinatae. Macrostomine has also been found in *P. arenarium* of the section Orthorhoeades, thus demonstrating that there is a chemical link between these two sections.

2.1.4 Sections Papaver (syn. Mecones Bernh.) and Glauca

Systematic investigation of the alkaloids of the Papaveraceae has shown that the section Papaver (syn. Mecones Bernh.) is rather heterogeneous in its chemical composition. The presence of the morphinan alkaloids, thebaine (25), codeine (26) and morphine (27), together with the secophthalideisoquinoline alkaloids narceine, nornaceine and narceine imide and the phthalideisoquinoline alkaloids narcotine and narcotoline, is characteristic of *P. somniferum* L. and *P. setigerum* DC. In contrast, rhoeadine and papaverrubine-type alkaloids predominate in *P. glaucum* Boiss. et Hausskn., *P. gracile* Auch. and *P. decaisnei* Hoechst et Steud. Traces of the rhoeadine-type alkaloids are also found in *P. somniferum* and *P. setigerum* whereas morphinan alkaloids have not been found in *P. glaucum* and *P. gracile.* There has been only one report of morphine from a species other than *P. somniferum* or *P. setigerum* and this has been as a minor alkaloid from the poppy heads of *P. decaisnei* (Slavík 1980b). The same plants contained papaverine as the major alkaloid and also rhoeadine as a minor alkaloid. Thebaine (25) is found in a whole series of plants from the sections Orthorhoeades, Pilosa, Miltantha and Macrantha but morphine (27) and codeine (26) appear to be unique to *P. somniferum* and *P. setigerum.* The phenolic *ortho-para* oxidation leading to the formation of thebaine appears to be a common phenomenon whereas the specific ability to demethylate the methoxyl groups in order to produce morphine or codeine appears to be a rare characteristic (Fig. 5).

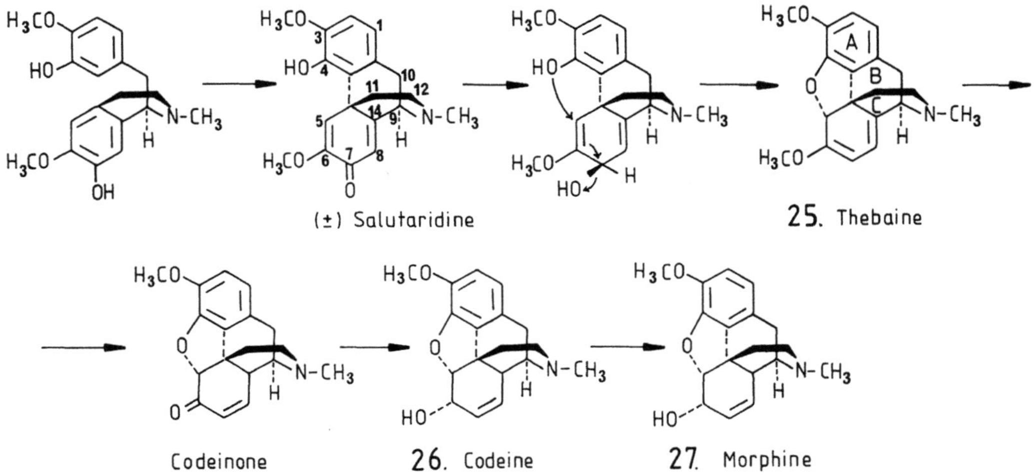

Fig. 5. Biosynthesis of the morphinan alkaloids thebaine (25), codeine (26) and morphine (27)

On the basis of these findings, it has been proposed that the original section Papaver (syn. Mecones) be divided into the section Papaver, to which *P. somniferum* and *P. setigerum* are categorized, and a new section Glauca J. Novák et V. Preininger (1980) which includes *P. glaucum, P. gracile* and *P. decaisnei.* This conclusion was substantiated by studies of the morphological features and their correlation with the basic chromosome numbers (Novák and Preininger 1980). Species of the section Papaver have a chromosome number of n = 11, whereas for species of the section Glauca, n = 7. In *P. somniferum* and *P. setigerum* there mostly occur diploid individuals (2n = 22) and very rarely tetraploid individuals (4n = 44). All species of the section Glauca are diploid. Plants from the new section Glauca differ in their habitat from species of the section Orthorhoeades, but even so, some authors (Stermitz 1968) consider that these plants are very closely related and certainly they do contain very similar alkaloids.

2.1.5 Section Miltantha

Detailed studies have been made on the alkaloids of *P. armeniacum* (L.) DC., *P. caucasicum* Marsch. Bieb., *P. fugax* Poir., *P. persicum* Lindl., *P. floribundum* Desf., *P. polychaetum* Schott et Kotschy and *P. triniaefolium* Boiss., cultivated from seeds obtained from different botanical gardens in Czechoslovakia and GDR. As a result of this work, we expressed the opinion that the chemotaxonomically significant alkaloids of the section are mecambrine (28), armepavine (29) and protopine (3) (Preininger and Šantavý 1970). The alkaloid armepavine has not been found in any other section of the genus Papaver or in other genera of the Papaveraceae or Fumariaceae. Studies carried out by Phillipson et al. (1981a) on the alkaloids of plants from the section Miltantha, which had been obtained from their natural habitats in Turkey, showed that some of the species are able to produce thebaine whereas some species produce rhoeadine alkaloids. *P. fugax* plants, which were cultivated in England from seeds collected in Turkey, continued to produce the same major alkaloid, thebaine (Sariyar and Phillipson 1980, Phillipson et al. 1981a). Other collections of *P. fugax* from Eastern Turkey yielded glaudine, glaucamine and rhoeadine while the latter alkaloid was also shown to be the major alkaloid of *P. armeniacum* collected in Turkey. Similarly, it was shown that the major alkaloids of different collections of *P. tauricolum* from Turkey were of the rhoeadine type. It is interesting to note the similarity between the alkaloids found in this plant material (i.e. rhoeadine, rhoeagenine, glaudine, sinactine) and those found in *P. rhoeas.* These are the first findings of rhoeadine type alkaloids, as major alkaloids, in the section Miltantha, although traces of rhoeadine have been demonstrated to be present in *P. caucasicum* (Preininger et al. 1967).

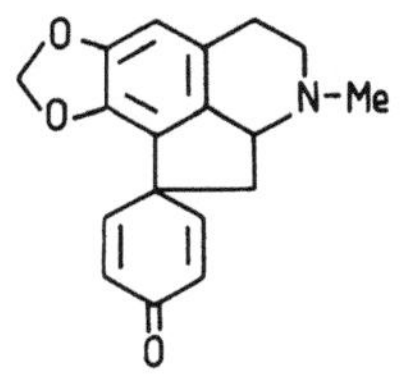

28.

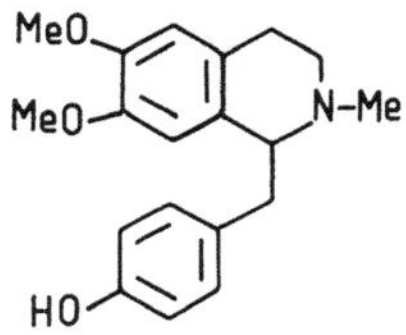

29.

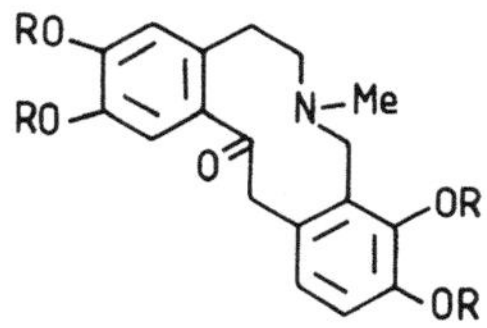

30.

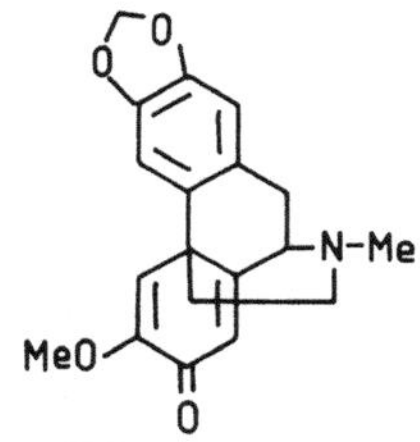

31.

Fig. 6

2.1.6 Section Pilosa

All of the species of the section Pilosa which have been investigated, contain rhoeadine- and papaverrubine-type alkaloids. Also noteworthy in species of this section are the alkaloids 13-oxoprotopine (30) and the glycoside latericine (20). Four species contain aporphine alkaloids and six species contain promorphinan and morphinan (thebaine) alkaloids. Böhm and Günther (1972) have indicated that the presence or absence of amurine (31) can be used in order to distinguish between *P. pilosum, P. atlanticum* and *P. lateritium.* Furthermore, it was noted that amurine was present in six species of section Pilosa, namely *P. oreophyllum, P. feddei, P. heldreichii, P. pilosum, P. strictum* and *P. spicatum* whereas it was absent from four other species, *P. rupifragum, P. atlanticum, P. lateritium* and *P. monanthum* (Preininger 1984). We believe that these conclusions, together with studies on the morphological features and chromosome studies, indicate the possibility of a subdivision within section Pilosa (Table 1).

2.1.7 Section Macrantha

Plants of the section Macrantha have been intensively investigated and particularly during the past 15 years. This was due, in part, to the finding that some of the varieties of *P. bracteatum* produce considerable yields of thebaine which is a poten-

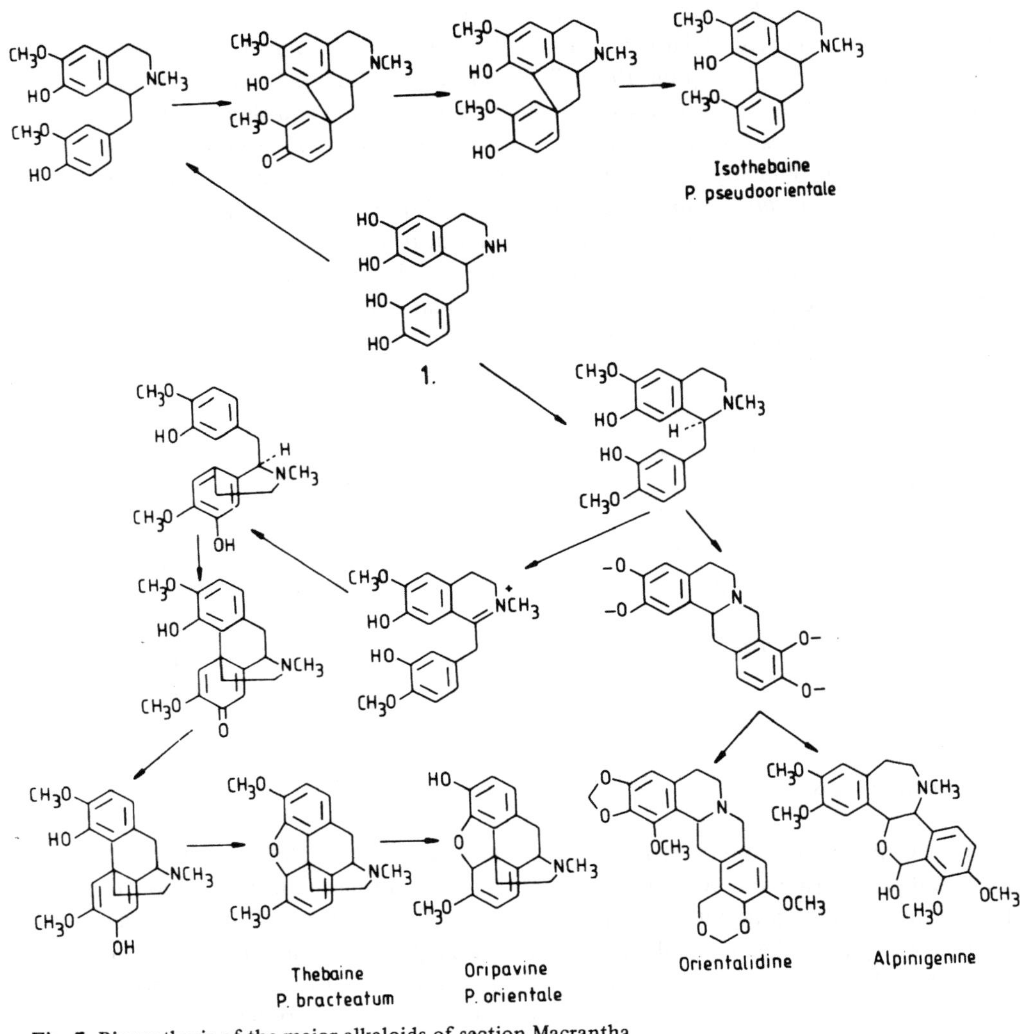

Fig. 7. Biosynthesis of the major alkaloids of section Macrantha

tial source of codeine. The biosynthetic relationships between the major alkaloids of section Macrantha are summarised in Fig. 7. *P. bracteatum* Halle III (Böhm 1967) and *P. bracteatum* of Iranian origin were found to contain thebaine as the major alkaloid and only small quantities of alpinigenine, orientalidine and isothebaine. In *P. bracteatum* of Turkish origin, the major alkaloids were thebaine and salutaridine (Sariyar 1975). It has been assumed that variety Halle III exists in two different races, one producing only thebaine and the other producing thebaine and alpinigenine. Analyses of well-defined *P. orientale* and *P. pseudo-orientale,* from natural localities, have shown that the dried latex of *P. orientale* contained 20% of oripavine and 9% of thebaine (Shafiee et al. 1975). Traces of isothebaine were found in plants collected from Khalkhala in north western Iran. Five different chemotypes of *P. orientale* have been reported but in each of them oripavine was the major alkaloid

(Shafiee et al. 1977). The dry latex of *P. pseudo-orientale* has yielded isothebaine as the major alkaloid in addition to orientalidine (32) and other alkaloids. Turkish samples of *P. pseudo-orientale*, on the other hand, have been shown to contain the promorphinan alkaloid salutaridine and macrantaline (33) as major alkaloids, together with the closely related minor alkaloid macrantoridine (Sariyar and Phillipson 1977). Thebaine and oripavine have been isolated as the dominant alkaloids from hybrids of *P. bracteatum* and *P. orientale*, whereas isothebaine, with lesser quantities of oripavine were isolated from hybrids of *P. orientale* and *P. pseudo-orientale*. Morphine, thebaine and alpinigenine have been isolated from the dried capsules of hybrids of *P. bracteatum* and *P. somniferum* (Böhm 1967, 1981).

Studies of the chromosome numbers have shown that, in most of the species of the genus *Papaver*, and also in plants of the section Macrantha, the basic chromosome number is n = 7 (Novák 1979, Phillipson et al. 1981b). It appears that the three species of the section are evidently determined on the basis of the somatic numbers of chromosomes and the karyologic conditions confirm the elaborated classification of the section. It can be concluded, on the available evidence to date, that the major alkaloid of *P. bracteatum* is thebaine, whereas that of *P. orientale* is oripavine, although mecambridine is the major alkaloid of at least one collection (Phillipson et al. 1981b). *P. pseudo-orientale* mainly contains isothebaine as the major alkaloid, although salutaridine has been isolated as the major alkaloid from three separate col-

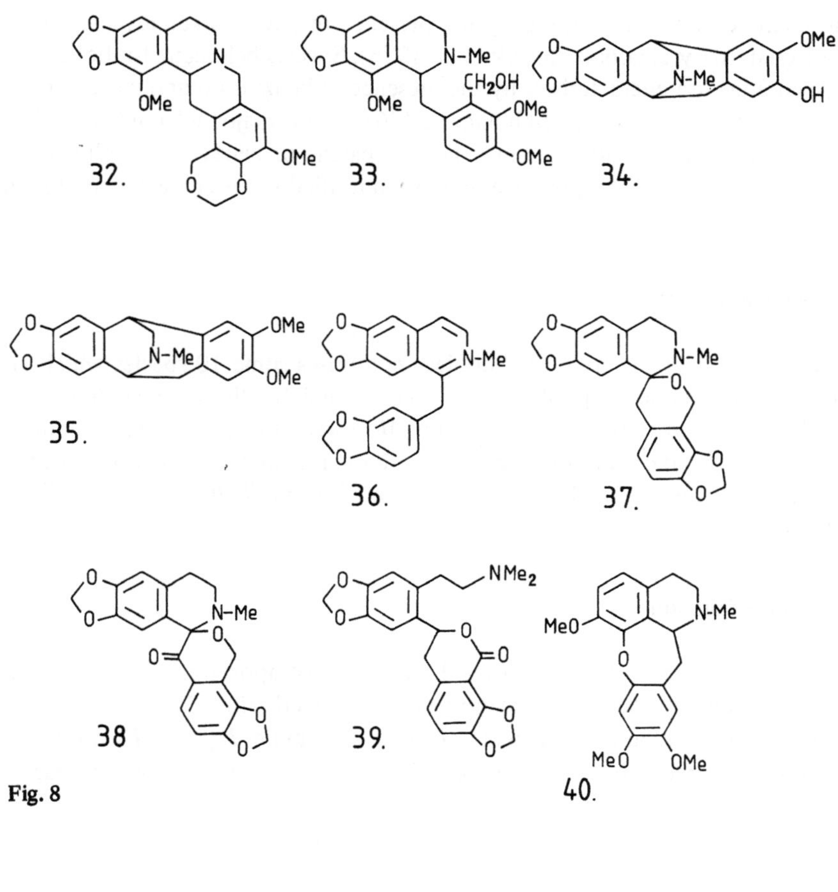

Fig. 8

lections (Phillipson et al. 1981b). Isothebaine and orientalidine do not appear to be present in other sections of the genus *Papaver* or in other genera of the Papaveraceae and Fumariaceae.

2.1.8 Section Scapiflora

The presence of the isopavine alkaloids amurensine (34) and amurensinine (35) can be considered to be of chemotaxonomic significance for the section Scapiflora. All the ssp. of *P. alpinum* contain rhoeadine-type alkaloids alpinine and alpinigenine, but rhoeadine itself has not been isolated. In contrast to these findings, rhoeadine is present in several of the varieties of *P. nudicaule.* The promorphinan alkaloid amurine (30) is present in all of the ssp. of *P. alpinum* and the majority of the varieties of *P. nudicaule.* Aporphine alkaloids have not yet been found in species of the section Scapiflora.

2.2 The Genus Eschscholzia

Fedde (1909) differentiated the genus *Eschscholzia* and its 123 species into two sections. This division is also consistent with the chemical characteristics of those species which have been investigated for their alkaloid content. *E. californica* Cham., *E. glauca* Greene and *E. douglasii* (Hook. et Arn.) Walp., which belong to the section Eurycraspedontae, are characterised by the presence of pavine alkaloids and a high content of quaternary alkaloids (Slavík et al. 1976). Species belonging to the section Stenocraspedontae are characterised by the presence of benzylisoquinoline alkaloids, e.g. *E. oregana* Greene contains escholamine (36) as the major alkaloid. *E. lobbii* Greene appears to be exceptional, since neither pavine or benzylisoquinoline alkaloids have been detected. The presence of aporphine alkaloids is characteristic for all species.

2.3 The Genus Bocconia

The genus *Bocconia* of the tribe Chelidonieae comprises some 10 species which are native to tropical and subtropical areas of Central and South America. *Bocconia,* which is closely related to the East Asian genus *Macleaya,* does not appear to contain alkaloids which are taxonomically significant although it may be noted that rhoeadine (18) and papaverrubine E have been isolated from *B. frutescens* L. (Slavík and Slavíková 1975).

2.4 The Genus Glaucium

Species of *Glaucium* are characterised by the presence of aporphine alkaloids in addition to the commonly occurring protopine (3), protoberberine (5) and benzophenanthridine (6) alkaloids. In these respects, *Glaucium* resembles *Papaver, Eschscholzia, Meconopsis* and *Roemeria* but it differs from the other genera of the Papaveraceae.

2.5 The Genus Meconopsis

Meconopsis comprises some 40 species which are mainly indigenous to China and the Himalayas. Protopine (3), protoberberine (5), benzophenanthridine (6), isopavine (13) and aporphine alkaloids (Fig. 2) have been isolated from a small number of species. Rhoeadine alkaloids, which are present in *Papaver* species, also occur in *Meconopsis* and their distribution appears to be limited to these two genera (Slavík and Slavíková 1977). The sole European representative, *M. cambrica* growing in England produces magnoflorine (4) as the major alkaloid, together with protoberberine (5), proaporphine (14) and promorphinan alkaloids (15) (Hemingway et al. 1981).

The promorphinan alkaloids isolated were (–)-amurine and (–)-flavinantine and not the (+)-isomers which have previously been isolated from other natural sources. It is not known whether this feature will prove to be of chemotaxonomic significance for the genus.

2.6 The Genus Argemone

Ten of the thirty species of *Argemone,* which have been studied to date, have yielded pavine (12) alkaloids and hence they resemble species of the genus *Eschscholzia.* These alkaloids have been obtained from only two genera of the family Papaveraceae (Stermitz 1968).

2.7 The Genus Hypecoum

Plants of the subfamily Hypecoideae differ chemotaxonomically from the plants of the subfamily Papaveroideae. Species of the genus *Hypecoum* contain secoberbine alkaloids (37–39) in addition to protopine (3), protoberberine (5) and benzophenanthridine (6) alkaloids. The presence of secoberbines (37–39) is indicative of a close relationship between *Hypecoum* and some species of Fumariaceae.

3 Chemotaxonomic Evaluation of the Family Fumariaceae DC

A close relationship exists between the alkaloids of the two families Papaveraceae and Fumariaceae. Their relationship, as well as their independence, is particularly well demonstrated by the external structure of the flowers and various anatomical characteristics of different organs (Novák and Preininger 1983). Characteristically, the Fumariaceae contain spirobenzylisoquinoline (7), indenobenzazepine (8), secophthalideisoquinoline (10) and phthalideisoquinoline (9) alkaloids, whereas rhoeadine (11), pavine (12) and isopavine (13) type alkaloids, which are characteristic of Papaveraceae, are absent. The secoberbines have been found in *C. incisa* (Thunb.) Pers. and in *C. ochotensis* Turcz. (37–39). The cularine alkaloids (40) which are characteristic of the genus *Corydalis* are also found in species of *Dicentra* which is a member of the Papaveraceae.

4 Conclusion

Although morphological characteristics are of practical importance for the determination of species, chemical characters can be used for establishing relationships between species, genera and families. The different isoquinoline alkaloids which are found in the genus *Papaver* have been applied as characters for assessing specific relationships within the genus. These chemical characters, which are independent of morphological and anatomical characteristics, have been discussed at specific and genera level within the families Papaveraceae and Fumariaceae. The variation in alkaloid content of many *Papaver* species clearly demonstrates that as many collections as possible should be analysed for their alkaloid content before chemotaxonomic conclusions can be made. Any proposals made on the basis of chemotaxonomic investigations should be related to morphological and anatomical characteristics and also to chromosome data.

References

Böhm H (1967) Über *Papaver bracteatum* Lindl. Planta Med 15:215–220

Böhm H (1981) *Papaver bracteatum* Lindl – Results and problem of the research on a potential medicinal plant. Planta Med 43:660–667

Böhm H, Günther KF (1972) Verteilung von Amurin innerhalb der *Papaver*-Sektion Pilosa Prantl. Pharmazie 27:125

Fedde F (1909) Papaveraceae-Hypecoideae et Papaveraceae-Papaveroideae. In: Engler A (ed) Das Pflanzenreich, part IV, no 40. Engelman, Leipzig, p 288

Grundon MF (1976–1983) The alkaloids, vol 6–13. Chem Soc, Burlington House, London

Günther KF (1975) Flora (Jena) Part 164/4

Hegnauer R (1961) Die Gliederung der Rhoeadales sensu Wettstein im Lichte der Inhaltstoffe. Planta Med 9:37–46

Hegnauer R (1962–1973) Chemotaxonomie der Pflanzen. Birkhäuser, Basel Stuttgart

Hemingway SR, Phillipson JD, Verpoorte R (1981) *Meconopsis cambrica* alkaloids, J Nat Prod 44:67–74

Hutchinson J (1921) The genera of Fumariaceae and their distribution. Kew Bull 97–115

Hutchinson J (1959) Families of flowering plants, vol II, 2nd edn. Oxford Univ Press, Oxford

Novák J (1979) Taxonomical Revision of the Section Macrantha, Genus *Papaver*, Preslia 51: 341–348

Novák J, Preininger V (1980) Glauca – a new section of the Genus *Papaver*. Preslia 52:97–101

Novák J, Preininger V (1983) Phytochemistry and Taxonomy of the Genus *Fumaria* L. Acta Univ Palacki Olomuc (Olomouc). Fac Rer Nat 78:39–45

Phillipson JD, Thomas OO, Gray AI, Sariyar G (1981a) Alkaloids from *Papaver armeniacum*, *P. fugax* and *P. tauricola*. Planta Med 41:105–118

Phillipson JD, Scutt A, Baytop A, Ozhatay N, Sariyar G (1981b) Alkaloids from Turkish samples of *Papaver orientale* and *P. pseudo-orientale*. Planta Med 43:261–271

Preininger V (1975) The pharmacology and toxicology of Papaveraceae alkaloids. In: Manske RHF (ed) The alkaloids, vol 15. Academic Press, London New York, p 207

Preininger V (1984) Chemotaxonomy of Plants of the Genus *Papaver* L., Acta Univ Olomuc (Olomouc). Fac Med 106:9–29

Preininger V, Šantavý F (1970) Isolierung der Alkaloide aus *Papaver bracteatum* Lindl, *P. fugax* Poir und *P. triniaefolium* Boiss und Identifizierung einiger früher isolierter Alkaloide aus Pflanzen der Sektionen Ortho-rhoeades, Mecones und Pilosa. Pharmazie 25:356–360

Preininger V, Appelt J, Slavíková L, Slavík J (1967) Alkaloide aus den Arten *Papaver persicum* Lindl. and *P. caucasicum* Marsch.-Bieb. Collect Czech Chem Commun 32:2682–2686

Preininger V, Novák J, Šantavý F (1981) Glauca – eine neue Sektion der Gattung Papaver. Planta Med 41:119–123

Šantavý F (1970, 1979) Papaveraceae Alkaloids. In: Manske RHF (ed) The alkaloids, vols 12, 17. Academic Press, London New York, pp 333, 385

Sariyar G (1975) Dissertation, Univ Istanbul

Sariyar G, Phillipson JD (1977) Macrantaline and Macrantoridine, New Alkaloids from a Turkish sample of *Papaver pseudo-orientale*. Phytochemistry 16:2009–2013

Sariyar G, Phillipson JD (1980) Alkaloids of Turkish *Papaver tauricola*. Phytochemistry 19: 2189–2192

Saxton JE (1971–1975) The alkaloids, vols 1–5. Chem Soc, Burlington House, London

Shafiee A, Lalezari I, Nasseri-Nouri P, Asgharian R (1975) Alkaloids of *Papaver orientale* and *P. pseudo-orientale*. J Pharm Sci 64:1570–1572

Shafiee A, Lalezari I, Assadi F, Khalafi F (1977) J Pharm Sci 66:1050

Shamma M (1972) The isoquinoline alkaloids. Academic Press, London New York

Shamma M, Moniot JL (1978) Isoquinoline alkaloids research, 1972–1977. Plenum Press, New York London

Slavík J (1980a) Chemistry and Distribution of Quaternary Alkaloids of Papaveraceae family. Acta Univ Olomouc Fac Med 93:5–21

Slavík J (1980b) Occurrence of morphine as a minor alkaloid in *Papaver decaisnei* Hochst. Collect Czech Chem Commun 45:2706–2709

Slavík J, Slavíková L (1975) On Alkaloids from the Leaves of *Bocconia frutescens* L. Collect Czech Chem Commun 40:3206–3210

Slavik J, Slavíková L (1977) Alkaloids of some Himalayan species of *Meconopsis* genus. Collect Czech Chem Commun 42:132–139

Slavík J, Novák V, Slavíková L (1976) Alkaloids from *Eschscholzia lobbii* Greene. Collect Czech Chem Commun 41:2429–2432

Stermitz FR (1968) Alkaloid Chemistry and Systematics of *Papaver* and *Argemone*. In: Mabry TJ, Alston RE, Runeckles VC (eds) Recent advances in Phytochemistry, vol I. Meredith Corporation, p 161

Takhtajan A (1959) Die evolution der angiospermen. Fischer, Jena

Structure Activities and Pharmacological Properties of the Opium Alkaloids

E. LINDNER [1]

1 Introduction

The early history of opium has been reviewed in Chapter 1. Since the isolation of morphine by Sertürner in 1803, some 40 alkaloids, representing different structural types of isoquinolines, have been isolated from opium. These alkaloids are combined with a number of acids including fumaric, lactic and the rare meconic acid (1). In addition, opium also contains a complex mixture of substances including protein, sugar, fat, resin, rubber, wax and mucilaginous substances. Because of the problems associated with the illegal use of opium, the United Nations have restricted the cultivation of the opium poppy, *Papaver somniferum* L. to Iran, India, Turkey, Bulgaria, Greece, Yugoslavia and Russia. There has been considerable interest in the pharmacology of all alkaloids obtained from the Papaveraceae (Preininger 1975) and this chapter concentrates on those alkaloids which have been obtained from opium.

2 Opium Alkaloids

2.1 Morphinans

Morphine itself (2) constitutes between 3% and 23% of opium and on average some 12%, while codeine (3) comprises between 0.2% and 3% and on average is about 1%. Neopine (4), which is isomeric with codeine, occurs as a minor alkaloid of opium together with 10-hydroxycodeine and 6-methylcodeine. Thebaine (5) is present at between 0.2% and 1% and on average occurs at 0.5%. Promorphinan alkaloids such as salutaridine (6) are also present in opium.

Morphine is marked by its strong analgaesic action and in humans 10 mg given orally will elevate the pain threshold by 50% and a 30 mg dose by 90%. At higher doses, morphine acts as a narcotic and sedative, 50–100 mg leading to deep sleep. In contrast, codeine and neopine have only one tenth of the analgaesic activity of morphine. Morphine also depresses respiration and this depression is pronounced when analgaesic doses are doubled. Although codeine inhibits respiration, it is much

---

1 Hoechst AG, Amselweg 8, 6000 Frankfurt/Main, FRG

The Chemistry and Biology of Isoquinoline Alkaloids, ed. by Phillipson et al.

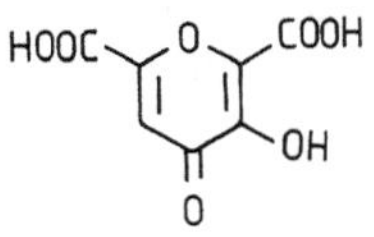

1.

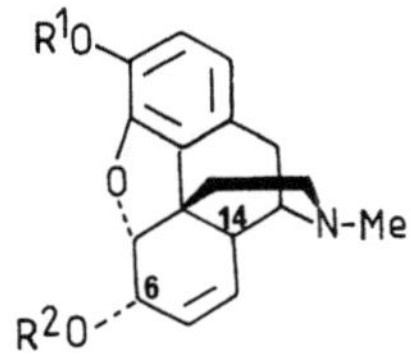

2. $R^1 = H, R^2 = H$
3. $R^1 = Me, R^2 = H$

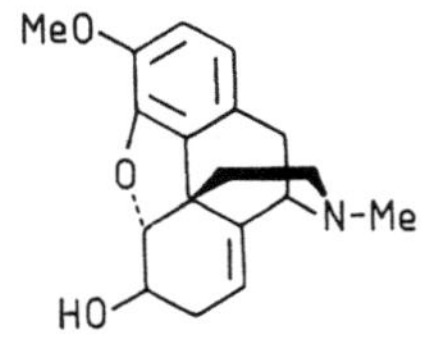

4.

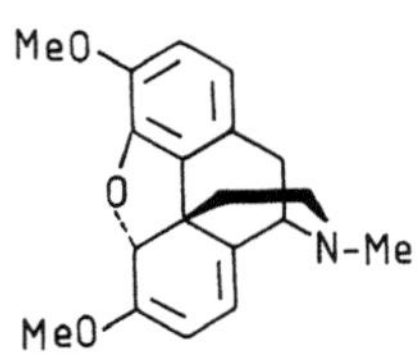

5.

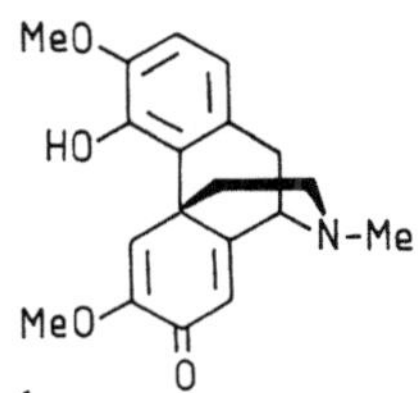

6.

Fig. 1

weaker in this respect than morphine and it takes a dose nine times higher to obtain the same level of depression of respiration. Both alkaloids have pronounced antitussive activity and in this respect morphine is more active than codeine. A dose of 2 mg of morphine is sufficient for cough suppression in humans, whereas 20 mg of codeine is necessary for the same effect. However, codeine is preferred as a cough depressant because its much weaker analgaesic action greatly reduces the danger of addiction. Even low doses of morphine can remove displeasure, anxiety, and the feeling of hopelessness, and can cause euphoria; psychic as well as physical dependence can develop. The reason for this physical dependence is due to the counter regulation of several parts of the central nervous system against the depressive action of morphine. The respiratory centre, however, does not respond in the same way, and this means that morphine is dangerous particularly when it is injected. When morphine is discontinued in humans, signs of deprivation develop in the form of restlessness, excitation, dysphoria, sweating and collapse.

Other essential effects of morphine are the elevation of muscle tone in the gastrointestinal tract due to the reduction of acetylcholine liberated from the parasympathetic nerve endings, thus leading to a depression of peristalsis and a reduction in the stretching of the g.i. tract. As a result there is a reduction in stomach emptying and defaecation. The tone of the gall bladder and also of the sphincter muscle of the urinary bladder is also enhanced.

The important action of morphine is undoubtedly its strong analgaesic effect. Natural morphine is laevorotatory and synthetic morphine-like compounds which have analgaesic activity are also laevorotatory. These findings led to the view that special receptors must be present in the nervous system and that when blocked there is an analgaesic effect. Eventually such receptors were located in nervous tissue and intestinal tissue not only for morphine but also for synthetic morphine-like analgaesics. The binding characteristics of such chematic receptors are repre-

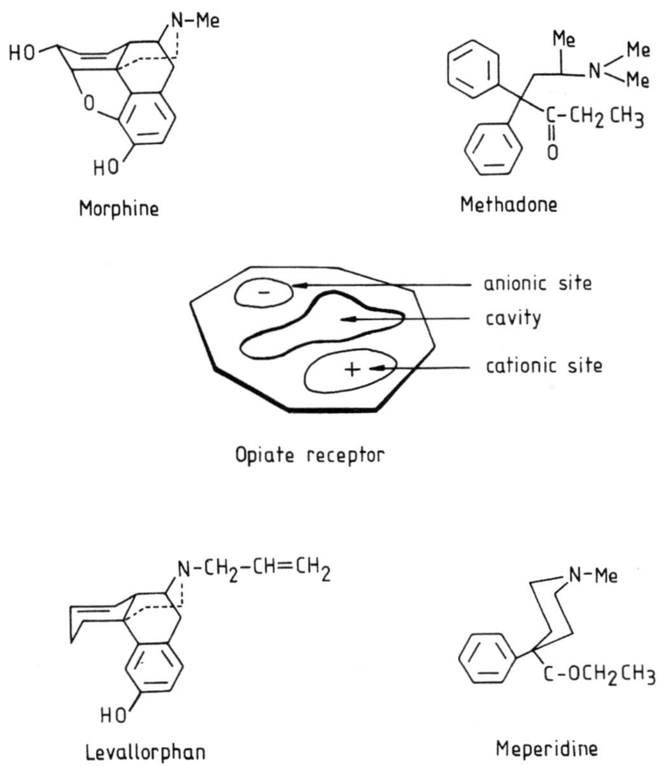

Fig. 2. Diagrammatic representation of the opiate receptor

sented in Fig. 2. Subsequently it became possible to distinguish between several different types of opiate receptors, e.g. μ-receptors which bind morphine, k- and δ-receptors which bind some of the semi-synthetic morphine derivatives. Some simple derivatives of morphine such as N-allyl-normorphine (nalorphine) have an antagonistic action on morphine because they can displace morphine from μ-receptors in a competitive manner. The finding of these receptors led to the assumption that such receptors must play a physiological role in the perception of pain. It seemed possible, therefore, that naturally occurring substances in animals could bind physiologically to receptors and effect analgaesia. In 1974 Kosterlitz and coworkers (Hughes et al. 1974) succeeded in finding two closely related peptides in the central nervous system and these two peptides, methionine-enkephalin and leucine-enkephalin, were capable of binding to opiate receptors of the intestine. Higher molecular weight peptides localised in the hypophysis were also isolated by hydrolysis of β-lipotropin and from these peptides it was possible to obtain β-endorphin and methionine-enkephalin. (Fig. 3) Endorphins and enkephalins have been located in the central nervous system and in the nervous plexus of the stomach and intestine.

The enkephalins contain the amino acid (−)-tyrosine which seems to be an important feature for binding to receptors (Fig. 4). When the tyrosine-6-glycine bond is split, the affinity for binding to the opiate receptor is lost. Morphine contains in its molecule a tyramine moiety which is the decarboxylation product of tyrosine,

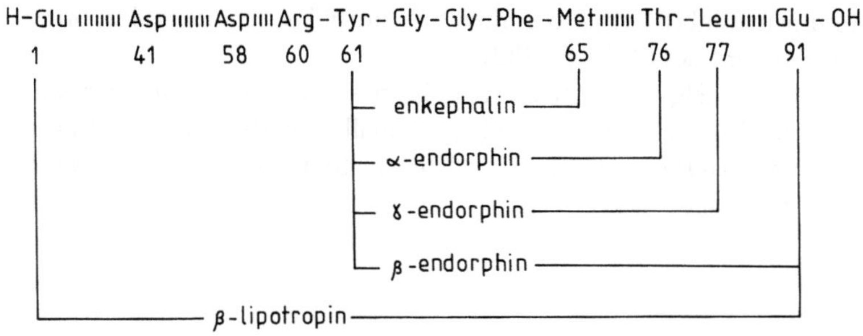

Fig. 3. Structures of enkephalin, endorphins and β-lipotropin

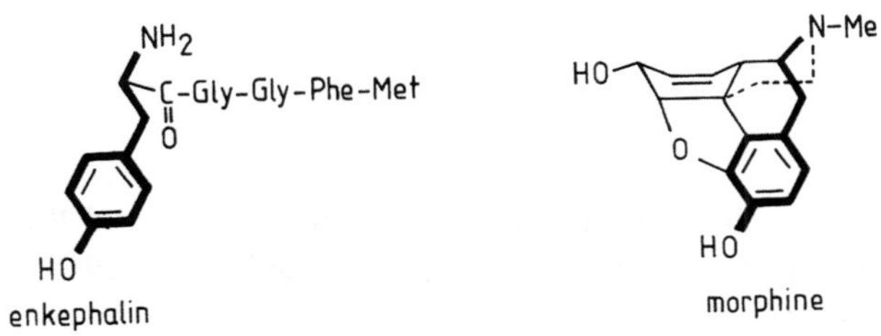

Fig. 4. "Tyramine moiety" of enkephalin and morphine

and this will fit the receptor site. The enkephalins are very active as analgaesics at μg levels when injected intravenously or into the side ventricle of rat brains. It was disappointing when it was realised that these natural analgaesics produced by animals also have addicting properties (Horn and Rodgers 1976, Kuschinsky 1979, Schulz 1979).

The occurrence of these active peptides in the body has given new insight into the presumed mechanism of pain perception and pain inhibition. There are nervous pathways of pain perception from the periphery to the spinal cord, to the brain stem and to the brain, and there are similar pathways for the inhibition of pain. It is possible to inhibit pain perception physiologically and it has long been known that an expected pain is not as painful as an unexpected pain, e.g. the ability to handle an object when it is known to be hot or the lack of pain while in a state of shock after a severe injury. Enkephalins are secreted at the terminals of pain-inhibiting neurones and they can modulate pain partly by the excitation of the inhibition pathways and partly by inhibition of the pain perception pathways. This can occur at several levels such as the spinal cord, the thalamus and the main brain. Electrical stimulation of the nucleus paragiganto-celluaris in the rat, from which central pain perception-inhibiting pathways go down to the spinal posterior horn, depresses the transmission of pain impulses. The competitive morphine antagonist naloxone inhibits the analgaesic effect of electrical stimulation of pain-inhibiting pathways. The analgaesic effects brought about by acupuncture may well be explained by the secretion of enkephalins, but it should be noted that these are not the only compounds secreted, as norepinephrine is also released by neurones of the pain-inhibiting pathways. Under normal circumstances, the secretion and action of endorphins is limited and therefore pain inhibition is also limited but morphine can be used in order to substitute the natural peptides (Takagi 1980).

Thebaine is devoid of analgaesic activity and is in some respects antagonistic and can also cause convulsions. It is not used medicinally by itself and, although it is present in opium preparations, the low levels mean that it only exerts a weak antagonistic effect on morphine.

2.2 Protopines

There are several protopine-type alkaloids including protopine (7), cryptopine, α-allocryptopine and 13-oxocryptopine. These alkaloids are of little importance and have no analgaesic activity. Protopine does have a sedative effect on laboratory animals.

2.3 Rhoeadines

There are a series of rhoeadine-type alkaloids including rhoeadine itself (8) and the nor-alkaloids known as the papaverrubines (Pfeiffer 1966, Dolejs and Hanus 1967). These alkaloids are of no medical importance.

2.4 Protoberberines

There are numerous protoberberine- and tetrahydroprotoberberine-type alkaloids including berberine (9), coptisine, scoulerine, isocorypalmine and coreximine. Berberine contracts uterine muscle and has been used to stop uterine bleeding and to help in childbirth. Its action is relatively weak and uncertain and because it has somewhat toxic properties it is no longer used for these purposes. Berberine also exerts some antiinflammatory activity (Otsuka et al. 1981) and it does depress the formation of connective tissue. When berberine is infused in a dose of 0.2 mg kg^{-1} it has a positive inotropic effect on dog heart, reducing heart rate and decreasing total peripheral resistance of blood circulation (Maroko et al. 1982). (–)-Coreximine (10), also found in opium, is closely related to berberine and it also acts with a positive inotropic effect on dog heart and reduces heart rate when it is infused with 0.2 mg kg^{-1}. Coptisine (11) has an inhibitory effect on the formation of connective tissue (Otsuka et al. 1981).

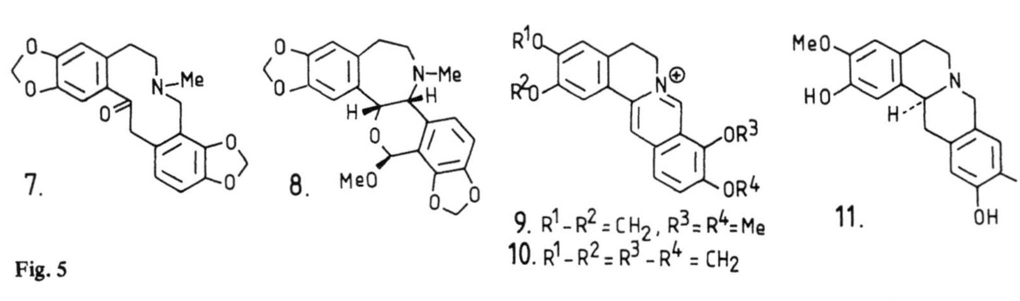

Fig. 5

2.5 Benzylisoquinolines

Some of the benzylisoquinolines, e.g. papaverine (12) and benzyltetrahydroisoquinolines, e.g. noscapine (13), have gained importance as clinical agents. Papaverine is found in opium in yields of 0.5% to 1.3% and on average is present at 1%. The content of (–)-noscapine and its dextrorotatory isomer in opium fluctuates from between 2% and 10%, with an average content of 5%. The closely related phenolic alkaloid narcotoline (14) is present in opium at about 0.03% (Wrede 1937). Meconine (15), a breakdown product of noscapine and of narcotoline is also present in opium. Narceine (16) is not a benzyltetrahydroisoquinoline, but the structure of this amphoteric compound is very similar to that of the phthalideisoquinoline alkaloid noscapine (13). Papaveraldine (17), an oxygenated derivative of papaverine, is also found in opium, together with the benzyltetrahydroisoquinoline alkaloid laudanine which is laevorotatory, but which also occurs as the dextrorotatory isomer (18) and laudanosine (19). The latter two alkaloids are not used medicinally since laudanosine may lead to convulsions while laudanine is also toxic. At high levels of 4 mg kg^{-1} i.v. in cats, laudanosine depresses blood pressure (Chapple and Clark 1983). Reticuline (20), which is not used as a drug, is a very closely related opium alkaloid. Other minor alkaloids which occur in opium but which have no medical importance include codamine, palaudine, somniferine and oxynoscapine.

Papaverine (12) is an excellent spasmolytic drug which relaxes the smooth muscle of blood vessels. It especially inhibits contractions brought about by direct stimulation of smooth muscle of the g.i. tract. Activity against spasms produced by the transmitter of parasympathetic impulses, acetylcholine, is only possible when much higher doses are used. An advantage of papaverine is that spasms are removed with small doses but without influencing the normal tone and peristaltic activity of the g.i. tract. Papaverine is combined with atropine in many medicinal preparations used against colic. Coronary arteries are dilated by papaverine as well as peripheral vessels, especially those of the g.i. tract. Thus, blood pressure can be suddenly

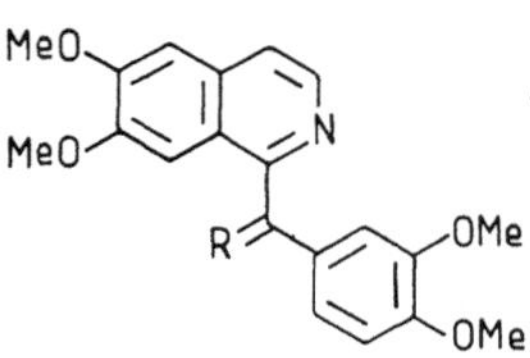

12. R = H,H
17. R = O

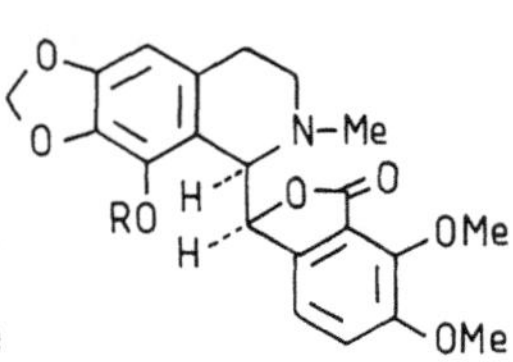

13. R = Me
14. R = H

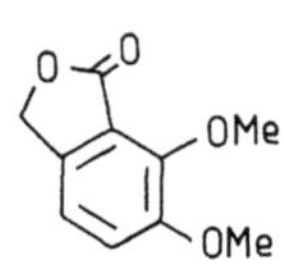

15.

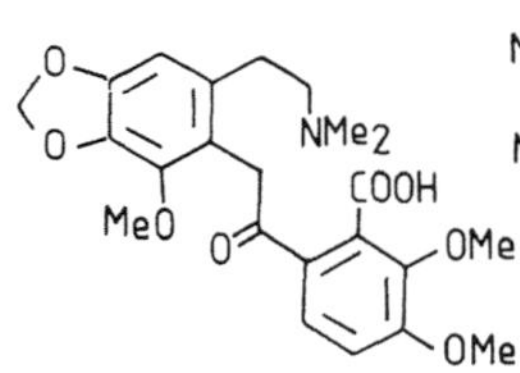

16.

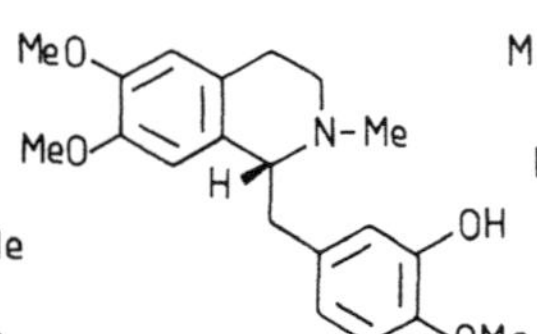

18

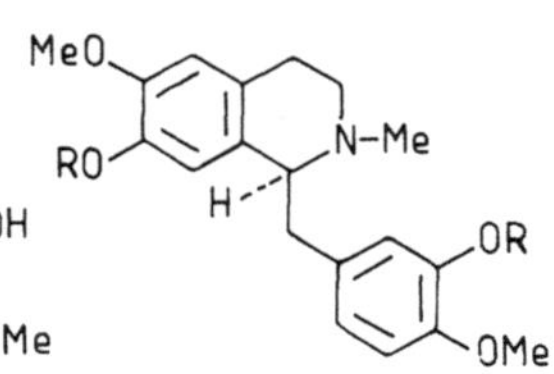

19. R = Me
20. R = H

Fig. 6

lowered by rapid intravenous injection of the drug. In addition to these activities, papaverine also acts as an antiarrhythmic drug against extrasystoles of the heart. Its central effects are very weak and only with very high doses can a small analgaesic effect be demonstrated. Papaverine is usually given in doses of 100–200 mg several times per day and is well absorbed when given orally.

Noscapine (13) also has spasmolytic effects, but the propulsion of the stomach contents and intestine are somewhat retarded. Normal doses do not give any analgaesic effects. The drug also has a central cough-inhibiting effect and in this respect noscapine is not much less effective than codeine, but it also has a negative effect on the contraction of the heart muscle. In small doses, noscapine stimulates respiration and is non-addictive. Narcotoline (14) also stimulates respiration, but it requires two to three times higher doses. The drug also acts as a spasmolytic, but again it requires larger doses than papaverine. Blood pressure is also depressed when the doses are fourfold higher than papaverine (Zimmermann 1937).

2.6 Aporphines

Examples of the aporphine alkaloids include corytuberine (21), magnoflorine (22) and isoboldine (23). The closely related alkaloid boldine (24) from boldo leaves has stimulating action on the gastric and bile secretions. It also increases the elimination of uric acid, has a sedative action and is used as an anthelmintic. The quaternary alkaloid, magnoflorine (22), acts as a neuromuscular blocking agent and lowers blood pressure in rats and in guinea pigs (Aiyar et al. 1979).

2.7 Benzophenanthridines

Sanguinarine (26), also found in low amounts in opium, elevates intraocular pressure at a dose of 2.5 mg kg^{-1} when given intravenously. The alkaloid inhibits NaK-ATPase similarly to the cardiac glycosides although it is completely different in its chemical constitution. It acts on the sulphydryl group of NaK-ATPase and it has a positive

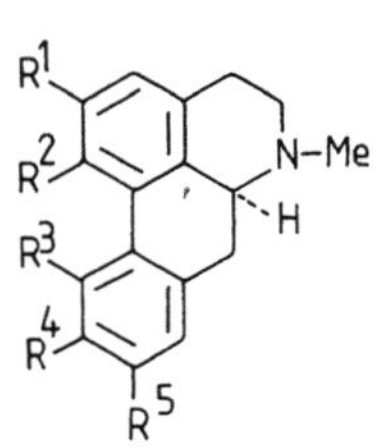

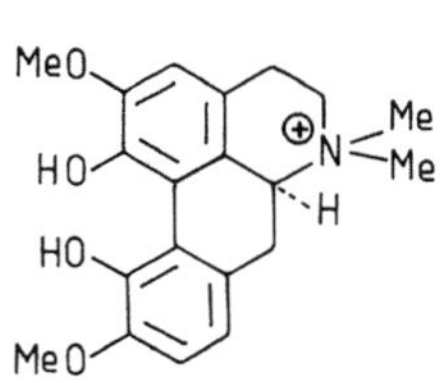

22.

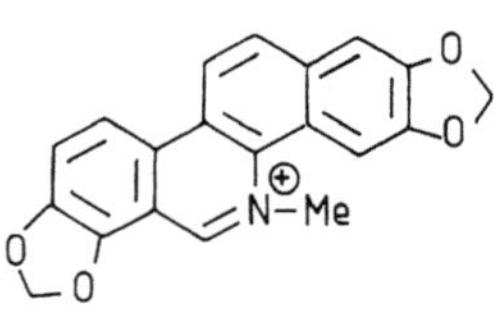

26.

	R^1	R^2	R^3	R^4	R^5
21.	OMe	OH	OH	OMe	H
23.	OMe	OH	H	OMe	OH
24.	OH	OMe	H	OMe	OH
25.	$-OCH_2O-$		H	H	H

Fig. 7

inotropic effect on guinea pig heart (Cohen et al. 1978). Several benzophenanthridine alkaloids have antibacterial, cytotoxic or antineoplastic activities (Stermitz et al. 1973 and 1975).

3 Aspects of Opium Pharmacology

The pharmacological effects of this complex mixture of alkaloids in opium may, in some respects, be more beneficial than the effects of an individual alkaloid. The effect of opium in the treatment of diarrhoea is better than that of either morphine or codeine given individually. It is possible to calm the gastrointestinal tract with a dose which contains only half of the dose which would be necessary for morphine alone. The combination of the increase in stretch reflex caused by morphine and the inhibition of increased peristaltic action by papaverine leads to a better therapeutic effect. The increase in muscle tone due to morphine is reduced by papaverine, and other benzylisoquinolines also participate in this tranquillisation of the g.i. tract. In the normal 50 mg dosage of opium used for the treatment of diarrhoea there is about 5 mg of morphine, 0.5 mg of papaverine and 3 mg of noscapine which contrasts markedly with a dose of 100–200 mg of papaverine if this were given alone. The analgaesic activity of opium is due mainly to the morphine present, as the levels of codeine, neopine and papaverine are relatively small. The respiratory depression effect of opium is also due to the morphine content and the antagonistic effects of noscapine and narcotoline are not so pronounced. Thus the main use of opium in medicine is for the treatment of diarrhoea.

There is a similarity between the use of opium with its complex mixture of alkaloids for the treatment of diarrhoea and the use of *Rauwolfia serpentina* extracts for the treatment of hypertension. The alkaloid combination of the *Rauwolfia* extract is more effective and better tolerated than reserpine alone because the effects of the centrally acting reserpine are facilitated by other alkaloids which are capable of causing dilation of peripheral blood vessels.

Thus it appears in these two particular cases of opium and of *Rauwolfia* extract that the whole is better than its parts.

References

Aiyar VN, Benn MH, Hanna T, Jacyno J, Roth SH, Wilkens JL (1979) The principal toxin of *Delphinium brownii* and its mode of action. Experientia 35:1367–1368

Chapple DJ, Clark JS (1983) Pharmacological action of breakdown products of atracurium and related substances. Br J Anaesth 55 Suppl 1: 11S

Cohen HG, Seifen EE, Straub KD, Tiefenback C, Stermitz FR (1978) Structural specificity of the NaK-ATPase inhibition by sanguinarine. Biochem Pharmacol 27:2555

Dolejs L, Hanus V (1967) Mass spectrometry of rhoeadine type alkaloids. Tetrahedron 23:2997

Horn AS, Rodgers JR (1976) Structural and conformational relationships between the opiates. Nature (London) 260:795–797

Hughes J, Smith TW, Kosterlitz HW, Fothergill LA, Morgan BA, Morris HR (1974) Identification of two related pentapeptides from the brain with potent opiate agonist activity. Nature (London) 258:577–579

Kuschinsky K (1979) Zur Physiologie und Pharmakologie von Endorphinen. Klin Wochenschr 57:701

Maroko PR, Zalewski A, Krol R, Cheung W (1982) Protoberberine alkaloids – a new family of inotropic agents. Clin Res 30:675A

Otsuka H, Fujimara H, Sawada T, Goto M (1981) Studies on antiinflammatory agents. Yakugaku Zasshi 101:883

Pfeiffer S (1966) The structure of porphyroxine (papaverrubine D). J Pharm Pharmac 18:133

Preininger V (1975) The pharmacology and toxicology of the Papaveraceae alkaloids. In: Manske RHF (ed) The alkaloids, vol 15. Academic Press, London New York, pp 207–271

Schulz R (1979) Endorphine. Umschau 79:431

Stermitz FE, Larson KA, Kim DK (1973) Some structural relationships among cytotoxic and antitumour benzophenanthridine alkaloid derivatives. J Med Chem 16:939–940

Stermitz FR, Gillespie JP, Amoros LG, Romero R, Stermitz TA (1975) Synthesis and biological activity of some antitumour benzophenanthridinium salts. J Med Chem 18:708–713

Takagi H (1980) The nucleus reticularis paragigantocellularis as a site of analgaesic action of morphine and enkephalin. Trends Pharmacol Sci 1:182–184

Wrede F (1937) Narkotolin, ein neues Alkaloid aus Mohn *(Papaver somniferum)*. Arch Exp Pharmakol Pathol 184:331–335

Zimmermann K (1937) Die pharmakologische Wirkung des Narkotolin, eines neuen Alkaloids aus *Papaver somniferum*. Arch Exp Pharmakol Pathol 184:336

The Occurrence of Simple Isoquinolines in Plants

J. LUNDSTRÖM [1]

1 Introduction

Simple isoquinoline alkaloids may be defined as those containing only one aromatic nucleus and no other cyclic structure except a methylenedioxy substituent. In general, they are derived from only one aromatic amino acid residue e.g. a phenylalanine or tyrosine moiety.

Compounds with the general structures shown in Fig. 1 may be included in this group of alkaloids. The tetrahydroisoquinolines constitute the major group, about 50 different structures are known to occur in nature (Fig. 2) (Lundström 1983). Ring-closing units (carbon-1 and its substituents) usually are one- or two-carbon units, but a five-carbon unit is also known. A few examples of 3,4-dihydroisoquinolines and one fully aromatic compound are known. Isoquinoline derivatives, where the carboxyl group of the amino acid precursor have been retained, constitute a small group of alkaloids.

Tetrahydro-isoquinolines | Isoquinoline amino acids | Isoquinolones

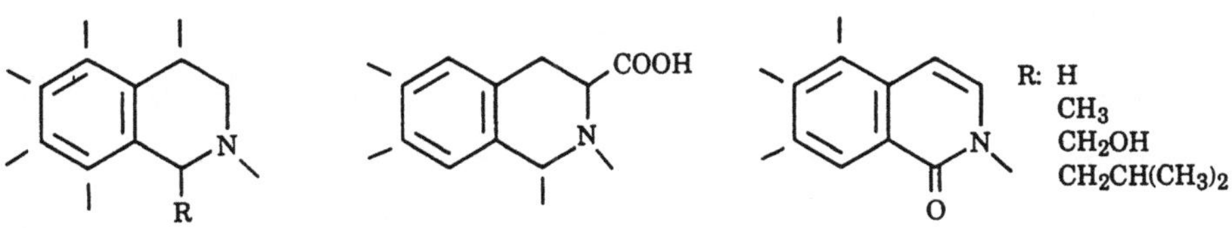

Fig. 1

Isoquinolones of the simple type may also be included in this general group of alkaloids, because of the structural similarity. Biosynthetically, however, these compounds are probably degradation products of more complex isoquinoline alkaloids, as has been pointed out in recent reviews (Lundström 1983, Krane and Shamma 1982).

In Table 1, the families of the species containing simple isoquinolines and isoquinolones are grouped in orders and superorders according to Dahlgren (1975). Species containing only simple isoquinolines are primarily found in the Cactaceae and the Chenopodiaceae of the Caryophyllales and in the Fabaceae (Leguminosae). The species of these three families contain simple isoquinolines as major alkaloids and these often occur together with β-phenethylamines.

1 Astra Läkemedel AB, S-151 85 Södertälje, Sweden

The Chemistry and Biology of Isoquinoline Alkaloids, ed. by Phillipson et al.

In the more primitive superorders, Magnolianae and Ranunculanae, species containing both simple isoquinolines and isoquinolones are found. Generally, these species contain simple isoquinolines and isoquinolones as minor constituents and mainly more elaborated isoquinoline alkaloids.

In the following sections, the plants and their alkaloid contents are presented according to their botanical distribution.

Fig. 2. Formulas of simple isoquinoline alkaloids. (Self-explanatory names as N-methyl and O-methyl derivatives and N-oxides are not included)

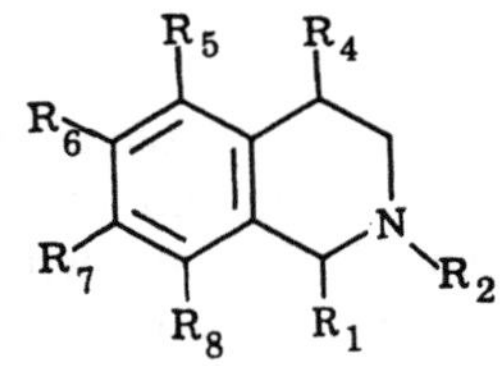

	R_1	R_2	R_4	R_5	R_6	R_7	R_8
Anhalamine	H	H	H	H	CH_3O	CH_3O	HO
Anhalidine	H	CH_3	H	H	CH_3O	CH_3O	HO
Anhalinine	H	H	H	H	CH_3O	CH_3O	CH_3O
Anhalonidine	CH_3	H	H	H	CH_3O	CH_3O	HO
Anhalonine	CH_3	H	H	H	CH_3O	$O - CH_2 - O$	
Anhalotine	H	$(CH_3)_2$	H	H	CH_3O	CH_3O	HO
Arizonine	CH_3	H	H	H	H	CH_3O	HO
Calycotomine	CH_2OH	H	H	H	CH_3O	CH_3O	H
Carnegine	CH_3	CH_3	H	H	CH_3O	CH_3O	H
Corypalline	H	CH_3	H	H	CH_3O	HO	H
Deglucopterocereine	CH_2OH	CH_3	H	HO	CH_3O	CH_3O	H
Dehydroheliamine[a]	H	H	H	H	CH_3O	CH_3O	H
Dehydrosalsolidine [a]	CH_3	H	H	H	CH_3O	CH_3O	H
Gigantine	CH_3	CH_3	H	HO	CH_3O	CH_3O	H
Heliamine	H	CH_3	H	H	CH_3O	CH_3O	H
Hydrocotarnine	CH_3	CH_3	H	H	$O - CH_2 - O$		CH_3O
Isoanhalamine	H	H	H	H	HO	CH_3O	CH_3O
Isoanhalidine	H	CH_3	H	H	HO	CH_3O	CH_3O
Isoanhalonidine	CH_3	H	H	H	HO	CH_3O	CH_3O

Fig. 2. (cont.)

	R_1	R_2	R_4	R_5	R_6	R_7	R_8
Isopellotine	CH_3	CH_3	H	H	HO	CH_3O	CH_3O
Isosalsoline	CH_3	H	H	H	HO	CH_3O	H
Lemairecereine	H	H	H	H	H	CH_3O	CH_3O
Longimammamine	H	CH_3	HO	H	H	H	HO
Longimammidine	H	CH_3	H	H	H	H	HO
Longimammatine	H	H	H	H	CH_3O	H	H
Longimammosine	H	CH_3	H	H	HO	H	H
Lophocerine	$CH_2CH(CH_3)_2$	CH_3	H	H	CH_3O	HO	H
Lophophorine	CH_3	CH_3	H	H	CH_3O	O - CH_2 - O	
Lophotine	CH_3	$(CH_3)_2$	H	H	CH_3O	O - CH_2 - O	
Nortehaunine	H	H	H	CH_3O	CH_3O	CH_3O	H
Pellotine	CH_3	CH_3	H	H	CH_3O	CH_3O	HO
Peyophorine	CH_3	CH_2CH_3	H	H	CH_3O	O - CH_2 - O	
Peyotine	CH_3	$(CH_3)_2$	H	H	CH_3O	CH_3O	HO
Peyoruvic acid	$<^{CH_3}_{CO_2H}$	H	H	H	CH_3O	CH_3O	HO
Peyoxylic acid	CO_2H	H	H	H	CH_3O	CH_3O	HO
Pycnarrhine[a]	H	CH_3	H	H	CH_3O	CH_3O	H
Salsoline	CH_3	H	H	H	HO	CH_3O	H
Salsolidine	CH_3	H	H	H	CH_3O	CH_3O	H
Salsolinol	CH_3	H	H	H	HO	HO	H
Tehaunine	H	H	H	CH_3O	CH_3O	CH_3O	H
Tepenine	CH_3	CH_3	H	H	H	CH_3O	CH_3O
Weberine	H	CH_3	H	CH_3O	CH_3O	CH_3O	CH_3O
Weberidine	H	H	H	H	H	CH_3O	H
Uberine	H	H	H	CH_3O	H	CH_3O	H

[a] 1,2-dehydro

Table 1. Number of plant species containing simple isoquinoline alkaloids

Superorder Order	Family	Isoquino-lines	Isoquino-lones	(Total)
Magnolianae				
Magnoliales:	Annonaceae	1	1	(2)
Laurales:	Atherospermataceae	1	1	(1)
	Lauraceae		1	(1)
	Hernandiaceae		1	(1)
Ranunculanae				
Nelumbonales:	Nelumbonaceae	1		(1)
Ranunculales:	Berberidaceae		1	(1)
	Menispermaceae	2	1	(3)
	Ranunculaceae	7	11	(16)
Papaverales:	Fumariaceae	6	2	(8)
	Papaveraceae	2	4	(6)
Dillenianae				
Euphorbiales:	Euphorbiacae	1[a]		(1)
Caryophyllanae				
Caryophyllales:	Cactaceae	20		(20)
	Chenopodiaceae	4		(4)
Fabanae				
Fabales:	Caesalpiniaceae		1	(1)
	Fabaceae	7[b]		(7)
Saxifraganae				
Cornales:	Alangiaceae	1		(1)
Celastranae				
Rhamnales:	Rhamnaceae	1		(1)

[a] Isoquinoline amino acid

[b] Including 3 species containing isoquinoline amino acids

2 Plants Containing Simple Isoquinolines as Main Alkaloids

2.1 Cactaceae Species

The Cactaceae family comprises some 3000 species in about 230 genera (Backeberg 1966). In the recent comprehensive tabular summary of the Cactaceae alkaloids by Mata and McLaughlin (1982), 153 alkaloid-bearing cactus species are listed. Most of these species contain β-phenethylamine alkaloids and only 21 species contain simple isoquinolines. In most of the species producing isoquinolines, phenethylamine derivatives are also found, and it is generally agreed that the tetrahydroisoquinolines evolve by cyclization of variously substituted phenethylamines with suitable ring-closing units. Alkaloids are mainly found in the Cereeae tribe of the Cereoideae subfamily, and the simple tetrahydroisoquinolines mainly in the semi-Tribe Boreocereeae (Table 2).

Table 2. Isoquinoline containing Cactaceae species of the Cereeae tribe and their alkaloid contents

Species	Alkaloids
Semitribe Austrocereeae	
Subtribe Austrocereinae	
Stetsonia coryne (S.D.) Br. & R.	Anhalidine, anhalonidine
Trichocereus pachanoi Br. & R.	Anhalonidine
Subtribe Austrocactinae	
Islaya minor Backeb.	Corypalline
Semitribe Boreocereeae	
Subtribe Boreocereinae	
Backebergia militaris (Andot) Bravo ex Sanches Mejorada	Heliamine, lemaireocereine
Carnegia gigantea (Engel) Br. & R.	Arizonine, carnegine, dehydroheliamine, dehydrosalsolidine, gigantine heliamine, salsolidine
Echinocereus merkerii Hildm.	Salsoline
Lophocereus australis (K. Brand) Borg	Pilocereine
Lophocereus gatesii M.E. Jones	Pilocereine
Lophocereus schottii (Engel) Br. & R.	Lophocerine, pilocereine
Pachycereus marginatus (D.C.) Br. & R.	Lophocerine, pilocereine
Pachycereus pecten-aboriginum Br. & R.	Arizonine, carnegine, heliamine, isosalsoline, salsoline, salsolidine
Pachycereus pringlei (S. Wats) Br. & R.	Heliamine, lemaireocereine, tehaunine, weberidine, tehaunine N-oxide

Table 2. (cont.)

Species	Alkaloids
Pachycereus tehuantepecanus Mac Dough & H. Bravo	Tepenine, tehaunine
Pachycereus weberi (Coult.) Backeb.	Anhalonidine, heliamine, lemaireocereine, N-methylheliamine, northehaunine, pellotine, tehaunine, weberidine, weberine
Pilosocereus guerreronis (Backeb.) Byl. et. Rowl.	N-Methylheliamine
Pterocereus gaumeri (Br. & R.) Mac Dough & Mir.	Pterocereine, deglucopterocereine Deglucopterocereine N-oxide
Subtribe Boreocactinae	
Dolichothele longimamma (D.C.) Br. & R.	Longimammamine, longimammatine, longimammidine, longimammosine
Dolichothele uberiformis (Zucc.) Br. & R.	Longimammamine, longimammatine, uberine
Lophophora diffusa (Croizat.) H. Bravo	Anhalamine, anhalonidine, lophophorine, pellotine, O-methylpellotine
Lophophora williamsii (Lem.) Coulter	Anhalamine (+N-acetyl and N-formyl), anhalidine, anhalinine (+N-formyl), anhalonidine (+N-formyl), anhalonine (+N-acetyl and N-formyl), anhalotine, isoanhalamine, isoanhalidine, isoanhalonidine, isopellotine, lophophorine, lophotine, O-methyl-anhalonidine, pellotine, peyophorine, dehydroanhalamine, dehydroanhalidine, dehydroanhalonidine, dehydropellotine, peyoruvic acid, O-methyl-peyoruvic acid, peyoxylic acid, O-methylpeyoxylic acid
Pelecyphora asselliformis Ehrenberg	Anhalidine, pellotine

For specific references see Lundström 1983

2.1.1 Semi-Tribe Austrocereeae

The three species of this semi-tribe contain mainly β-phenetyhlamine derivatives and only traces of tetrahydroisoquinoline. Most well known is the "San Pedro" cactus *Trichocereus pachanoi,* for its content of the hallucinogenic phenethylamine mescaline (3,4,5-trimethoxyphenethylamine) (Lundström 1971).

2.1.2 Semi-Tribe Boreocereeae, Subtribe Boreocereinae

Most of the isoquinoline-containing cactus species belong to this subtribe. These species are all large tree-like cacti except *Echinocereus merkerii.* Several of the *Pachycereus* species and *Carnegia gigantea* may reach heights of over 10 m.

A famous and well-known cactus is the giant cactus Saguaro or *C. gigantea,* growing in Sonora in northern Mexico and in Arizona (Bruhn 1971). This cactus is rich in alkaloids (0.6% of fresh weight) und apart from phenethylamines contains a variety of tetrahydroisoquinolines. Most alkaloids are 6,7-dioxygenated like the first isolated carnegine, but also the less common 7,8-dioxy and 5,6,7-trioxy substitution patterns as in arizonine and gigantine have been found (Bruhn and Lundström 1976). The two dihydroisoquinolines, dehydrosalsolidine and dehydroheliamine may be of biogenetic importance (Ordaz et al. 1983).

The alkaloids of the "sinita" cactus *Lophocereus schottii* and other *Lophocereus* species are unique in having an isobutyl substituent at carbon-1. These plants mainly contain the trimeric pilocereine, but the monomeric lophocerine has also been isolated (Djerassi et al. 1954).

There are no reports on phenethylamine derivatives from *Lophocereus* species; however, these cacti are rich in dopamine, a possible progenitor of the various isoquinolines found (Lundström unpublished).

The tallest Mexican columnar cactus *Pachycereus weberi* contains a great variety of differently substituted tetrahydroisoquinolines, mono-, di-, tri- and tetraoxygenated in the aromatic nucleus. Apparently this cactus has a highly evolved enzyme system for ring oxidation and O-methylation (Mata and McLaughlin 1982).

Pterocereus gaumeri contains the unusual glucoalkaloid pterocereine (Mohamed et al. 1979). The alkaloid may be hydrolyzed by acid or β-glucosidase to the aglycone deglucopterocereine. A hydroxymethyl group at carbon-1 was previously only known in the Fabaceae alkaloid calycotomine. Also the N-oxide of deglucopterocereine has been found (Pummangura et al. 1982).

2.1.3 Semi-Tribe Boreocereeae, Subtribe Boreocactinae

The two *Dolichothele* species, *D. longimamma* and *D. uberiformis,* contain simple isoquinolines that are monooxygenated in the 6- or 8-position. Probably, these compounds are biosynthesized by cyclisation of *meta*-tyramines. In *D. uberiformis* a 5,7-dioxygenated compound, uberine, was also found (Ranieri and McLaughlin 1976, 1977).

The Mexican peyote cactus, *Lophophora williamsii*, is the cactus species appearing to be richest in alkaloids. This little spineless sage-green cactus has over the years enjoyed widespread interest, mainly because of its content of the hallucinogenic phenethylamine mescaline. Its natural habitat is northern Mexico and the southern parts of Texas, USA (Anonymous 1959).

The whole fresh plant contains 0.4–0.5% of alkaloids, most of it residing in the green aerial part (0.9%). Dried slices, so called Mescal buttons, contain 8–9% of alkaloids (Bruhn and Holmstedt 1974).

More than 50 different alkaloidal compounds have so far been identified in peyote (Kapadia and Fayez 1970, 1973). The plant contains a large veriety of phenethylamines, mono-, di- or tri-oxygenated in the nucleus. The simple isoquinolines are all trioxygenated at the 6, 7 and 8 positions and this fact led early to the assumption that they evolve by cyclisation of 3,4,5-trioxygenated phenethylamines. The major alkaloids are the phenolic anhalamine, anhalidine, anhalonidine and pellotine and the non-phenolic anhalinine, anhalonine, lophophorine and O-methylanhalonidine. The minor non-phenolic alkaloid peyophorine contains an unusual N-ethyl substituent. The isomeric phenolic compounds isoanhalamine, isoanhalidine, isoanhalonidine and isopellotine are trace constituents that are probably formed by cyclisation of the same precursor 4,5-dimethoxy-3-hydroxyphenethylamine as their counterparts, the major phenolic alkaloids, but with cyclisation taking place para to the phenolic group instead of ortho.

The dehydroderivatives of the main phenolic alkaloids and the glyoxylic and pyruvic acid condensation products, peyoruvic acid and peyoxylic acid, are interesting in the light of biosynthetic considerations, as these compounds may be intermediates in the biogenetic transformation of phenethylamines into tetrahydroisoquinolines (cf. Lundström 1983).

The non-phenolic alkaloids including a carbon-1 substituent all possess the S-configuration. Curiously, their phenolic counterparts possess the R-confirguration (cf. Lundström 1983)

Lophophora diffusa is a species closely related to *L.williamsii.* This cactus mainly contains phenolic tetrahydroisoquinolines and very little of non-phenolic tetrahydroisoquinolines and phenethylamine derivatives (Bruhn and Holmstedt 1974).

2.2 Chenopodiaceae Species

The Chenopodiaceae family is closely related to the Cactaceae, both belonging to the order of Caryophyllales. Four species of this family are known to contain simple tetrahydroisoquinoline alkaloids of the same kind as found in cactus species (Table 3). *Salsola arbuscula* was the first known natural source of salsoline and salsolidine (Orekhov and Proskurnina 1934). No phenethylamine derivatives are reported from the Chenopodiaceae species containing tetrahydroisoquinolines. However, another species of this family, *Haloxylon salicornum* (Moq.-Tand.) Boiss. is rich in substituted phenethylamines (Michel and Sandberg 1968) which further points at the close relationship with the Cactaceae family.

Table 3. Isoquinoline containing species of the Chenopodiaceae

Species	Alkaloids
Corispermum leptopyrum L.	Salsoline, salsolidine
Haloxylon articulatum (Cav.) Bge.	Carnegine, N-methylisosalsoline
Salsola arbuscula Pall.	Salsoline, salsolidine
Salsola kali L.	Salsoline, salsolidine

2.3 Fabaceae Species

A few species of the Fabaceae family (Table 4) contain the tetrahydroisoquinoline alkaloids salsoline and salsolidine that are also found in the Cactaceae and the Chenopodiaceae. It is interesting to note that the *Alhagi* and *Desmodium* species investigated by Ghosal et al. (1973a, b, 1974) also contain a variety of substituted β-phenethylamines. The Fabaceae family belongs to the order of Fabales, that is deeply

Table 4. Isoquinoline containing species of the Fabaceae

Species	Alkaloids
Alhagi pseudalhagi (Bieb.) Desv.	Salsolidine
Calycotome spinosa Link	Calycotomine
Cassia siamea Lam.	Siamine (isoquinolone)
Desmodium cephalotes Wall.	Salsolidine
Desmodium tiliaefolium G. Don.	Salsoline
Genista purgans L.	Salsoline, salsolidine

diverging from the Caryophyllales, and the presence of β-phenethylamines and simple tetrahydroisoquinolines in a few species of the Fabaceae is probably a result of convergent evolution. *Desmodium tiliaefolium* is a unique species containing, apart from β-phenethylamines and simple tetrahydroisoquinolines, also indole-3-alkylamines. Calycotomine isolated from *Calycotome spinosa* was the first known simple tetrahydroisoquinoline with a hydroxymethyl substituent at carbon-1 (White 1944).

3 Plants Containing Isoquinoline Amino Acids

Three 3-carboxyl substituted simple tetrahydroisoquinoline alkaloids are known from natural sources (Table 5). The compound found in *Euphorbia myrsinites* has been shown to originate from m-tyrosine (Müller and Schütte 1971).

The seeds of the Fabaceae species of Table 5 contain large amounts of L-dopa (Bell et al. 1971, Daxenbichler et al. 1972) and biosynthetic experiments in a callus culture of *Stizolobium hassjoo* have shown that this amino acid is a direct precursor of the cyclized compounds. The origin of the ring closing units (carbon-1 and its substituent) remains unclear. Pyruvate did not serve at all as a precursor and acetate was only poorly incorporated in this two-carbon unit (Obata-Sasamoto and Komamine 1981).

Table 5. Plants containing isoquinoline amino acids

Family/species	Alkaloid
Euphorbiaceae	
Euphorbia myrinitis L.	3-Carboxy-6-hydroxy-1-methyltetrahydro-isoquinoline
Fabaceae	
Mucuna deeringiana (Bort.) Merr	3-Carboxy-6,7-dihydroxy-1-methyltetrahydro-isoquinoline
Mucuna mutisiana D.C.	3-Carboxy-6,7-dihydroxytetrahydroisoquinoline
Stitzolobium hassjoo (callus culture)	3-Carboxy-6,7-dihydroxytetrahydrosioquinoline, 3-Carboxy-6,7-dihydroxy-1-methyltetrahydro-isoquinoline

4 Plants Containing Isoquinolones and Simple Isoquinolines as Minor Constituents

The isoquinolones form a distinct group of alkaloids. Thirteen different compounds of the monomeric type (Fig. 3) are presently known (Krane and Shamma 1982, Lundström 1983). Three dimeric structures are also known, but these are not further treated in this presentation.

Several species belonging in the superorders Magnolianae and Ranunculanae contain isoquinolones as minor alkaloids (Table 6). In particular, several *Thalictrum* species of the Ranunculaceae family contain isoquinolones. Also simple isoquinolines occur as minor alkaloids in species of these orders and often isoquinolones and simple isoquinolines co-occur in the same species. Major alkaloids in these plants are benzylisoquinolines or more complex alkaloids derived from benzylisoquinoline units.

The first example of a naturally occurring isoquinolone was doryanine isolated by Gharbo ,et al. (1965) from the Australian sassafras tree *Doryphora sassafras,* and workers of the same laboratory later isolated doryphornine from the same species (Chen et al. 1974). The authors assumed that isoquinolones may be formed in the plant by oxidation of benzylisoquinolines or similar alkaloids, and this view now appears to be generally accepted (Krane and Shamma 1982). Indirect support for this hypothesis may be the presence of the bisbenzylisoquinoline alkaloid thalicarpine and its two oxidation products hernandaline and N-methyl-6,7-dimethoxy-1-isoquinolone in *Hernandia ovigera* (Cava and Bessho 1966).

It has been suggested (Urzúa and Cassels 1978) that the simple isoquinolines in this group of plants may be formed by reduction of isoquinolones, e.g. corypalline by reduction of doryfornine. Another possibility would be that these plants

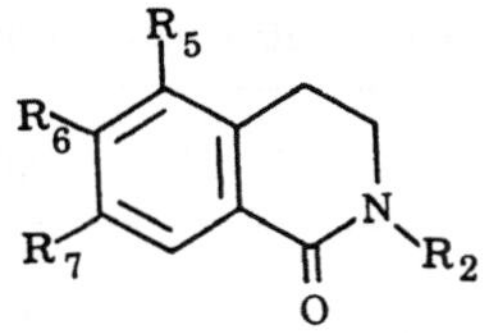

Fig. 3. Formulas of isoquinolone alkaloids

	R_2	R_5	R_6	R_7
Corydaldine	H	H	CH_3O	CH_3O
N-Methylcorydaldine	CH_3	H	CH_3O	CH_3O
N-Methylthalidaldine	CH_3	CH_3O	CH_3O	CH_3O
Noroxyhydrastinine	H	H	O - CH_2 - O	
Oxyhydrastinine	CH_3	H	O - CH_2 - O	
Thalflavine	CH_3	CH_3O	O - CH_2 - O	
Thalifoline	CH_3	H	CH_3O	HO

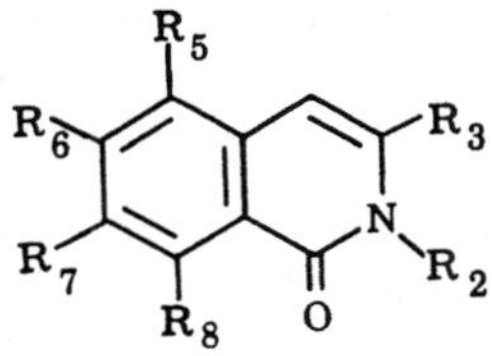

	R_2	R_3	R_5	R_6	R_7	R_8
Doryaninine	CH_3	H	H	O - CH_2 - O		H
Doryfornine	CH_3	H	H	CH_3O	HO	H
6,7-Methylenedioxy-isoquinolone	H	H	H	O - CH_2 - O		H
N-Methyl-6,7-di-methoxyisoquinolone	CH_3	H	H	CH_3O	CH_3O	H
Siamine	H	CH_3	H	HO	H	HO
Thalactamine	CH_3	H	CH_3O	CH_3O	CH_3O	H

possess metabolic pathways leading directly to simple isoquinolines by condensation of a phenethylamine derivative with a suitable ring-closing unit.

The experiments by Canonica et al. (1981) suggest a third plausible formation of simple isoquinolines with no carbon-1 substituent in plants mainly producing alkaloids of the benzylisoquinoline type. Treatment of the benzylisoquinoline N-methylcoclaurine with ascorbic acid oxidase, an enzyme widespread in plants, resulted in benzylic fission giving about 40% yield of corypalline, its quaternary 1,2-dehydroderivative and benzyl alcohol. Thus, both isoquinolones and simple isoquinolines may be formed directly by the action of oxidative enzymes in benzylisoquinoline-producing plants.

Table 6. Plants containing isoquinolones and simple isoquinolines as minor constituents

Family/Species	Alkaloid ([a] isoquinolone)
Alangiaceae	
Alangium lamarchii Thw.	Salsoline
Annonaceae	
Annona reticulata L.	Salsolinol (Forgacs et al. 1981)
Enantia polycarpa Engl. et Diels.	Corydaldine[a]
Atherospermataceae	
Doryphora sassafras Endlicher	Corypalline, doryanine[a], doryfornine[a]
Berberidaceae	
Berberis baluchistanica Ahrendt	Corydaldine[a] (Shamma et al. 1974)
Fumariaceae	
Corydalis ambigua Cham et Schlecht.	N-Methylsalsoline
Corydalis aurea Willd.	Corypalline
Corydalis ophiocarpa Hook. et Thoms.	Corypalline, dehydrocorypalline, noroxyhydrastinine (Iwasa and Takao 1982) (Tani et al. 1978)
Corydalis pallida Pers.	Corypalline
Corydalis speciosa Maxim.	Corypalline (Tani et al. 1975)
Corydalis stricta Steph. ex DC.	N-Methylcorypalline, corypalline, pycnarrhine (Irgashev et al. 1983)
Fumaria parviflora Lam.	Noroxyhydrastinine[a] (Hussain et al. 1983[a])
Fumaria schleicheri Soy.-Will.	Oxyhydrastinine[a]
Hernandiaceae	
Hernandia ovigera L.	N-Methyl-6,7-dimethoxyisoquinolone[a], N-Methylcorydaldine[a]
Lauraceae	
Cryptocaria longifolia Kostermans	Thalifoline[a] (Bick et al. 1981)
Menispermaceae	
Archangelisia flava (L.) Merr.	Pycnarrhine (Verpoorte et al. 1982)
Pycnarrhena longifolia (Decne. ex Miq.) Beccari	Pycnarrhine
Stephania sasakii Hayata	N-Methyl-6,7-dimethoxyisoquinolone[a] (Kunitomo et al. 1981)
Nelumbonaceae	
Nelumbo nucifera Gaertn.	N-Methylheliamine
Papaveraceae	
Argemone mexicana L.	Oxyhydrastinine[a]
Papaver bracteatum Lindl. var	Corypalline, O-methylcorypalline
Arya I	N-Methylcorydaldine[a] (Theuns et al. 1983)
Papaver dubium L. var. glabrum	Oxyhydrastinine[a] (Hussain et al. 1983[a])
Papaver somniferum L.	Hydrocotarnine
Papaver urbanianum Fedde	N-Methylcorydaldine[a]

Table 6 (cont.)

Family Species	Alkaloid ([a]isoquinolone)
Ranunculaceae	
Aconitum carmichaeli Debx.	Salsolinol (D. Chen and Liang 1982)
Thalictrum alpinum L.	Noroxyhydrastinine[a], N-Methyl-6,7-dimethoxy-isoquinolone[a]
Thalictrum dasycarpum Fish et Lall	Corypalline (Kupchan and Yoshitake 1969)
Thalictrum dioicum L.	O-Methylcorypalline (Shamma and Rothenburg 1978)
Thalictrum flavum Cav.	Thaliflavine[a]
Thalictrum fendleri Engelm. ex Gray	N-Methylcorydaldine[a] N-Methylthalidaldine[a]
Thalictrum foliolosum Hook. et Thoms.	Noroxyhydrastinine[a]
Thalictrum glaucum Desf.	Noroxyhydrastinine[a] 6,7-Methylenedioxyisoquinolone[a]
Thalictrum isopyroides C.A. Mey	N-Methyl-6,7-dimethoxyisoquinolone[a]
Thalictrum minus L.	Thalactamine[a] Noroxyhydrastinine[a]
Thalictrum minus var. *adiantifolium* Hort.	Noroxyhydrastinine[a], Thalifoline[a]
Thalictrum minus var. *microphyllum* Ait.	Noroxyhydrastinine[a] Thalactamine[a] (Baser 1982)
Thalictrum polygamum Muhl.	Corypalline, noroxyhydrastinine[a] (Shamma et al. 1978)
Thalictrum revolutum D.C.	N-Methyl-6,7-dimethoxyisoquinolinium chloride
Thalictrum rugosum Ait.	Corypalline, noroxyhydrastinine[a] 6,7-Methylenedioxyisoquinolone[a]
Thalictrum uschiyami Nakai	Corypalline (Iee and Lee 1982)

[a]For specific references not included see Lundström 1983

The isoquinolone siamine found in the Caesalpiniaceae plant *Cassia siamea* has an oxygenation pattern clearly different from other isoquinolones and also carries a methyl substituent at carbon-3. This compound is probably not derived from an aromatic amino acid but rather from structurally related polyketides that occur in the same plant (Ahn et al. 1978).

References

Ahn BC, Degen U, Lienjayetz C, Pachaly P, Zymalkowski F (1978) Inhaltstoffe von *Cassia siamea.* Arch Pharm 311:569–578

Anonymous (1959) Peyotl. Bull Narcot UN Dep Social Affairs 11:16–29

Backeberg C (1966) Das Kakteenlexikon. Fischer, Jena

Baser KH (1982) Isolation and identification of anisaldehyde and three alkaloids from leaves of *Thalictrum minus* var. *microphyllum.* J Nat Prod 45:704–706

Bell EA, Nulu JR, Cone C (1971) L-Dopa and L-3-carboxy-6,7-dihydroxy-1,2,3,4-tetrahydro-isoquinoline, a new amino acid from seeds of *Mucuna mutisiana.* Phytochemistry 10:2191–2194

Bick IRC, Sévenet T, Sinchai W, Skelton BW, White AH (1981) Alkaloids of *Cryptocarya longifolia:* X-ray crystal structure of thalifoline and longifoline. Aust J Chem 34:195–207
Bruhn JG (1971) *Carnegia gigantea:* The Saguaro and its uses. Econ Bot 25:320–329
Bruhn JG, Holmstedt B (1974) Early peyote research. An interdisciplinary study. Econ Bot 28: 353–390
Bruhn JG, Lundström J (1976) Alkaloids of *Carnegia gigantea.* Arizonine, a new tetrahydroisoquinoline alkaloid. Lloydia 39:197–203
Canonica L, Galliani G, Lesma G, Tollari S (1981) Enzymatic transformations of alkaloids utilizing pure enzymes. Proc 1st Int Conf Chem Biotechnol Biol Act Nat Prod 1st vol. 2:522–527
Cava MP, Bessho K (1966) Hernandia alkaloids II. Hernandaline, a new elaborated aporphine structural type. Tetrahedron Lett 36:4279–4282
Chen CR, Beal JL, Doskotch RW, Mitscher LA, Svoboda GH (1974) A phytochemical study of *Doryphora sassafras* II. Isolation of eleven crystalline alkaloids from the bark. Lloydia 37: 493–500
Chen D, Liang X (1982) Studies on the constituents of lateral root of *Aconitum carmichaeli* Debt (Fu Zi), a traditional Chinese medicine I. Isolation and structural determination of salsolinol. Xaoxue Xuebao 17:793–794, CA 98:40465k
Dahlgren R (1975) A system of classification of the angiosperms to be used to demonstrate the distribution of characters. Bot Not 128:119–147
Daxenbichler ME, Kleiman R, Wiesleder D, VanEtten CH, Carlson KD (1972) A new amino acid, (–)-1-methyl-3-carboxy-6,7-dihydroxy-1,2,3,4-tetrahydroisoquinoline, from velvet beans. Tetrahedron Lett 18:1801–1802
Djerassi C, Smith CR, Marfey SP, McDonald RN, Lemin AJ, Figdor SK, Estrada H (1954) Alkaloid studies III. Isolation of pilocereine and anhalonidine from four cactus species. J Am Chem Soc 76:3215–3217
Forgacs P, Desconclois JF, Mansard D, Provost J, Tiberghien R, Tocquer J, Touché A (1981) Dopamine et alcaloïdes tétrahydroisoquinoléiques *d'Annona reticulata* L. Annonacées. Plant Med Phytother XV:10–15
Gharbo SA, Beal JL, Schlessinger RH, Cava MP, Svoboda GH (1965) A phytochemical study of *Doryphora sassafras.* I. Isolation of eight crystalline alkaloids from the leaves. Lloydia 28:237–244
Ghosal S, Metha R (1974) β-Phenethylamine and tetrahydroisoquinoline alkaloids of *Desmodium cephalotes.* Phytochemistry 13:1628–1629
Ghosal S, Srivastava RS (1973a) Chemical investigation of *Alhagi pseudalhagi* (Bieb.) Desv.: β-Phenethylamine and tetrahydroisoquinoline alkaloids. J Pharm Sci 62:1555–1556
Ghosal S, Srivastava RS (1973b) β-Phenethylamine, tetrahydroisoquinoline and indole alkaloids of *Desmodium tiliaefolium.* Phytochemistry 12:193–197
Hussain SF, Minard RD, Freyer AJ, Shamma M (1983a) New alkaloids from *Fumaria parviflora.* J Nat Prod 44:169–178
Hussain SF, Nakkady S, Khan L, Shamma M (1983b) Oxyhydrastinine, an isoquinolone alkaloid from the Papaveraceae. Phytochemistry 22:319–320
Iee IR, Lee MM (1982) Studies of chemical constituents of the genus *Thalictrum* in Korea I. Alkaloids from the root of *Thalictrum uchiyamai* Nakai. Saegyate Hakhoe Chi 13:132–135, CA 98:221674y
Irgashev T, Israilov IA, Batsuren D, Yunosov MS (1983) *Corydalis stricta* alkaloids. Khim Prir Soedin 1983:490–493, CA 100:82695a
Iwasa K, Takao N (1982) Formation of alkaloids in *Corydalis ophiocarpa* callus culture. Phytochemistry 21:611–614
Kapadia G, Fayez MBE (1970) Peyote constituents: Chemistry, biogenesis and biological effects. J Pharm Sci 59:1699–1726
Kapadia G, Fayez MBE (1973) The chemistry of peyote alkaloids. Lloydia 36:9–35
Krane BD, Shamma M (1982) The isoquinolone alkaloids. J Nat Prod 45:377–384
Kunitomo J-I, Murakami Y, Oshikata M, Shingu T, Lu S-T, Chen I-S, Akasu M (1981) The alkaloids of *Stephania sasakii* Hayata (13). On the tertiary minor bases. Yakugaku Zasshi 101:431–436

Kupchan SM, Yoshitake A (1969) Thalictrum alkaloids IX. The isolation, structural elucidation and synthesis of thalisopavine. J Org Chem 34:1062–1065

Lundström J (1971) Biosynthetic studies on mescaline and related cactus alkaloids. Acta Pharm Suec 8:275–302

Lundström J (1983) Simple isoquinoline alkaloids. In: Brossi A (ed) The Alkaloids, vol. XXI. Academic Press, London New York, pp 255–327

Mata R, McLaughlin JL (1982) Cactus alkaloids 50. A comprehensive tabular summary. Rev Latinoam Quím 12:95–117

Michel K-H, Sandberg F (1968) Occurrence of tyramine and N-methyltyramine in *Haloxylon salicornum.* Acta Pharm Suec 5:67–70

Mohamed YAH, Chang C-j, McLaughlin JL (1979) Cactus alkaloids XXXIX. A glucotetrahydroisoquinoline from the Mexican cactus *Pterocereus gaumeri.* J Nat Prod 42:197–202

Müller P, Schütte HR (1971) Biosynthesis and metabolism of phenolic latex amino acids. Biochem Physiol Pflanz 162:234–244

Obata-Sasamoto H, Komamine A (1981) Biosynthesis of 3-carboxy-67,-dihydroxy-1,2,3,4-tetrahydroisoquinoline in a callus culture of *Stizolobium hassjoo.* Z Naturforsch 36:921–924

Ordaz C, Ferrigni NR, McLaughlin JL (1983) Dehydroheliamine, a trace alkaloid from the saguaro, *Carnegia gigantea* (Cactaceae). Phytochemistry 22:2101–2102

Orekhov A, Proskurnina N (1934) Alkaloids of *Salsola richteri* II. Constitution of salsoline. Ber 67B:878–884

Pummangura S, Mohamed YAH, Chang C-j, McLaughlin JL (1982) Two simple tetrahydroisoquinoline alkaloid N-oxides from cacti. Phytochemistry 21:2375–2377

Ranieri RL, McLaughlin JL (1976) β-Phenethylamine and tetrahydroisoquinoline alkaloids from the Mexian cactus *Dolichothele longimamma.* J Org Chem 41:319–323

Ranieri RL, McLaughlin JL (1977) Cactus alkaloids XXXI. β-Phenethylamines and tetrahydroisoquinolines from the Mexican cactus *Dolichothele uberiformis.* Lloydia 40:173–177

Shamma M, Rothenberg AS (1978) The alkaloids of *Thalictrum dioicum.* Lloydia 41:169–178

Shamma M, Foy JE, Miana GA (1974) Baluchistanamine. Novel type dimeric isoquinoline alkaloid. J Am Chem Soc 96:7809–7810

Tani C, Nagakura N, Sugiyama N (1975) Studies on the alkaloids of Papaveraceous plants XXI. Alkaloids of *Corydalis speciosa* Maxim. The isolation of the tertiary bases by the multibuffered DCCC Yakugaku Zasshi 95:838–842

Tani C, Nagakura N, Kuriyama C (1978) Studies on the alkaloids of papaveraceous plants XXXI. Separation of the alkaloids by droplet countercurrent chromatography. The alkaloids of *Corydalis ophiocarpa* Hook. et Thoms. Yakugaku Zasshi 98:1243–1251

Theuns HG, Vlietstra EJ, Salemink CA (1983) Corypalline and O-methylcorypalline, two alkaloids from *Papaver bracteatum.* Phytochemistry 22:247–250

Urzúa A, Cassels BK (1978) Alkaloid chemosystematics, chemotaxonomy and biogenesis in the Atherospermataceae. Lloydia 41:98–113

Verpoorte R, Siwon J, van Essen GFA, Tieken M, Baerheim-Svendsen A (1982) Studies on the Indonesian medicinal plants VII. Alkaloids of *Archangelisia flava.* J Nat Prod 45:582–584

White EP (1944) Alkaloids of the Leguminosae. N Z J Sci Tech 25B:137–162

Erythrina Alkaloids

A. H. JACKSON[1]

1 Introduction

The *Erythrina* genus consists of a range of some 108 species of magnificent orange or red-flowered trees, shrubs and herbaceous plants. These are widely distributed throughout the tropical and semi-tropical regions of the world, and several of the more common species are used for decorative purposes in gardens and city streets. Largely through the efforts of the late Dr. B.A. Krukoff, the genus has been fully classified into 5 sub-genera and 26 sub-sections on the basis of the characters of the flowers, fruits and inflorescence. (Krukoff and Barnaby 1974).

The close botanical relationships between all the members of the *Erythrina* genus is a paralleled by the presence of a series of spirocyclic isoquinoline alkaloids which are found in all parts of the plants. There are three main types of *Erythrina* alkaloids usually referred to as dienoid (1) alkenoid (2) and lactonic (3) alkaloids. The dienoid alkaloids are characterised by a conjugated diene system in the A and B rings, whilst the alkenoids have a 1,6-double bond in the A-ring; α- and β-erythroidine (3a) and (3b) have a lactone ring in place of the benzene ring. The erythrinane skeleton is almost unique to *Erythrina* genus and the only other examples are in the *Cocculus* genus (cf. Dyke and Quessy 1981) and in *Pachygone ovata* (Menispermaceae) (Bhat et al. 1980) although alkaloids with the homoerythrinane skeleton can be found in *Cephalotaxus* species.

An important factor in stimulating the early structural studies of *Erythrina* alkaloids during the 1940's was the discovery of their curariform activity (cf. Craig 1955); this appears to be associated with the spiroamine moiety, and in contrast with the curare alkaloids, which are quaternary salts, the *Erythrina* alkaloids are all tertiary amines. Many other types of pharmacological activity have also been discovered with *Erythrina* alkaloids, e.g. sedative, hypotensive, CNS depressing, laxative and diuretic properties.

1 Department of Chemistry, University College, Cardiff, United Kingdom

The Chemistry and Biology of Isoquinoline
Alkaloids, ed. by Phillipson et al.

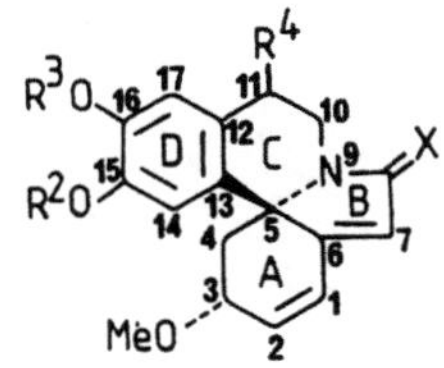

1

DIENOID

Alkaloid (1)	R²	R³	R⁴	X
a) Erysodine	Me	H	H	H_2
b) Erysovine	H	Me	H	H_2
c) Erysopine	H	H	H	H_2
d) Erythraline	$-CH_2-$		H	H_2
e) Erythrabine	Me	Me	H	=0*
f) Erysotramidine	Me	Me	H	=0
g) 11-β-methoxyerysodine	Me	H	β-OMe	H_2
h) Erythristemine	Me	Me	β-OMe	H_2
i) 11-α-Hydroxyerysodine	Me	H	α-OH	H_2
j) 11-β-Hydroxyerysodine	Me	H	β-OH	H_2
k) 11-β-Methoxyerythraline	$-CH_2-$		β-OMe	H_2
l) 8-Oxo-11-β-methoxyerythroline	$-CH_2-$		β-OMe	=0
m) 8-Oxoerythraline	$-CH_2-$		H	=0
n) Crystamidine	$-CH_2-$		H	=0

* 11,11-double bond

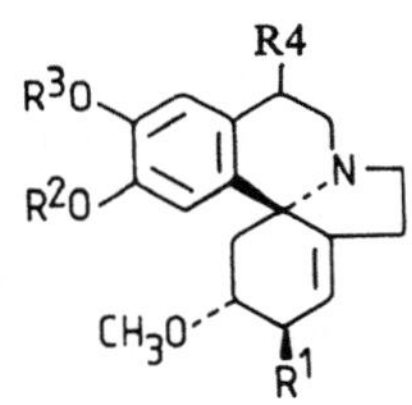

2

ALKENOID

Alkaloid (2)	R_1	R_2	R_3	R_4
a) Erythratidine	α-OH	Me	Me	H
b) *epi*-Erythratidine	β-OH	Me	Me	H
c) Erythratine	OH	$-CH_2-$		H
d) 11–Hydroxoyerysotine	OH	Me	H	OH

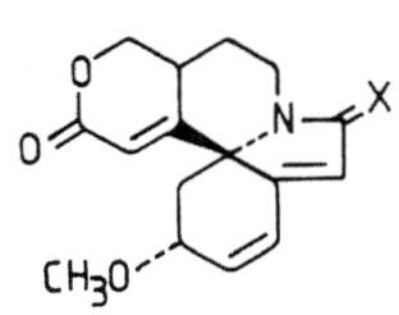

3

ERYTHROIDINES

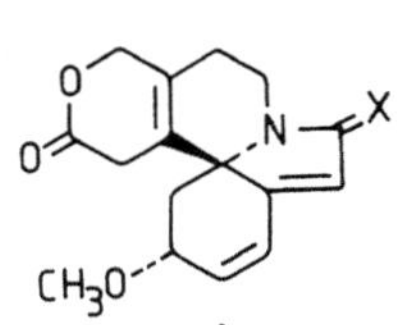

4

α-Erythroidine	β-Erythroidine
(3a) $X=H_2$	(3b) $X=H_2$
8-Oxo-α-erythroidine	8-Oxo-β-erythroidine
(4a) X=0	(4b) X=0

Fig. 1

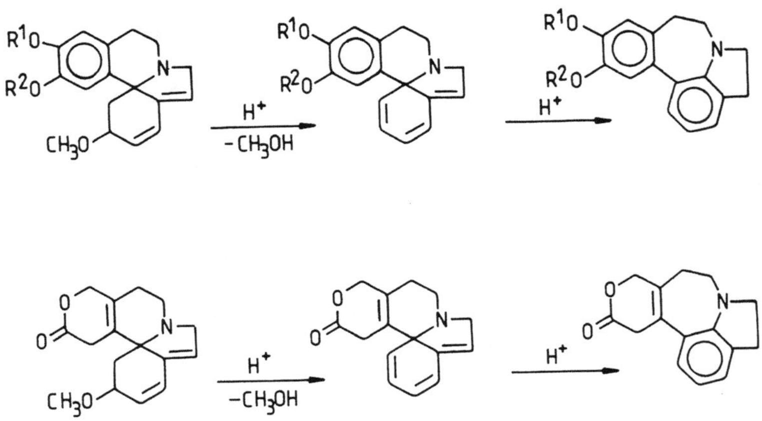

Fig. 2. Apo-rearrangement of *Erythrina* alkaloids

2 Structure Determination of *Erythrina* Alkaloids

2.1 Hydrogenation and Acid-Catalysed Degradation

Key reactions in the early structural studies were hydrogenation and acid-catalysed degradation. Thus hydrogenation of "dienoid" alkaloids led to a marked diminution in the UV absorption at 235 nm. due to saturation of the double bonds, whilst mild acid treatment afforded a strong UV absporption at 313 nm due to elimination of methanol and formation of a conjugated triene system (Carmack et al. 1951, Boekelheide et al. 1953, Mondon 1970). On heating with acid, rearrangement occurred to an "apo" derivative (Fig. 2) by a process somewhat akin to the well-known morphine/apomorphine rearrangement (cf. Boekelheide 1960).

These results led to the postulation of the spirocyclic erythrinane skeleton and this was confirmed by synthesis of the parent nucleus by Belleau (1953). The structures of the two lactonic alkaloids (3) were also elucidated by Boekelheide (1960) and coworkers, who recognised their close structural identity to the other "aromatic" erythrina alkaloids (1); the lactonic alkaloids also underwent an "apo"-type rearrangement under drastic acidic conditions. Subsequently the structures of both the aromatic and lactonic alkaloids were confirmed by X-ray crystallography of erythraline hydrobromide (Nowacki and Bonsma 1958) and of the erythroidines (Hanson 1963). The spiro centre was shown to have the same configuration in all the erythrina alkaloids by use of optical rotatory measurements (Weiss and Ziffer 1963, Beecham 1971).

2.2 Mass Spectra

In recent years mass and nmr spectral methods have been increasingly utilised in the structure determination of *Erythrina* alkaloids. Mass spectra have proved particularly useful in preliminary screening of *Erythrina* species for their alkaloid content especially in conjunction with gas chromatography (Games et al. 1974, Hargreaves et al.

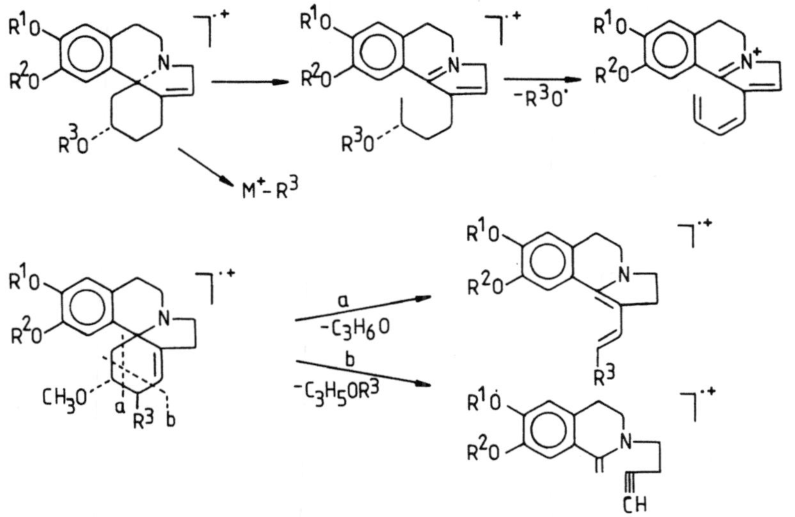

Fig. 3. Mass spectral fragmentation of dienoid and alkenoid *Erythrina* alkaloids

1974, Millington et al. 1974). The alkaloids usually occur as the free bases, although some phenolic alkaloids may occur as their glycosides; the "free" alkaloid fraction, and the "liberated" alkaloid fraction (obtained by hydrolysis of glycosides) can be analysed by GC-MS after conversion of any free hydroxyl groups to their trimethyl silyl derivatives. A detailed analysis of the mass spectra of a range of *Erythrina* alkaloids (Boar and Widdowson 1970) showed that the predominant fragmentation of the dienoid and lactonic alkaloids corresponded to cleavage of methyl or methoxyl from the C-3 methoxyl unit whereas although the alkenoid alkaloids also showed similar fragmentations, another prominent peak at M^+-58 corresponding to a retro-Diels Alder fragmentation with loss of C-3, C-4 and the C-3 methoxyl group also occurred (Fig. 3). In some instances mass spectrometry allows one to distinguish between epimers, e.g. in the case of erythratidine (2a) and epierythratidine (2b) (Fig. 4). The derivatized alkaloids containing $OSi(CH_3)_3$ groups also show very similar fragmentation to the parent alkaloids, but there is usually a strong peak at m/z 73 corresponding to cleavage on the silyl residue, and derivatives with a secondary $OSi(CH_3)_3$ group also fragment with loss of 89 and 90 mass units. Mass spectrometry is particularly useful for microscale analysis of very small amounts of material (μg or less).

2.3 NMR Spectra

NMR spectra can give detailed information on structure and stereochemistry, especially with the availability of very high field instruments, even on quantities as small as 10 μg or less. The dienoid alkaloids are readily distinguished by the presence of an ABX system corresponding to the three olefinic protons, whereas the alkenoid alkaloids only have a single olefinic resonance and the lactonic alkaloids lack the two aromatic singlet resonances.

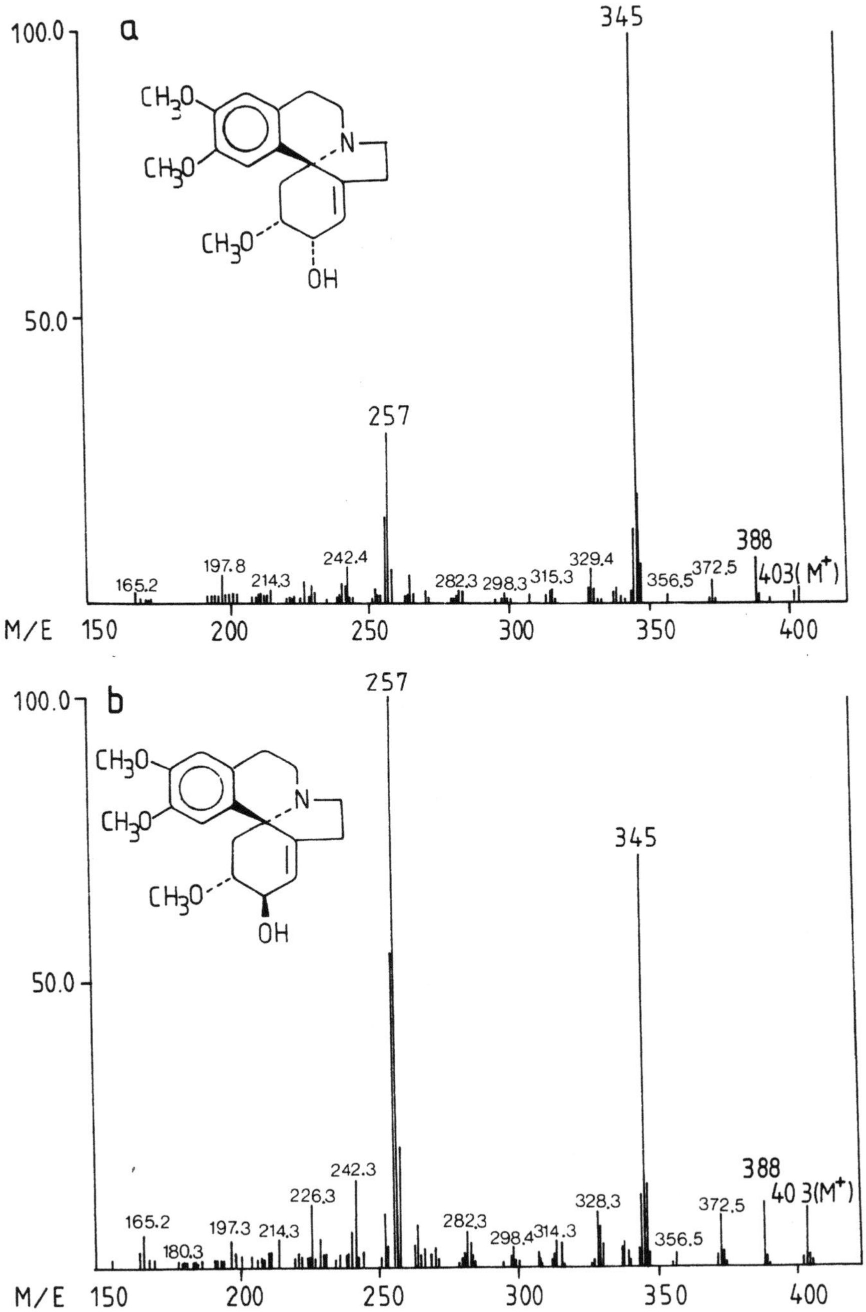

Fig. 4 a, b. Mass Spectra of the TMS Derivatives of **a** erythratidine and **b** epi-erythratidine

Early examples of the use of NMR spectroscopy in structure determination were to distinguish between erysodine (1a) and erysovine (1b) (Barton et al. 1966) and to establish the structures of erythratine (2c) (Barton et al. 1966) and of 11-methoxyerythraline (lk) (Letcher 1971). The epimeric alkaloids erythratidine (2a) and epi-erythratidine (2b) may easily be distinguished from each other by NMR spectroscopy. INDOR and NOE methods promise to be very useful techniques in structural studies, e.g. dihydroerysodine (Juichi et al. 1978), erythraline (ld), erysotramidine (lf) and 11-hydroxyerysotine (2d) (Ito et al. 1973).

3 Biosynthesis

Early experiments on ***Erythrina*** alkaloid biosynthesis carried out by Leete and Ahmad (1966) confirmed that tyrosine is a precursor of the erythroidines and that a symmetrical intermediate is involved. Several speculative schemes for ***Erythrina***

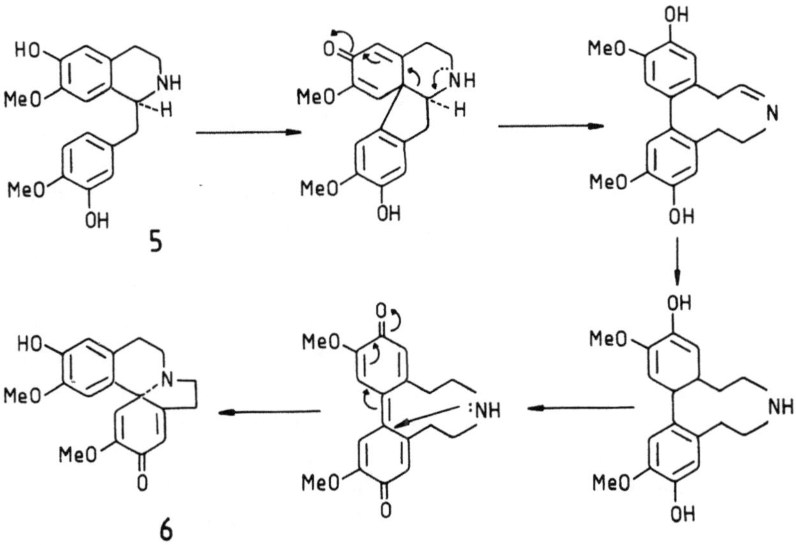

Fig. 5. Biosynthesis of *Erythrina* Alkaloids

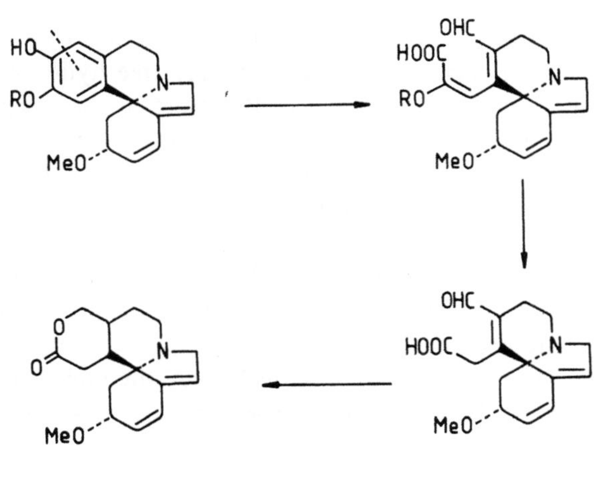

Fig. 6. Biosynthesis of Erythroidines

alkaloid biosynthesis based on phenolic oxidative coupling were put forward in the 1960's (cf. e.g. Barton and Cohen 1957), but a series of experiments carried out by Barton et al. 1970a, 1974) led to the conclusion that N-norprotosinomenine (5) was the specific precursor rather than any of its close relatives or an open chain intermediate. The pathway now accepted involves phenolic oxidative coupling and ring opening to form a dibenzazonine, which subsequently ring closes to erysodienone (6) (Fig. 5). It is thought that formation of lactonic alkaloids (Fig. 6) occurs at a relatively late stage in the pathway (cf. Boekelheide and Grundon 1953) as tracer studies show that C_{16} was lost in the oxidative cleavage process (Barton et al. 1974). Isolation of the N-methyl dibenzazonine alkaloid, erybidine (Ito et al. 1973) and the detection of erysodienone and N-norprotosinomenine (Ghosal et al. 1971) provided additional evidence for the general pathway shown in Fig. 5.

4 Chemotaxonomic Studies

Our own work in Cardiff has largely been concerned with chemotaxonomic studies of *Erythrina* alkaloids using GC-MS as the primary analytical technique for screening a wide range of *Erythrina* species. (Games et al. 1974, Barakat et al. 1977, Abdullah et al. 1979, Jackson and Chawla 1982, Jackson et al. 1982). The alkaloids occur in all parts of the plants and earlier studies in other laboratories have been carried out with roots, bark, leaves and seeds as well as whole plants. Most of our studies have been carried out with seeds, and to a lesser extent leaves. The alkaloid content has varied from 0.5–1% but is sometimes much less than 0.1%. Initial studies were usually carried out with 5–10 g samples of seeds which were extracted for alkaloids with methanol, and separated into "free" alkaloids or alkaloid glycoside fractions; hot acidic hydrolysis of the latter then afforded a "liberated" alkaloid fraction. In our more recent work an initial extraction of the seeds has also been carried out with hexane to remove lipid material prior to the methanol extraction; acid extraction of the hexane extract then afforded a "hexane" fraction with a similar pattern of alkaloids to the "free" alkaloids. In this way a somewhat higher recovery of alkaloids was obtained from the samples in preparative scale isolations, than by methanol extraction alone.

The crude alkaloid extracts ("hexane", "free", or "liberated" fractions) were each treated with N,O-bistrimethylsilylacetamide to derivatize any free hydroxyl groups, and then subjected firstly to GC and then to GC-MS. Most of the work so far has been carried out with packed columns coated with OV-17 stationary phase, but during the last year or two encouraging results have been obtained with capillary columns, which appear to give better resolution, particularly of isomeric alkaloids.

The crude alkaloid extracts can also be examined directly by field desorption (or field ionisation) mass spectrometry (Games et al. 1973). This has proved a particularly useful method for analysing mixtures of similar compounds because little or no fragmentation occurs. Thus one observes essentially only the molecular ions, and this provides a useful crosscheck with the GC-MS results. High resolution measurements of the field ionisation spectra also provide information about the composition of the molecular ions.

5 Recent Studies on *Erythrina* Alkaloids

5.1 *E.lysistemon*

Following the preliminary screening studies further large scale extraction (100–500 g amounts if available) of each particular species is then carried out, and the "free" and "liberated" alkaloid fractions examined chromatographically, preferably by semi preparative HPLC rather than by ordinary open column methods. Taking *E.lysistemon* as an example, a re-investigation of the alkaloid content of the seeds has shown the presence of three new 11-oxygenated and two 8-oxygenated alkaloids (Chunchatprasert 1983). The structures of these new alkaloids were deduced very largely from their mass and NMR spectra. Thus the molecular formula of 11-β-methoxyerysodine (lg) ($C_{19}H_{23}NO_4$) was first established by high resolution mass spectrometry, and the fragmentation pattern was clearly that of a dienoid alkaloid with prominent peaks at M^+ (329), (M^+ - 15) and (M^+ - 31); there were also additional ions at M^+ - 32 (loss of MeOH) M^+ - 47 (loss of Me and MeOH) M^+ - 63 (M^+ - OMe and MeOH) presumably corresponding to the elimination of the 11-methoxyl group and hydrogen from the 10-position. The TMS derivative gave a molecular ion at m/z 401 confirming the presence of only one free hydroxyl function. Analysis of the 400 MHz NMR spectrum (Fig. 6) and comparisons with that of erysodine itself, clearly showed that the new methoxyl group was at the 11-position, and this was confirmed by a series of decoupling experiments. Final proof of structure was obtained by methylation of the phenolic hydroxyl group to afford the known alkaloid erythristemine (3h) (Barton et al. 1970b) the NMR spectrum of which is also shown in Fig. 7.

11-α-and 11-β-Hydroxyerysodine (li) and (lj) were also isolated from *E.lysistemon* and their structures likewise determined by NMR spectroscopic comparisons with each other, with erysodine (la) and with 11-β-methoxyerysodine (lg). The known alkaloid 11-β-methoxyerythraline (lk) was also identified both in the preliminary GC-MS experiments, and by isolation by HPLC; even more interestingly, however, 8-oxo-11-β-methoxyerythraline (II) and 8-oxo-erythraline (lm) were also found in *E.lysistemon.* This is the first known occurrence of an 8,11-dioxygenated *Erythrina* alkaloid, although we first isolated 8-oxoerythraline from *E.chiriquensis* seeds (Chunchatprasert 1983).

5.2 *E. chiriquensis*

GC-MS studies of *E.chiriquensis* showed that erythraline (le) was the main component of the "free" alkaloid fraction, although traces of erysodine and erysotine were found; the "liberated" fraction contained erysopine, erysodine and erysovine.

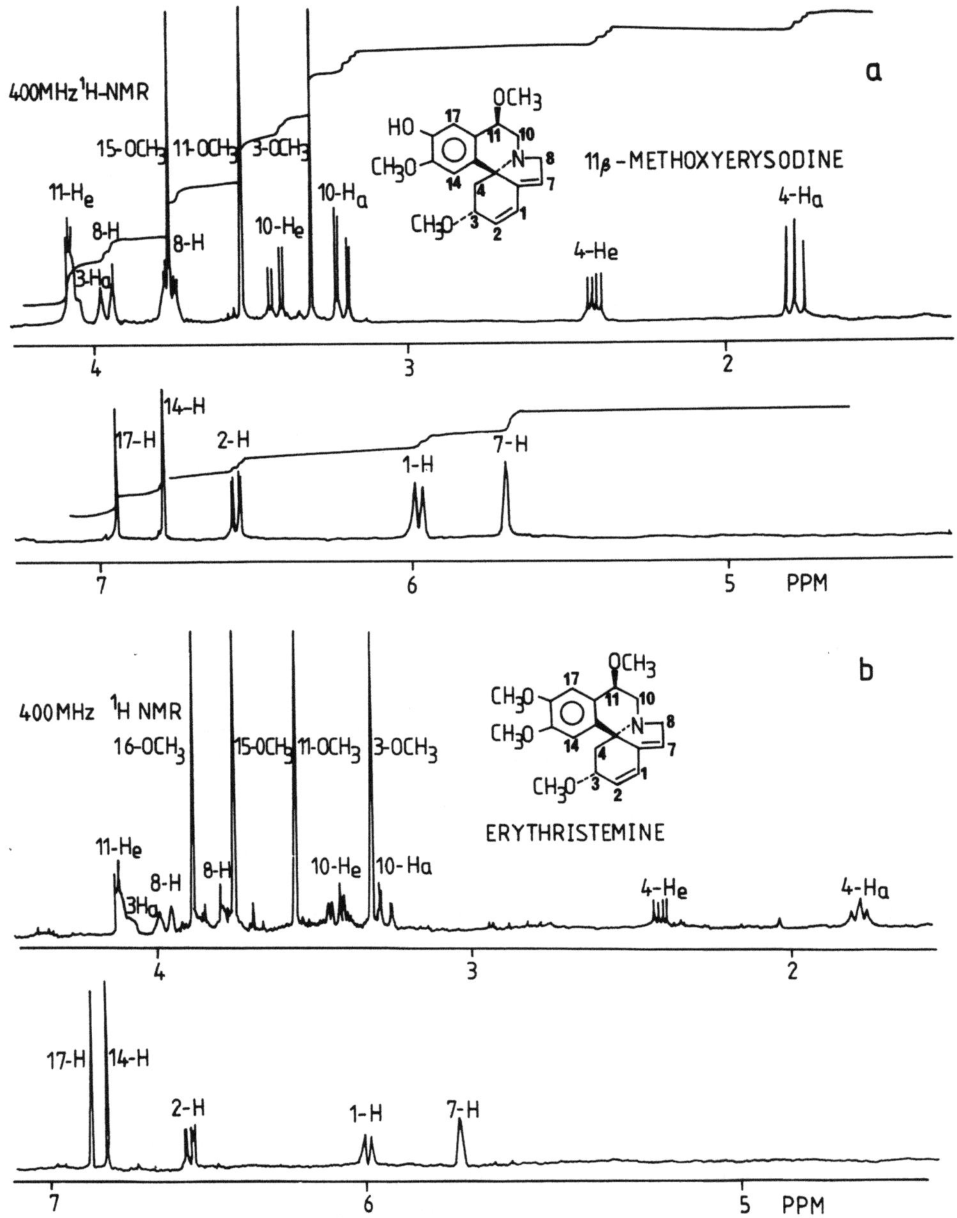

Fig. 7 a, b. 400 MHz proton NMR Spectra of **a** 11-β-methoxyerysodine and **b** erythristemine

However, semipreparative HPLC analysis of the free alkaloid fraction (Fig. 8a) revealed the presence of another alkaloid as well as erythraline. The mass spectrum (Fig. 8b) showed a molecular ion at m/z311 with major fragments at M^+-15 and M^+-31 confirming that the new alkaloid was of a dienoid type. Comparisons of the ^1H NMR spectrum with that of erythraline (ld) showed considerable similarities but there were no resonances corresponding to an 8-methylene group and most of the other reso-

Fig. 8 a,b. Identification of 8-oxoerythraline in *E.chiriquensis* seeds **a** HPLC separation of 8-oxoerythraline *(1)* and erythraline *(2)* **b** Mass spectrum of 8-oxoerythraline

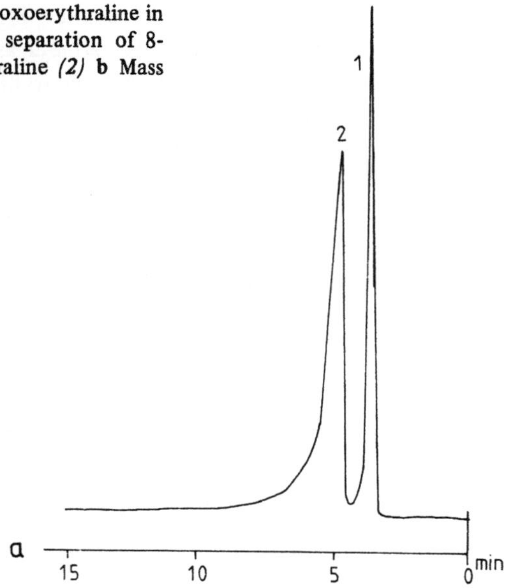

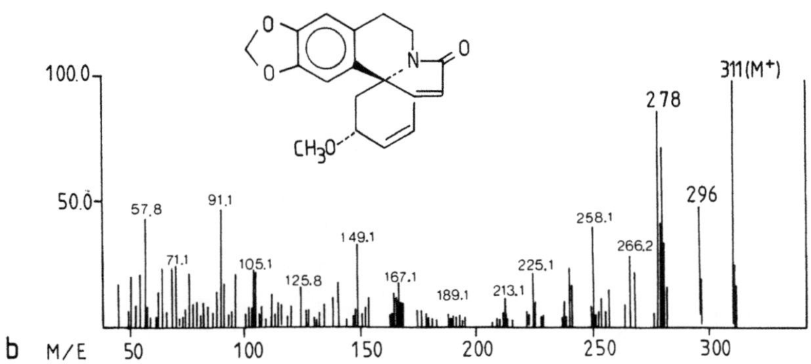

nances were shifted somewhat downfield. These results together with the mass spectral data were consistent with the structure, 8-oxoerythraline (lm) and this was confirmed by a series of decoupling experiments; the ^{13}C NMR spectrum (cf. Chawla et al. 1983) also clearly showed the presence of the 8-oxo-group by the very low field shift of C-8 to 171.2 ppm, a value closely in accord with that observed for the corresponding carbon resonance of the known 8-oxo-alkaloid erysotramidine (lf) (170.8 ppm). Apart from *E.lysistemon* and *E.chiriquensis* we have also found 8-oxo-erythraline in *E.abyssinica* and *E.tahitensis.*

The isolation of these new 8-oxo-alkaloids adds two further examples to a growing family of 8-oxo-erythrina alkaloids, other examples being erythrabine (ld) and erysotramidine (lf) (Ito et al. 1973) isolated from *E.arborescens* and crystamidine (ln) from *E.crysta-galli* (Ito et al. 1973) whilst two other oxo-alkaloids have been found in *Cocculus laurifolius* (Uprety and Bhakuni 1975), (Bhakuni and Jain 1980).

In our own studies in Cardiff we have also isolated 8-oxo-α- and β-erythroidine (4a) and (4b) (Chawla et al. 1982) from *E.berteroana* as well as 8-oxo-erysodine from *E.tahitensis* (Chunchatprasert 1983); it thus seems likely that 8-oxo- analogues of many of the other dienoid alkaloids are also likely to be found accompanying the parent alkaloid, albeit in small amounts. However, the relatively high polarity makes them less amenable to GC-MS and only 8-oxo-β-erythroidine (4b) has been detected by GC, presumably because the derivatization of the lactone residue by the silylating reagent makes it much more volatile. (The TMS derivative of β-erythroidine itself has the shortest retention time of all the *Erythrina* alkaloids we have studied so far).

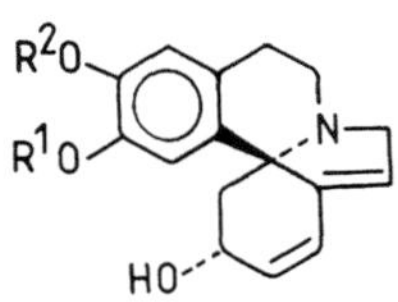

7

Fig. 9.

Alkaloid (7)	R^1	R^2
a) Erysoline	H	Me
b) Erysonine	Me	H
c) Erythrocarine	$-CH_2-$	

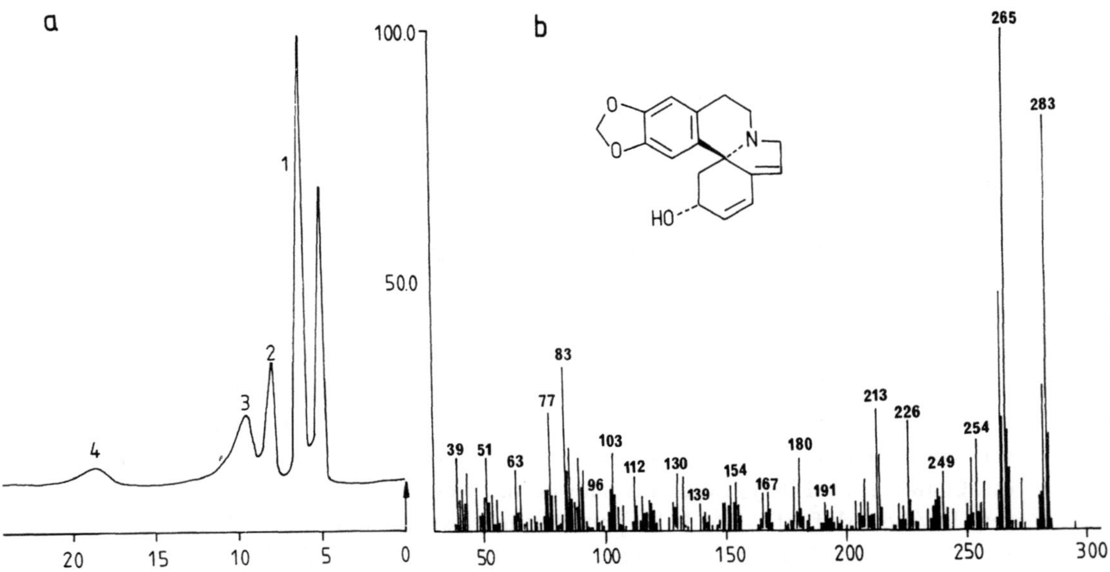

Fig. 10. **a** HPLC separation of "free" alkaloids from *E. caribea* seeds. *1* erysotrine; *2* erythraline; *3* erysovine and erysovine; *4* erythrocarine (a new alkaloid). **b** Mass spectrum of erythrocarine

5.3 *E. caribea*

Recently we have examined the seeds of a Mexican species, *E.caribea*, which has not previously been examined before. This proved to be relatively rich in alkaloids (Redha 1983), the "free" fraction containing a range of dienoid alkaloids as well as the alkenoid erythratidine (2a). Erysopine (1c), erysodine (1a) and erysovine (1b) were also found in substantial amounts in the "liberated" fraction. A small amount of the new alkaloid erythrocarine was isolated by HPLC, from the hexane fraction (Fig. 10) and its mass spectrum showed significant peaks at m/z 283, 282 (M^+ - H), 266 (M^+ - OH) and 264 (M^+ H_2O - H); this was consistent with a dienoid type of alkaloid, but bearing an hydroxyl group at C_3 rather than a methoxyl group as in erysodine (1a) and erysovine (1b). Based on molecular weight considerations the structure was assigned as the methylenedioxy analogue (7c) of erysoline (7a) and erysonine (7b); this was confirmed by the NMR spectrum which showed a signal at δ 5.85 integrating for the two protons of the methylene dioxy group in addition to signals in the expected positions for the two aromatic protons (δ 6.59 and 6.72), olefinic protons and the remaining aliphatic protons (Chawla et al. 1984).

5.4 *E. melanacantha* and *E. merilliana*

Another example to be discussed in this chapter is the isolation and structure determination of two novel alkaloids from *E.melanacantha* and *E.merilliana*. GC-MS studies of the alkaloids in *E.melanacantha,* an African species of tall trees, revealed the presence of several known alkaloids in the "free" fraction (i.e. erysopine, erysodine, erysonine, erysotine and erythratidine) whereas the "liberated" alkaloid fraction contained erysonine and/or erysoline, erysodine and traces of erysopine (Redha 1983). The novel feature of the GC traces, however, was the presence of another major component (with a much longer retention time than any of the other alkaloids) in the "free" alkaloid fraction. The molecular ion of this new alkaloid was at m/z 312 shown by high resolution measurements on material isolated by HPLC to correspond to $C_{18}H_{20}N_2O_3$. A minor component accompanying the new alkaloid in the HPLC experiments had the molecular formula $C_{18}H_{18}N_2O_4$, and was therefore thought to be an oxo-analogue. The fragmentation patterns of the two new alkaloids were of the dienoid type with major fragments at M^+-15 and M^+-31 (100%). The high field ^{1}H-NMR spectrum (Fig. 11) of the new alkaloid was very similar to those of the well-known alkaloids erysodine (1a) and erysovine (1b), but the two aromatic proton resonances were both at much lower field. The IR spectrum showed an ester carbonyl absorption at 1730 cm^{-1} and so the methoxyl resonance at δ 3.96 was assigned to a methoxy carbonyl group (rather than an aromatic methoxyl group). The ^{13}C NMR spectrum showed considerable similarities to those of the dienoid alkaloids, but lacked one of the aromatic carbon resonances, whilst a peak at δ 165.8 confirmed the presence of the carbonyl group. These results and comparisons with the ^{1}H and ^{13}C chemical shifts of pyridine derivatives led to the conclusions that the structure of the new alkaloid was (8a) i.e. the normal dioxygenated aromatic ring present in the dienoid alkaloid had been replaced by an α-methoxy carbonyl pyridine residue. The minor component was assigned the

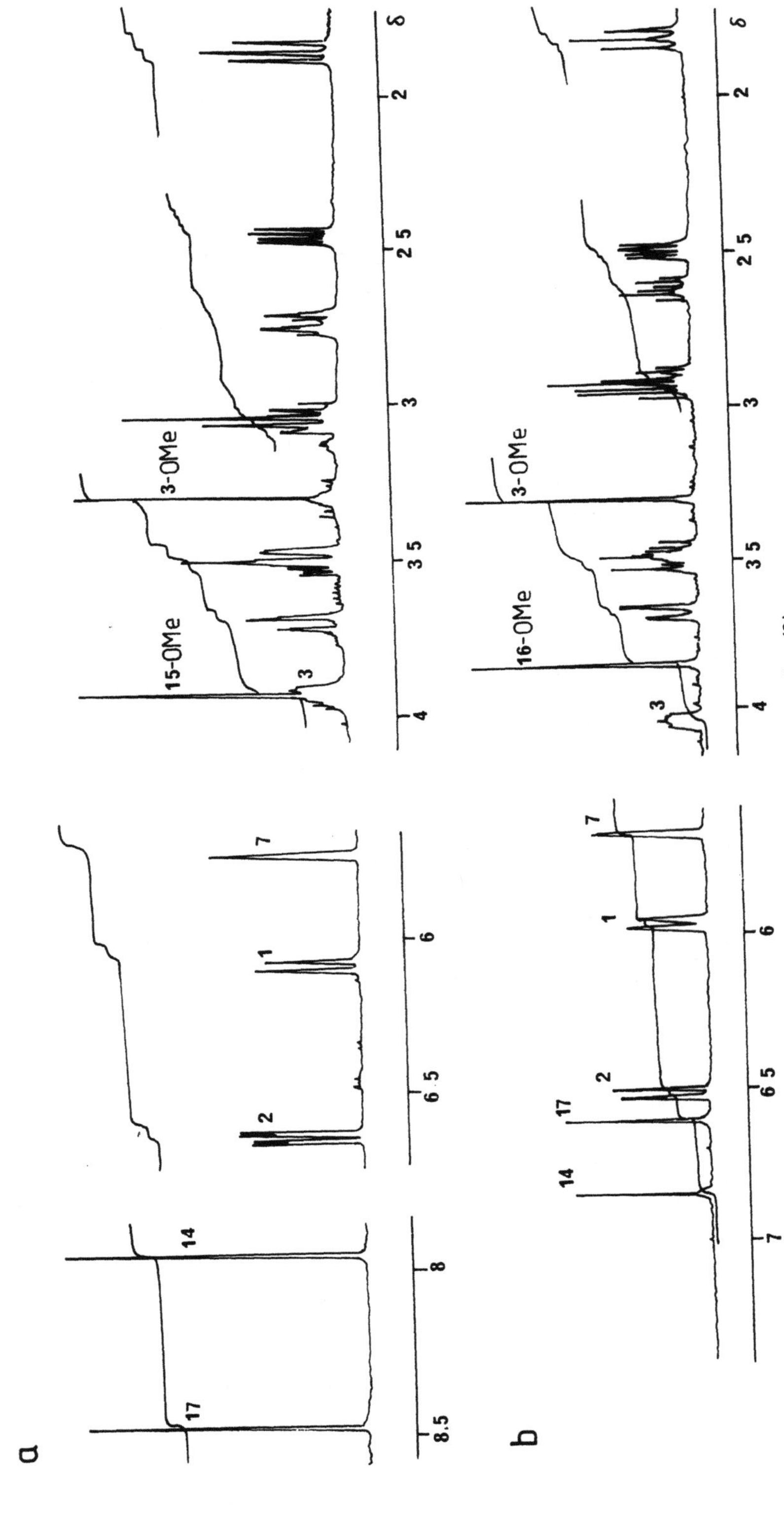

Fig. 11 a, b. 360 MHz Proton NMR Spectra of a erymelanthine (8a) and b erysovine (2)

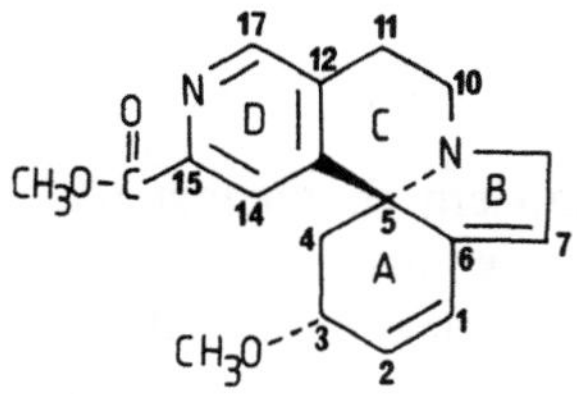

8

Fig. 12.

Alkaloid (8)	
a) Erymelanthine	$X=H_2$
b) Melanacanthine	X=O

8-oxo structure by analogy with other 8-oxo dienoid alkaloids, and this was confirmed by its ^1H and ^{13}C NMR spectra. NOE difference NMR spectra showed enhancement of the 14-H signal of the major new alkaloid when the methyl of the methoxycarbonyl group was irradiated (and vice versa). When the 11-H proton multiplet (δ 3.13) in the oxo-derivative was irradiated a small enhancement of the 17-H resonance δ 8.60 was observed, and similarly irradiation of the 14-H singlet at δ 7.97 gave rise to a small enhancement of the 3-H resonance at δ 3.80. These and other NOE and decoupling experiments confirmed the structures of the two new alkaloids, and particularly that the nitrogen and CO_2Me groups were at positions 15 and 16 respectively (not vice versa). Their absolute configurations at the spiro-centre C_5 were shown to be the same as all the other *Erythrina* alkaloids by the similarity of their CD spectra. Shortly after this work was completed, Dagne and Steglich (1983) also described the isolation of the major new alkaloid (named erymelanthine) from *E.melanacantha* although not the 8-oxo derivative which we now name melanacanthine (8b).

We have also isolated erymelanthine from *E.merilliana* seeds but much less of the plant material was available in this case and we did not also find melanacanthine. Like most of the other 8-oxo-dienoid alkaloids, melanacanthine was apparently too polar to be eluted in gas chromatographic studies.

5.5 *E.variegata*

In connection with these findings it is interesting to note that the GC-MS analysis of several samples of *E.variegata* from California, Guyana, India, Indonesia, and Nigeria showed the presence of a related alkaloid (M^+ 254) with a short retention time and a typical dienoid fragmentation pattern. (I. Barakat, F. Redha and A.H. Jackson unpublished work). High resolution measurements showed that the molecular formula was $C_{16}H_{18}N_2O$, and this led to the conclusion that it was probably the des-methoxycarbonyl analogue (9) of erymelanthine. The FD mass spectrum of the crude mixture of alkaloids obtained from one of the *E.variegata* samples also showed an ion at m/z 312, indicating the possibility that erymelanthine was also present.

The biosynthesis of these novel pyridine analogues of the dienoid *Erythrina* alkaloids is of considerable interest and it seems likely that they arise by a variant of the pathway leading to the lactonic alkaloids α- and β-erythroidine (cf. Fig. 5). Thus oxidative ring openings of the C_{16}-C_{17} bond of erysovine (1b) followed by amination could afford erymelanthine directly.

6 Conclusions

The results described in this paper show the power of modern chromatographic and spectroscopic techniques for the rapid isolation and structural identification of *Erythrina* alkaloids. The dienoid alkaloids and the erythroidines are usually the major alkaloids, although in some species substantial amounts of alkenoid alkaloids also occur. Increasing numbers of alkaloids oxygenated at 8- and 11-positions are now being found, and it seems likely that all the possible variants of the dienoid alkaloids with 8-oxo-, 11-hydroxy and 11-methoxy groups (and perhaps also 11-oxo groups) will be found, albeit in small amounts in one or more of the *Erythrina* species. The presence of 11-oxygenated alkaloids seems more characteristic of the Old World species, than of those found in the Americas. (Games et al. 1974). The discovery of the pyridine analogues erymelanthine (8a) and melanacanthine (8b) was, however, a complete surprise. Their biosynthesis is certainly worthy of further study, as well as of the abnormal *Erythrina* alkaloids occurring in *Cocculus* species, some of which lack an oxygen function at C_{16}, while others bear an amide or ester residue at C_{15}.

Acknowledgements. This work could not have been possible without the generous support of the late Dr. B.A. Krukoff; he provided all the seeds used in these studies and his enthusiasm for taxonomic investigations of *Erythrina* alkaloids was a lifelong passion. Grateful acknowledgement is also made to the Science and Engineering Research Council for support in the provision of spectroscopic and chromatographic equipment used in this work. Finally, it is a pleasure to acknowledge the dedicated contributions of a number of co-workers who have played an essential role in all of this work over the last few years, D.S. Millington, N. Khan, I. Barakat, V.U. Ratnayake, M. Abdullah, P. Ludgate, V. Mavraganis, A.S. Chawla, S. Chunchatprasert, F. Redha and D. Scallabrin, some of whose work is described in this article.

References

Abdulla MI, Barakat IE, Games DE, Ludgate P, Marraganis VG, Ratnayake VU, Jackson AH (1979) Studies of *Erythrina* Alkaloids Pt. III. GC-MS Investigations of Alkaloids in the seeds of a further fourteen species. Ann Mo Bot Gard 66:533–540

Barakat I, Jackson AH, Abdulla MI (1977) Further studies of *Erythrina* alkaloids. Lloydia 40:471–475

Barton DHR, Cohen T (1957) Some Biogenetic aspects of phenol oxidation. Festschrift A Stoll, Birkhäuser, Basel, pp 117–140

Barton DHR, James R, Kirby GW, Turner DW, Widdowson DA (1966) Constitutions of Erythratine and Erysodine and the Biosynthesis of *Erythrina* alkaloids. Chem Commun 294–295

Barton DHR, Boar RB, Widdowson DA (1970a) Phenol Oxidation and Biosynthesis Part XXI. The Biosynthesis of the *Erythrina* alkaloids. J Chem Soc (c):1213–1219

Barton DHR, Jenkins PN, Letcher R, Widdowson DA (1970b) Erythristemine, a new alkaloid from *Erythrina lysistemon:* a spectroscopic and crystallographic study. Chem Commun 391–392

Barton DHR, Potter CJ, Widdowson DA (1974) Phenol oxidation and biosynthesis, part XXIII. On the benzyltetrahydroisoquinoline origins of the *Erythrina* alkaloids. PCS Perkin I; 347–352

Beecham AF (1971) The influence of allylic oxygen on the CD of certain chromophores. Tetrahedron Lett 27:5207–5210

Belleau B (1953) The synthesis of Erythrinane. J Am Chem Soc 75:5765–5769
Bhakuni DS, Jain S (1980) Alkaloids of *Cocculus laurifolius* DC. Tetrahedron Lett 3107–3113
Bhat SV, Donnauer H, DeSouza J (1980) Structure of Pachygonine. A new quaternary alkaloid from *Pachygone ovata*. Lloydia 43:588–591
Boar RB, Widdowson DA (1970) Mass spectra of the *Erythrina* Alkaloids – A novel fragmentation of the spiran system. J Chem Soc (B):1591–1595
Boekelheide V (1960) In: Manske RHF, Holmes HL (eds) The alkaloids, chap 11. *Erythrina* alkaloids. Academic Press, London New York, pp 201–227
Boekelheide V, Grundon MF (1953) A characterisation of α-Erythroidine. J Am Chem Soc 75: 2563–2568
Boekelheide V, Weinstock J, Grundon MF, Sauvage GL, Agnello EJ (1953) The structure of β-Erythroidine and its derivatives. J Am Chem Soc 75:2550–2554
Carmack M, McKusick BC, Prelog V (1951) *Erythrina* Alkaloide: Über das Apo-Erysodin und das Apo-Erythralin. Helv Chim Acta 34:1601–1609
Chawla AS, Jackson AH, Ludgate P (1982) *Erythrina* alkaloids Pt 6 Isolation and characterisation of alkaloids from *Erythrina berteroana* seeds and leaves: formation of oxoerythroidines. J Chem Soc Perkin I:2903–2907
Chawla AS, Chunchatprasert S, Jackson AH (1983) Studies of *Erythrina* alkaloids VII ^{13}C-NMR Spectral Studies of some *Erythrina* alkaloids. Org Mag Res 21:39–41
Chawla AS, Redha FMJ, Jackson AH (1985) Alkaloids in the seeds of *E.brucai, E.cochleata, E.thollonia* and *E.caribea:* Isolation of a new Dienoid Alkaloid, Erythrocarine. Phytochemistry (in press)
Chunchatprasert S (1983) Chemical and spectroscopic studies of *Erythrina* alkaloids. Ph D thesis, Cardiff
Craig LE (1955) 2. *Erythrina* alkaloids (curare like effects). In: Manske RHF, Holmes HL (eds) The alkaloids vol V. Academic Press, London New York, pp 281–287
Dagne E, Steglich W (1983) Erymelanthine, A new type of *Erythrina* alkaloid containing a 16-azaerythrinane skeleton. Tetrahedron Lett:5067–5070
Dyke SF, Quessy SN (1981) Erythrina and related alkaloids. In: Manske RHF (ed) The alkaloids, vol 18. Academic Press, London New York, pp 1–93
Games DE, Jackson AH, Millington DS (1973) Application of field ionization Mass Spectrometry in the analysis of organic mixtures. Tetrahedron Lett 32:3063–3066
Games DE, Jackson AH, Khan NA, Millington DS (1974) Alkaloids of some African, Asian, Polynesian and Australian species of *Erythrina* Lloydia 37:581–588
Ghosal S, Majumdas, Chakraborti (1971) *Erythrina* alkaloids III. Occurrence of (+)-N-Norprotosinomenine and other alkaloids in *Erythrina lithosperma* (Leguminosae). Aust J Chem 24:2733–2738
Hanson AW (1963) The crystal structure of dihydro β-erythroidine hydrobromide. Acta Crystallogr 16:939–942
Hargreaves RT, Johnson RD, Millington DS, Mindal MN, Beavers W, Becker L, Young C, Rinehart KL Jr (1974) Alkaloids of American species of *Erythrina*. Lloydia 37:569–580
Ito K, Furukawa H, Tanaka H (1973) *Erythrina* Alkaloids 1. Alkaloids of *Erythrina bidwillii* 1. Isolation and charąterisation of bases. Yakugaku Zasshi 93:1211–1214
Jackson AH, Chawla AS (1982) Studies of *Erythrina* Alkaloids Pt IV. Gc/ms Investigations of Alkaloids in the leaves of *E.poeppigiana, E.macrophylla, E.berteroana,* and *E.saliviflora* (1982). Allertonia 3:39–45
Jackson AH, Ludgate P, Mavraganis V, Redha FMJ (1982) Studies of *Erythrina* Alkaloids Pt V. Gc/ms Investigations of Alkaloids in the seeds of *E.suburnbrans, E.lanata, E.rubinervia, E. acanthocarpa, E. variegata* and *E. melanacantha.* Allertonia 3:47–51
Juici M, Andoyoshi KO, Satoh A, Kunitomo J, Shingu T, Furukawa H (1978) Application of INDOR and NOE techniques to determination of the substitution pattern on aromatic ring of Erythrinan alkaloids. Chem Pharm Bull 26(2):563–570
Krukoff BA, Barneby RC (1974) Conspectus of species of the genus *Erythrina*. Lloydia 37: 332–459

Leete E, Ahmad A (1966) Biosynthesis of the *Erythrina* alkaloids. The incorporation of Tyrosine 2-C into Erythroidines. J Am Chem Soc 88:4722–4727

Letcher RM (1971) Alkaloids of *Erythrina lysistemon.* 11-Methoxyerythraline, a new Alkaloid. J Chem Soc (C):652–657

Millington S, Steinman H, Rinehart KL Jr (1974) Isolation, gas chromatography, mass spectrometry and structures of new alkaloids from *Erythrina folkersii* Krukoff and Moldenke and *Erythrina saliviflora* Krukoff and Barneby. J Am Chem Soc 96:1909–1914

Mondon A (1970) In: Pelletier SW (ed) Chemistry of the alkaloids, *Erythrina* alkaloids. Van Nostrand Reinhold, New York, pp 173–196

Nowacki W, Bonsma GF (1958) The crystal and molecular structure of Erythraline hydrobromide. Z Kristallogr 110:89–93

Redha FMJ (1983) Chromatographic and spectroscopic studies of *Erythrina* alkaloids, Ph D thesis, Cardiff

Uprety H, Bhakuni DS (1975) Laurifotine, Laurifine and Laurifinine. Three new Dibenz-(d,f)-azonine alkaloids from *Cocculus laurifolia* D.C. Tetrahedron Lett 14:1201–1204

Weiss U, Ziffer H (1963) Absolute configuration of the Spiro C atom of *Erythrina* alkaloids: evidence from optical rotary dispersion. Experientia 19:108–112

Annonaceae Alkaloids

A. CAVE [1]

1 Introduction

The Annonaceae is a large family comprising about 120 genera and more than 2000 species. About 150 species belonging to 41 genera have been studied chemically. Although much data on the alkaloids is not known, there is enough to get some idea about some chemotaxonomic relationships and to help to imagine some biogenetic pathways in the biosynthesis of isoquinoline alkaloids.

One important characteristic of this family is the presence of alkaloids, which exist in the greater part of the Annonaceae and almost all of them possess an isoquinoline-derived structure (Leboeuf et al. 1982a).

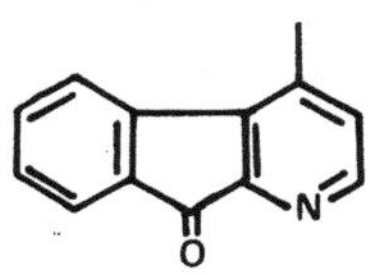

(1) onychine
(Onychopetalum amazonicum)

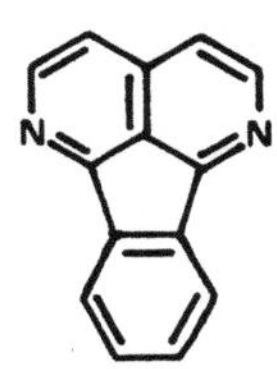

(2) eupolauridine
(Cananga odorata)

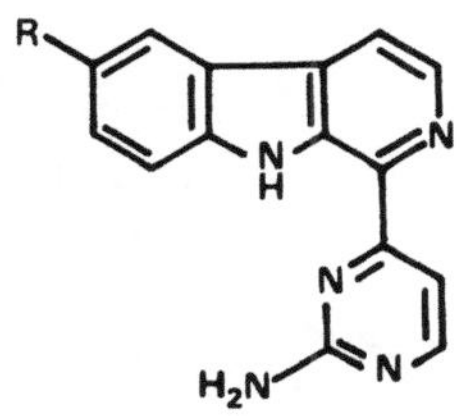

(3) R = H : annomontine
(4) R = OMe : methoxyannomontine
(Annona montana)

The few alkaloids devoid of an isoquinoline skeleton are onychine (1) from *Onychopetalum amazonicum;* eupolauridine (2), a naphthyridine alkaloid, although it is considered as a product of degradation of an oxoaporphine, liriodenine; annomontine (3) and methoxyannomontine (4), first examples of a new class of pyrimi-

---

1 Faculté de Pharmacie, Rue Jean Baptiste Clément, 92290 Chatenay-Malabry, France

The Chemistry and Biology of Isoquinoline Alkaloids, ed. by Phillipson et al.

dine-β-carboline alkaloids composed of a harman moiety linked to a 2-aminopyrimidine; two prenylated indoles (5), (6), isolated from *Monodora tenuifolia* and from *Uvaria elliotiana*, and the original class of indolosesquiterpenes (7) to (14). This last type of alkaloid has been found, until now, only in African species of *Polyalthia*. This characteristic was recognized by Waterman as a confirmation of Verdcourt's distinction of African species of Polyalthia as the separate genus *Greenwayodendron* (Hassan et al. 1982).

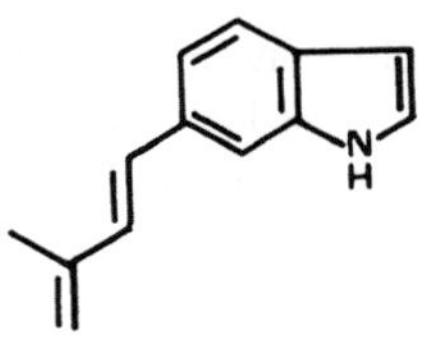

(5) (Monodora tenuifolia)

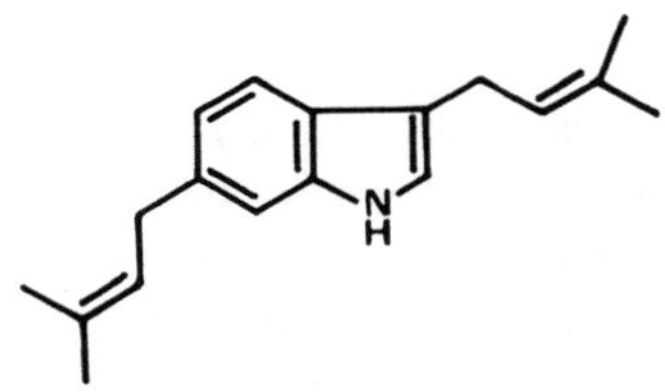

(6) (Uvaria eliotiana)

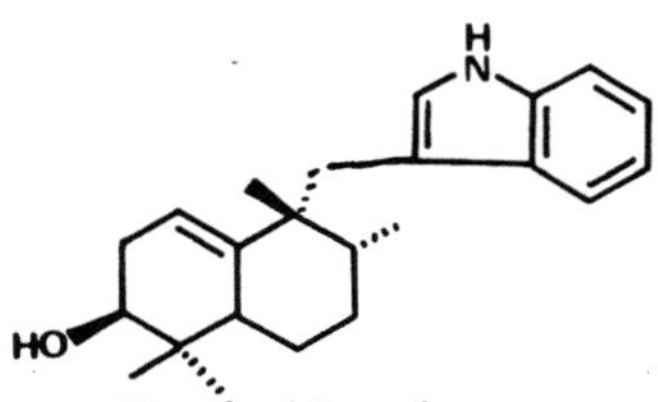

(7) polyalthenol
(Polyalthia oliveri ; P. suaveolens)

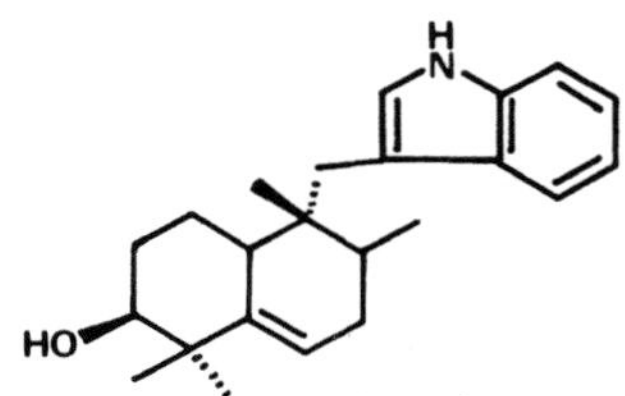

(8) isopolyalthenol
(P. suaveolens)

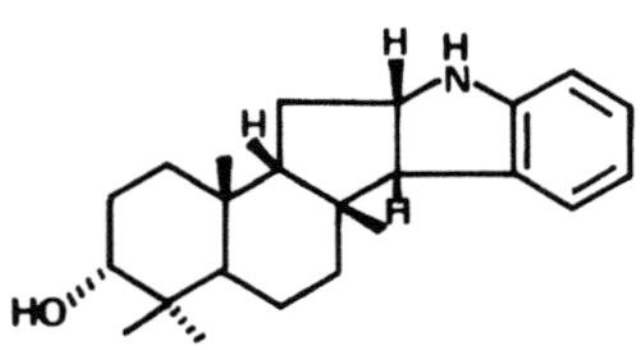

(9) polyveoline
(P. suayeolens)

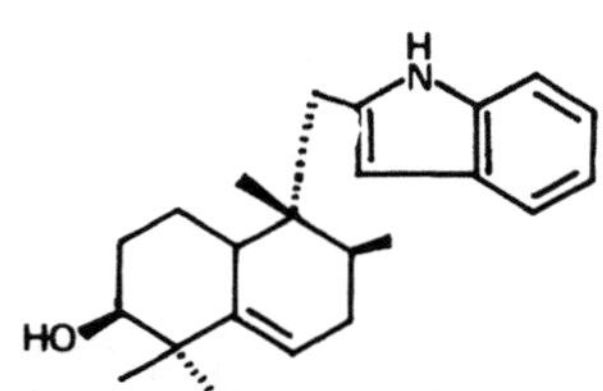

(10) neopolyalthenol
(P. suaveolens)

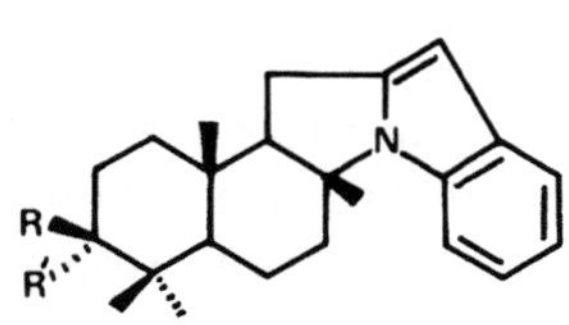

(11) R.R′= O greenwayodendrinone = polyavolensinone
(12) R = OAc , R′= H acetoxygreenwayodendrine = polyavolensine
(13) R = OH , R′= H hydroxy 3ß greenwayodendrine = polyavolensinol
(14) R = H , R′= OH hydroxy 3α greenwayodendrine

Fig. 1 (P. suaveolens) = Greenwayodendron suaveolens

All isoquinoline alkaloids are derived from benzylisoquinolines, with a few exceptions in the Cactaceae, for instance. The initial reaction in benzylisoquinoline biosynthesis (Fig. 2) is a condensation of two aromatic units; both derived from tyrosine, namely dopamine and 3,4-dihydroxyphenylacetaldehyde.

This scheme, first proposed several decades ago, was modified when it was postulated that DOPA was converted into the substituted pyruvic acid rather than the aldehyde, and that condensation with dopamine provided norlaudanosoline-1-carboxylic acid, which subsequently underwent decarboxylation with the eventual formation of norlaudanosoline, as is shown in route II (Fig. 2).

Recently, Zenk et al. proved by the discovery of a novel enzyme named norlaudanosoline synthase that route I is a real pathway of benzylisoquinoline biosynthesis, and that the previously postulated intermediate norlaudanosoline-1-carboxylic acid is probably an artefact (Schumacher et al. 1983).

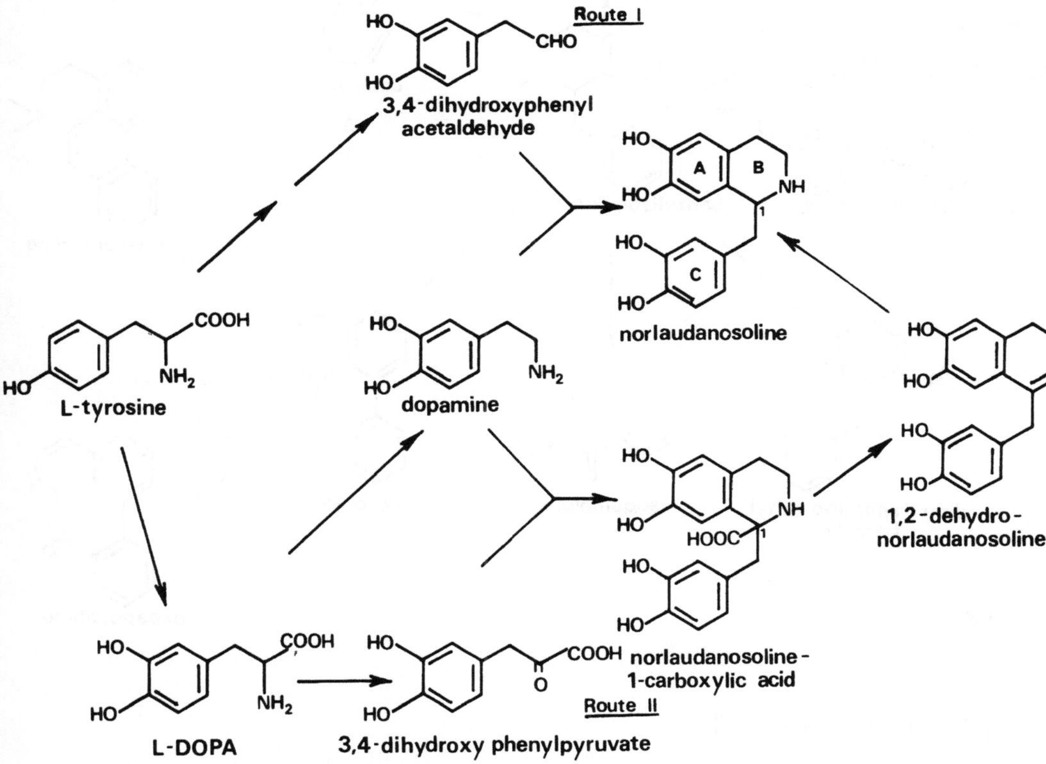

Fig. 2. Two putative biosynthetic routes to the benzylisoquinoline skeleton (Schumacher et al. 1983)

From the benzylisoquinolines are derived all the types of isoquinoline alkaloids found in the Annonaceae (Fig. 3): berberines, aporphines, oxoaporphines, phenanthrenes, cularines, morphinanedienones, isoquinolones, dimethylaminoethylbenziles and bisbenzylisoquinolines.

In this review I will not catalogue all the isoquinoline alkaloids which have been found in the Annonaceae, but instead will choose some examples of recent studies, particularly from my own laboratory.

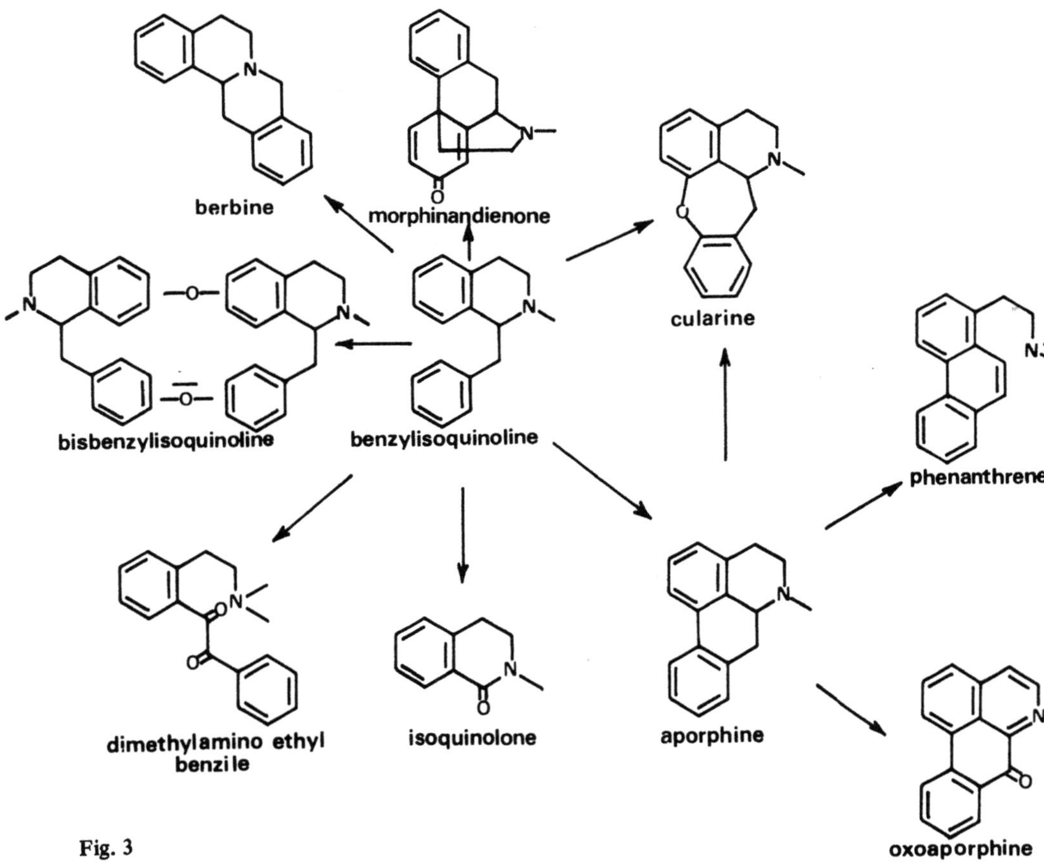

Fig. 3

2 Monomeric Alkaloids

2.1 *Guatteria* Alkaloids

The genus *Guatteria* is the largest Annonaceous genus, comprising about 250 species. Until 1980 (Table 1) the alkaloid composition of the six species which had been studied did not seem to show any homogeneity and little in the way of originality in structural types, since they were all bisbenzylisoquinolines, aporphines or oxoaporphines (Leboeuf et al. 1982a).

Table 1. Occurrence of alkaloids in *Guatteria* species until 1980. (Leboeuf et al. 1982b)

Species	Alkaloid	Alkaloidal type
Guatteria elata	Norlaureline	Aporphine
	Puterine	Aporphine
	Oxolaureline	Oxoaporphine
	Oxoputerine	Oxoaporphine
G. megalophylla	O,O-dimethylcurine	Bisbenzylisoquinoline
	Isochondodendrine	Bisbenzylisoquinoline
	O-methylcurine	Bisbenzylisoquinoline
G. melosma	Oxoanolobine	Oxoaporphine
G. modesta	Roemerine	Aporphine
	Liriodenine	Oxoaporphine
G. psilopus	Guatterine	7-OH-aporphine
	Atherospermidine	Oxoaporphine
G. subsessilis	O-methylmoschatoline	Oxoaporphine
	Subsessiline	Oxoaporphine

2.1.1. *Guatteria melosma*

In 1980 an alkaloid of a new type, melosmine (15), isolated from *Guatteria melosma*, was described by Schiff and Slatkin's group (Phoebe 1980, Zabel et al. 1982). In the proton NMR spectrum of this alkaloid, a six-proton singlet at δ 1.73 ppm ap-

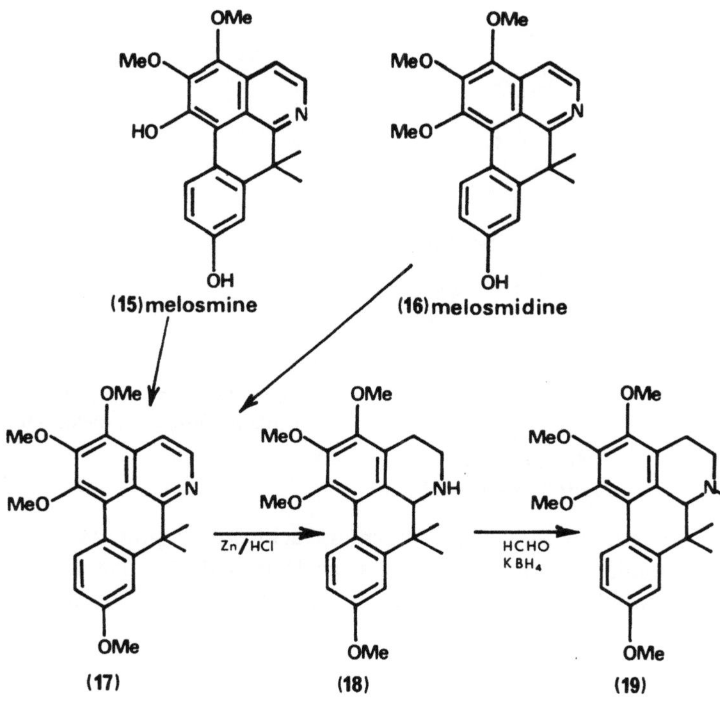

Fig. 4

peared, and this chemical shift, never seen previously in the spectra of benzylisoquinoline alkaloids, indicated the presence of a gem-dimethyl group in the molecule. As the proton NMR spectrum partially ressembled that of a 1,2,3,9-tetrasubstituted oxoaporphine alkaloid, the presence of the singlet at 1.73 ppm and the lack of spectral evidence for the presence of a 7-keto group of an oxoaporphine favoured the assignment of the gem-dimethyl group to carbon-7.

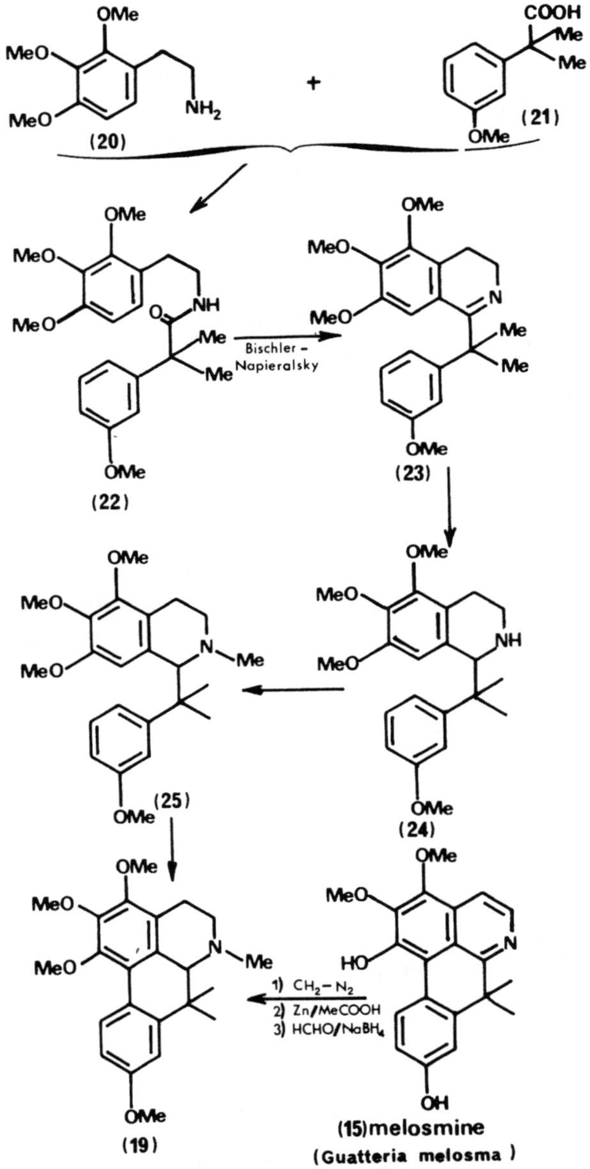

Fig. 5. Synthesis of N,O,O-trimethyltetrahydromelosmine

Melosmine (15) was reduced with zinc/hydrochloric acid and the product showed the addition of four atomic mass units in its mass spectrum, with the characteristic fragmentation of noraporphines, as well as a typical aporphine UV spectrum. When this reduction was performed on O,O-dimethylmelosmine (17) and was followed by N-methylation, it yielded N,O,O-trimethyltetrahydromelosmine (19) (Fig. 4) which was synthesized by the following route (Fig. 5). The amine (20) and the acid (21) were condensed using a Schoten-Bauman reaction and the resulting product (22) cyclised using the conditions of the Bischler-Napieralski reaction. The resulting imine (23) was reduced by sodium borohydride to afford the amine (24). This amine, N-methylated, yielded the benzylisoquinoline (25) subsequently cyclised using thallium trifluoroacetate to afford N,O,O,-trimethyltetrahydromelosmine (19) (Ammar et al. 1983).

2.1.2 *Guatteria scandens*

At the same time that Schiff et al. were working on *Guatteria melosma,* we had started the study of *Guatteria scandens* and had isolated two new, original alkaloids, guattescine (26) and guattescidine (27) (Hocquemiller et al. 1982a). These alkaloids also showed an upfield singlet in their ^{1}H NMR spectra which can be attributed to a quaternary methyl group, but this signal only integrated for three protons instead of six as in melosmine (15). Besides this, these alkaloids possess an imine function due to partial saturation of ring B, and reduction with potassium borohydride gives a noraporphine with two additional mass units in the mass spectrum. Finally, a 7-hydroxy-7-methyl-6,6a-dehydroaporphine structure was proposed for these alkaloids and confirmed by an X-ray crystallographic study of guattescine (26) (Chiaroni et al. 1983) (Fig. 6

2.1.3 *Guatteria discolor*

Shortly after this, while studying *Guatteria discolor,* two other 7-hydroxy-7-methylaporphine imines were isolated, guacoline (28) and guacolidine (29) (Hocquemiller et al. 1984a), and also other types of aporphines bearing two methyl substituents at C-7 and an imine function which could be reduced with potassium borohydride. These were named guadiscine (30), guadiscidine (31) and guadiscoline (32). They are 6,6a-dehydro-7-gemdimethylaporphines which are thus close to the melosmine structural type (Hocquemiller et al. 1982b) (Fig. 6).

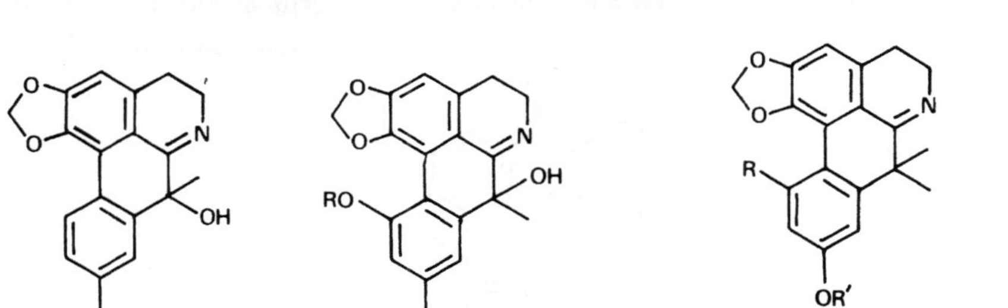

(26) R = Me guattescine (28) R = R' = Me guacoline (30) R = H R' = Me guadiscine
(27) R = H guattescidine (29) R = Me R' = H guacolidine (31) R = H R' = H guadiscidine
(32) R = OMe R' = Me guadiscoline

(Guatteria scandens) (Guatteria discolor) (Guatteria discolor)

Fig. 6

2.1.4 *Guatteria ouregou*

These two types of gem-dimethylaporphines have been found together in another *Guatteria* from Guiana, *Guatteria ouregou* (Leboeuf et al. 1982b). *Guatteria ouregou* contains melosmine (15), found previously in *Guatteria melosma,* and also a 6,6a-dehydroaporphine, dihydromelosmine (35), and the 7-hydroxy-7-methyldehydro-aporphines guattouregine (37) and guattouregidine (36) (Leboeuf et al. 1983). But this plant also contains a new type of gem-dimethylated alkaloid, a cularine named gouregine (38), the structure of which was confirmed by X-ray crystallographic analysis (Leboeuf et al. 1982c) (Fig. 7).

A priori, formation of cularines would seem to be from the cyclisation of 8-hydroxylated benzylisoquinolines. The *Guatteria ouregou* alkaloids, however, ex-

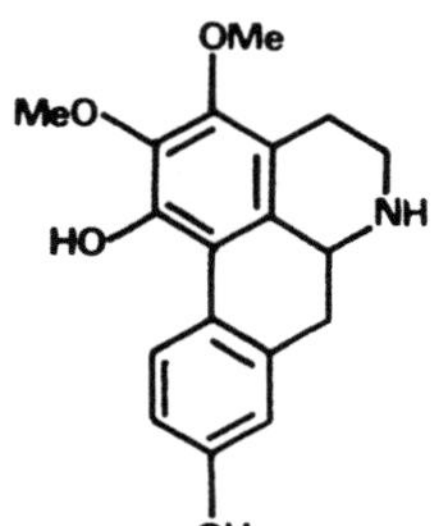

(33) oureguattidine

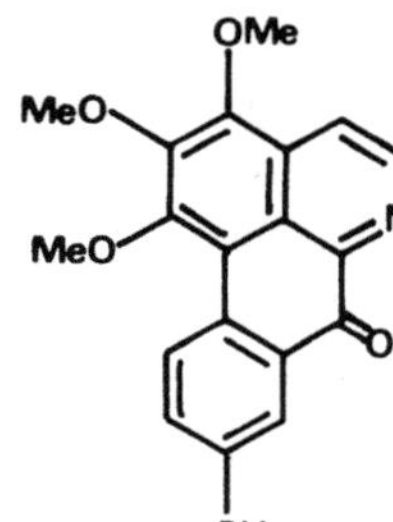

(34) subsessiline

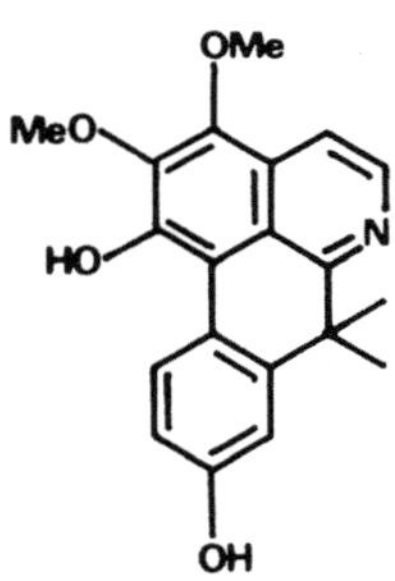

(15) melosmine

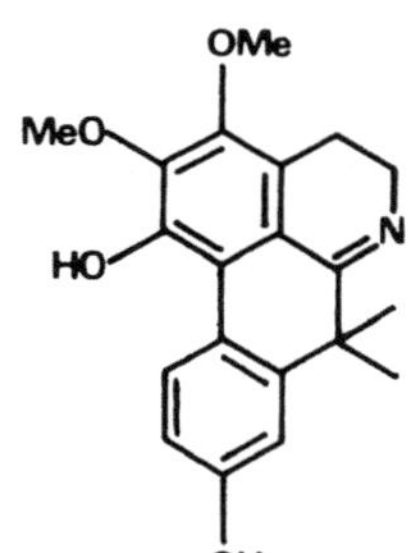

(35) dihydromelosmine

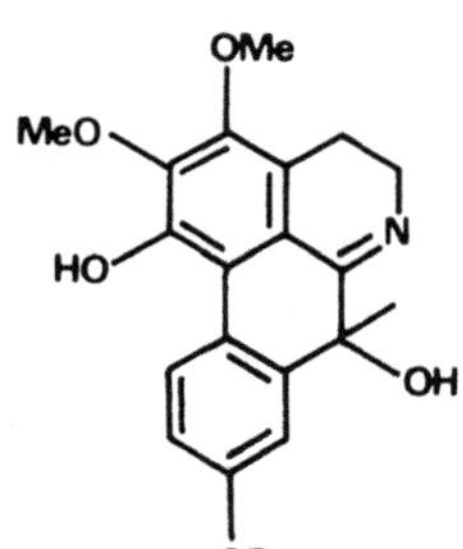

(36) guattouregidine R = H
(37) guattouregine R = Me

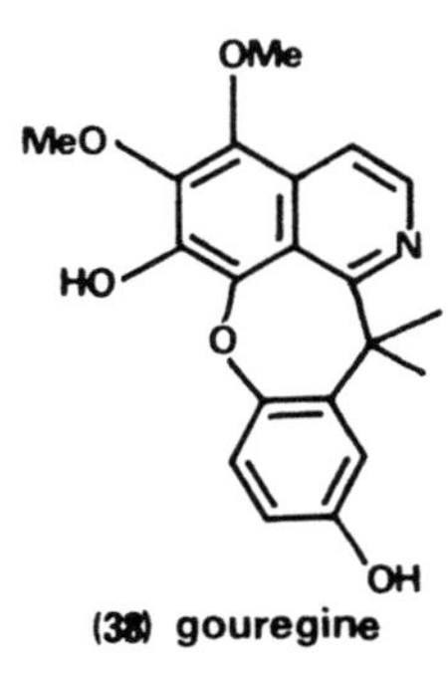

(38) gouregine

Fig. 7 (Guatteria ouregou)

hibited a striking similarity in the substitution patterns of some of the aporphines, as oureguattidine (33), oxoaporphines as subsessiline (34), dimethyltetradehydro-aporphines (15), dimethyldehydroaporphines (35), 7-hydroxy-7-methyldehydro-aporphines (36), and the cularine (38). In view of this, we thought that a biogenetic relationship between all these alkaloids was likely, and that gouregine might be an oxidized derivative of melosmine. This was confirmed by treating melosmine with Fenton's reagent, which gave gouregine in 90% yield (Fig. 8).

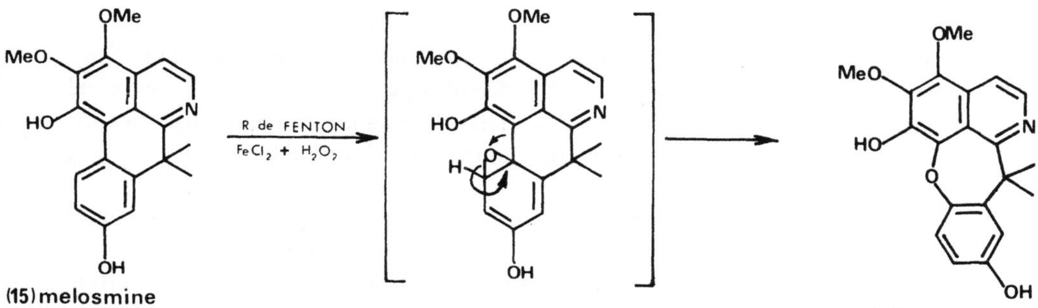

Fig. 8

If the formation of gouregine is explained in this way, it is interesting to consider the formation of the 7-methylated derivatives and a biogenetic sequence can be reasonably postulated (Fig. 9):

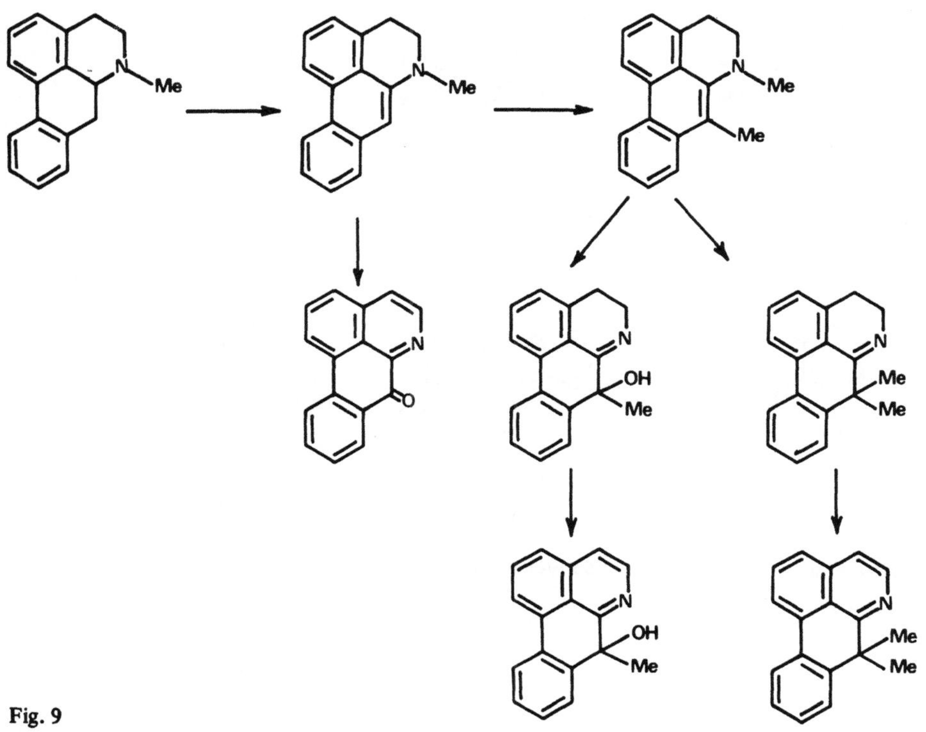

Fig. 9

Aporphines are oxidised easily to dehydroaporphines, which usually lead to oxoaporphines. If the necessary enzyme is present, however, one can envisage the methylation at carbon-7 as an alternative route. A 6a,7-dehydro-7-methyl derivative should be the logical intermediate in this sequence, and should exist in the genus *Guatteria*. We were lucky enough to find such a compound in *Guatteria schomburgkiana*, a Brazilian species (Cortes et al. 1984a, b).

2.1.5 *Guatteria schomburgkiana*

Together with aporphines, *Guatteria schomburgkiana* contains guattescine (26), guadiscine (30), and the postulated methyldehydroaporphine which we have named belemine (39). This plant contains another, related new alkaloid: dehydroguattescine (40), the ultimate product of oxidation, similar to melosmine (15) with its B ring unsaturated (Fig. 10).

From belemine (39) we obtained dehydroguattescine (40), using chloroperbenzoic acid. Chloroperbenzoic acid is not specially a biogenetic reagent, but we can think that a reaction of the same type can naturally afford dehydroguattescine (40) from belemine (39) in *Guatteria schomburgkiana*.

(Guatteria schomburgkiana)

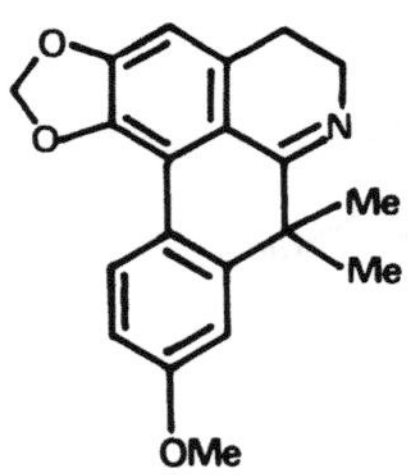

(30) guadiscine

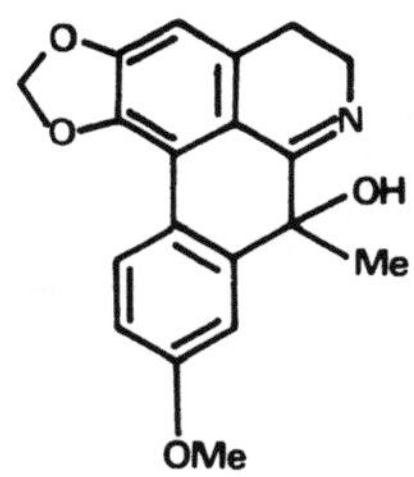

(26) guattescine

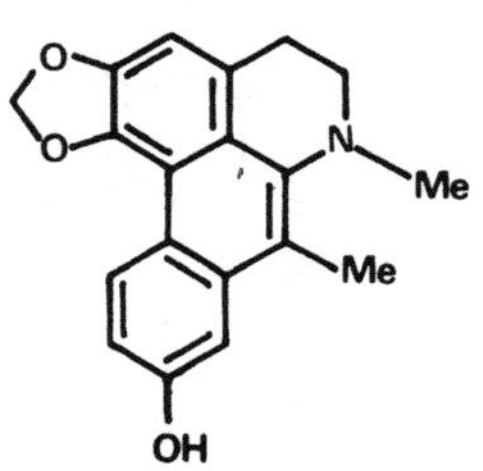

(39) belemine

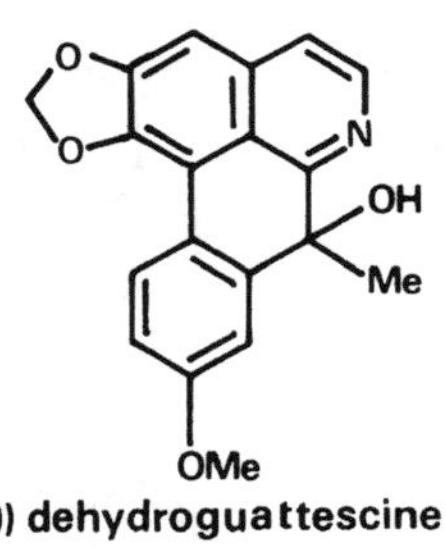

(40) dehydroguattescine

Fig. 10

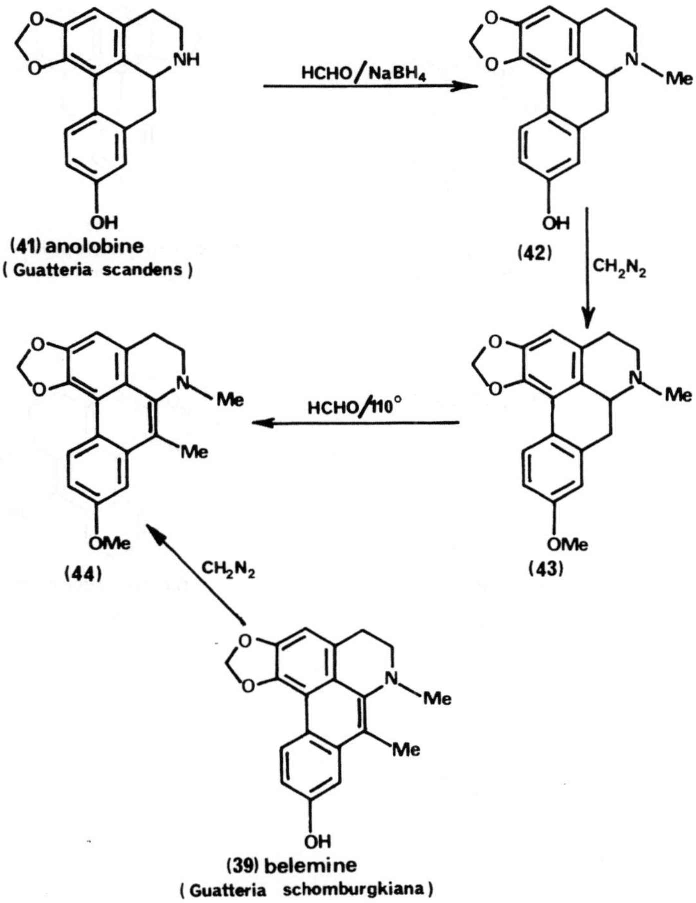

Fig. 11

O-methylbelemine (44) was prepared (Fig. 11) starting from anolobine (41), an aporphine found in abundance in *Guatteria scandens* (Hocquemiller et al. 1983). Anolobine (41) is N-methylated to (42) by formaldehyde/Na-borohydride, then O-methylated by diazomethane and the O,N-dimethylated product (43) treated with formaldehyde according to Mollov and Philipov (1979), at 110°C degrees under pressure. It gives in a poor yield the expected O-methylated belemine (44).

2.2 *Duguetia* Alkaloids

The presence of methyl substituents at C-7 of the aporphine nucleus now appears to be an important taxonomic marker which may lead to a subdivision of the genus *Guatteria*. It may also indicate the proximity of the genus *Duguetia* to *Guatteria*. In fact, a 7-alkyl substitution has been demonstrated in certain alkaloids of *Duguetia calycina* (Roblot et al. 1981). These are duguecalyne (45) and duguenaine (46),

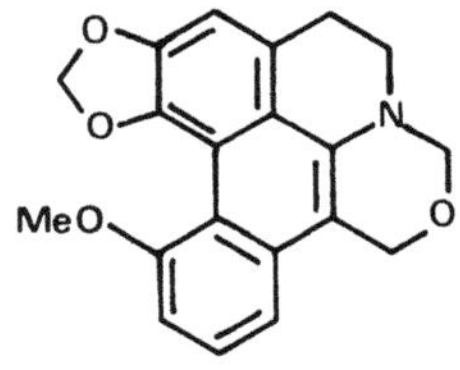

(45) duguecalyne
(Duguetia calycina)

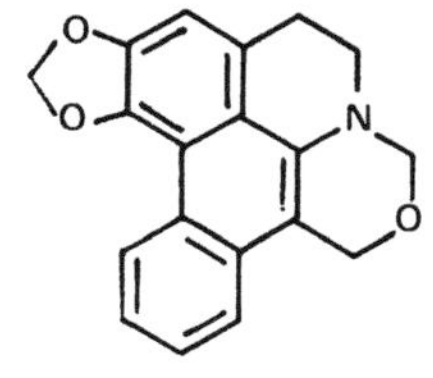

(46) duguenaine
(Duguetia calycina)

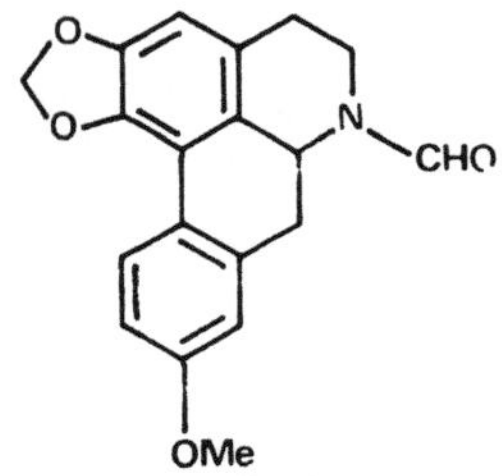

(47) N-formyl xylopine
(Duguetia obovata)

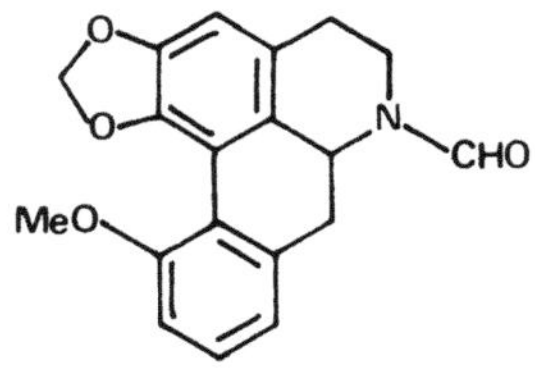

(48) N-formylputerine
(Duguetia calycina)
(Guatteria schomburgkiana)

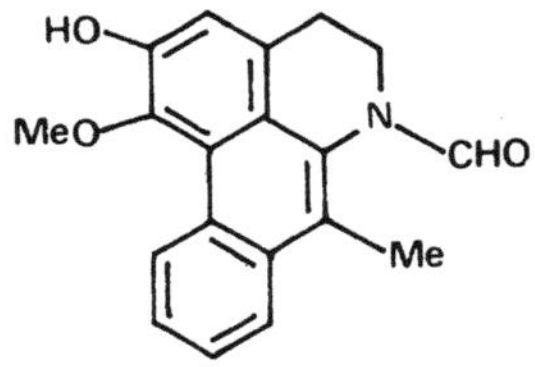

(Duguetia spixiana)
(49)

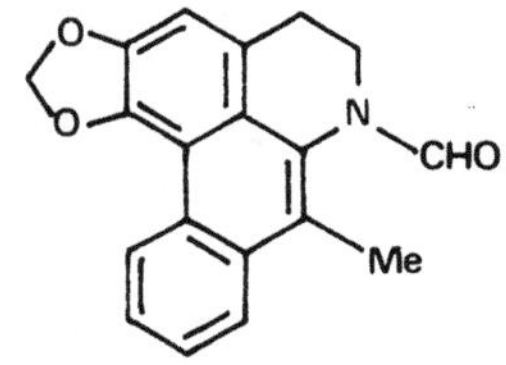

(Guatteria trichostemon)
(50)

Fig. 12

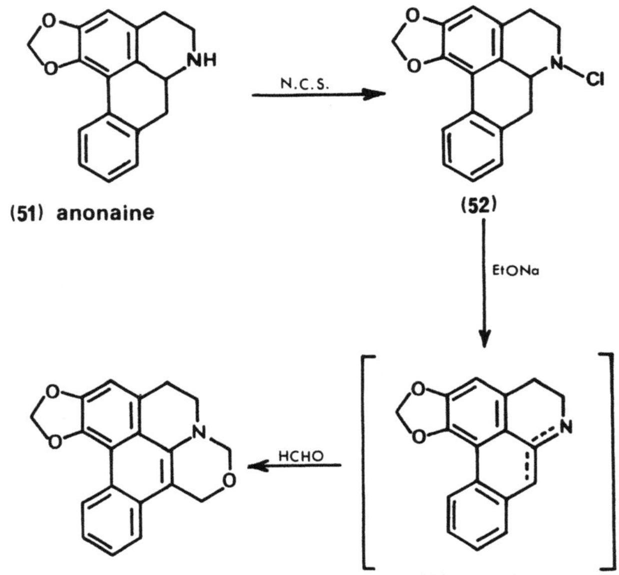

(51) anonaine
(52)
(46) duguenaine
(53 ⇌ 54)

Fig. 13

in which the 7-alkyl group is bonded in an oxazine ring formed between C-7 and the aporphine nitrogen atom. This heterocycle may arise from a 7-formylaporphine with a N-methyl group or from a 7-methylaporphine with a formyl group on the nitrogen. N-formylaporphines as (47) and (48) have already been found in *Duguetia* (Roblot et al. 1983), *Hexalobus* (Achenbach et al. 1982) and *Guatteria* (Cortes et al. 1984b).

Such a potential intermediate has been recently found in another related species, (49) in *Duguetia spixiana* (Debourges et al. 1984) and (50) in *Guatteria trichostemon* (Hocquemiller et al. 1984b).

These alkaloids gave us the idea of the hemisynthesis of duguenaine (46) and we recently suceeded (Fig. 13), (Debourges et al. 1984). It was likely that duguenaine (46) could be synthesized starting from a 6a,7-dehydronoraporphine which adds two units of formaldeyhde. Anonaine (51) treated by N-chlorosuccinimide gives the N-chloroanonaine (52) and this substance treated by sodium ethoxide gives the imine (53) probably in equilibrium with the 6a,7-dehydroaporphine (54). An excess of formaldehyde was added immediately to this reaction product and refluxed for 1 h. In this way, duguenaine (46) has been obtained.

metadisubstituted aporphines from Guatteria discolor

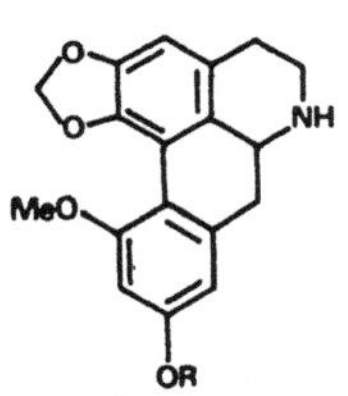
(55)R = H isocalycinine
(56)R = Me discoguattine

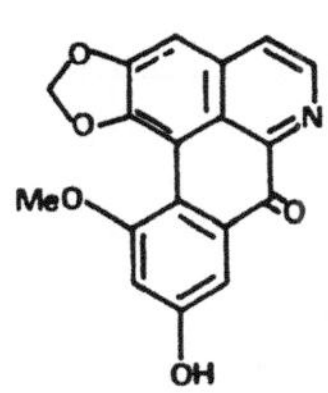
(57)oxoisocalycinine

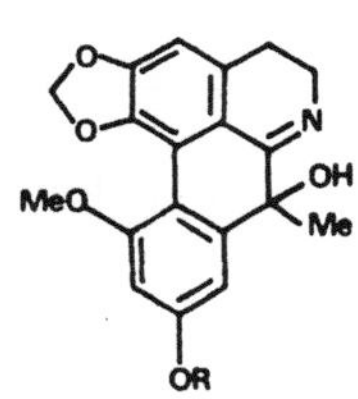
(58)R = H guacolidine
(59)R = Me guacoline

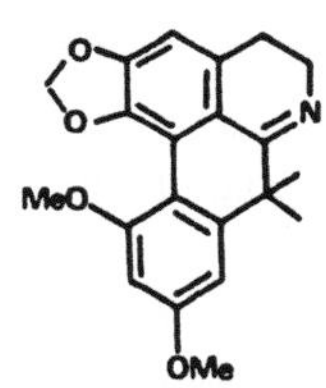
(60)guadiscoline

metadisubstituted aporphines from Duguetia

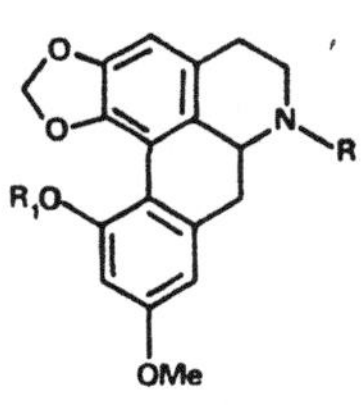
(61) R = R₁ = H calycinine
(62) R = Me, R₁ = H N-methyl calycinine
(D. calycina)
(D. obovata)

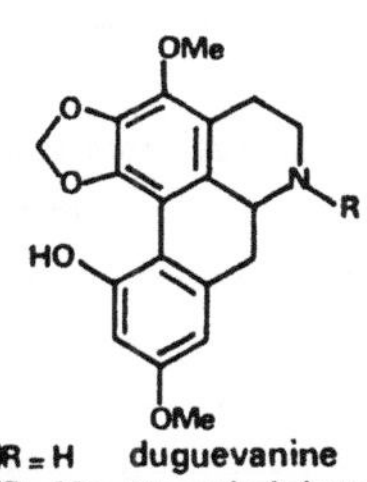
(63)R = H duguevanine
(64)R = Me N-methyl duguevanine
(65)R = CHO N-formyl duguevanine
(D. obovata)

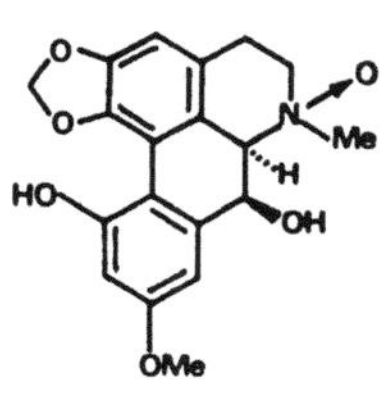
(66)N-oxy spixianine
(D. spixiana)

Fig. 14

Another feature which deserves pointing out also characterizes the *Duguetia* aporphines and is found in some *Guatteria* species, again suggesting a relationship between these two genera. This is meta-disubstitution of ring D of the aporphine (Fig. 14). This meta-disubstitution has been found in several alkaloids (55) to (60) isolated from *Guatteria discolor* (Hocquemiller et al. 1984a) and in alkaloids (61) to (66) isolated from several *Duguetia* species in which it seems to be a constant character (Roblot et al. 1978, 1983; Cavé et al. 1981). The phenol function is located at C-9 in the *Guatteria discolor* alkaloids, while it is always at C-11 in the alkaloids isolated from *Duguetia*. It would be interesting to know the biogenetic origin of this disubstitution.

3 Bisbenzylisoquinoline Alkaloids

Until 1981, only twelve bisbenzylisoquinolines had been isolated from Annonaceae, specifically from the genera *Crematosperma, Guatteria, Isolona, Phaeanthus, Popowia* and *Uvaria* (Table 2, Leboeuf et al. 1982a).

Table 2. Occurrence of bisbenzylisoquinolines and bisbenzyltetrahydroisoquinolines in Annonaceae until 1981. (Leboeuf et al. 1982a)

Chondrofoline	*Uvaria ovata*
Curine	*Isolona pilosa*
Cycleanine	*Isolona hexaloba*
Dauricine O-methyl ether	*Popowia* cv. *cyanocarpa*
O,O-Dimethylcurine	*Guatteria megalophylla*
Isochondodendrine	*Guatteria megalophylla*
	Isolona hexaloba
	Isolona pilosa
Limacine	*Phaeanthus* aff. *macropodus*
12′-O-Methylcurine	*Guatteria megalophylla*
Norcycleanine (7-O-methyl-isochondodendrine)	*Isolona hexaloba*
Phaeanthine	*Phaeanthus ebracteolatus*
	Phaeanthus aff. *macropodus*
Phlebicine	*Crematosperma polyphlebum*
Phaeantharine	*Phaeanthus ebracteolatus*

3.1 *Popowia* Alkaloids

Many new alkaloids of this type have recently been isolated from Annonaceous plants. *Popowia pisocarpa* has yielded three aporphines and one known bisbenzylisoquinoline, and also two parallel series of new bisbenzylisoquinolines (Jossang 1983a, Jossang et al. 1983b).

The first series consists of eight alkaloids (67) to (74) with the same structural type in which two coclaurine-derived molecules are joined by a C-11-O-C-12′ diaryl ether bridge. All of these bases bear five oxygen substituents at C-6, C-7, C-12, C-6′, and C-7′. All give O-methyldauricine (68) by methylation with diazomethane. All have the same absolute configuration, namely 1(R), 1′(R), proved by examination of their CD spectra and by cleavage with sodium in liquid ammonia, a method which has allowed the structures of all these alkaloids to be demonstrated beyond doubt. Only one of these alkaloids, O-methyldauricine (68), was known as a constituent of the Annonaceae, and it had only been isolated from the single species of *Popowia* studied previously, *Popowia cyanocarpa* (Leboeuf et al. 1982a).

The second series of *Popowia pisocarpa* alkaloids (75) to (80) consists of bisbenzylisoquinolines in which the two monomeric units are also derived from coclaurine, but are coupled to each other only through a C-11-C-11′-biphenyl carbon-carbon bond. These alkaloids bear six oxygen substituents at C-6, C-7, C-12, C-6′, C-7′, and C-12′ (Jossang et al. 1983b). Their absolute configuration has not been determined with certainty, but it is probably 1 (R), 1′(R). Their circular dichroism curves are superimposable on those of the foregoing alkaloids, but this cannot be taken as absolute proof of their configuration in spite of the conformational similarities that can be inferred from their UV and ^{1}H NMR spectra.

This type of alkaloid is new, not only in the Annonaceae, but in the plant kingdom. Although about 20 bisbenzylisoquinoline alkaloids have been described so far with a biphenyl bond joining their benzyl rings, all of them also possess one or two diaryl ether bridges between their isoquinoline moieties.

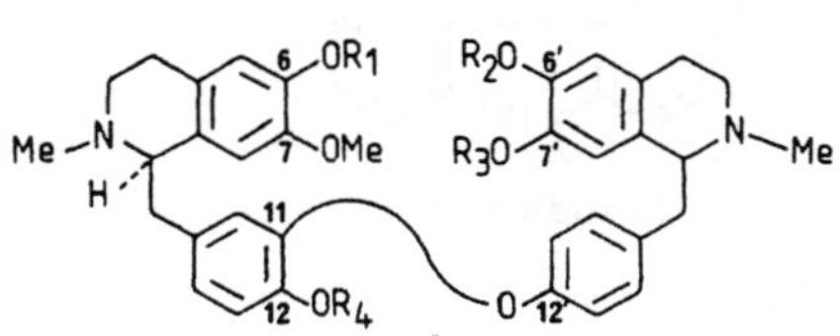

(67) dauricine $R_1 = R_2 = R_3 = Me,\ R_4 = H$
(68) OMe dauricine $R_1 = R_2 = R_3 = R_4 = Me$
(69) popisine $R_1 = R_2 = R_4 = Me,\ R_3 = H$
(70) popisidine $R_1 = R_3 = R_4 = Me,\ R_2 = H$
(71) popidine $R_2 = R_3 = R_4 = Me,\ R_1 = H$
(72) popisopine $R_2 = R_4 = Me,\ R_1 = R_3 = H$
(73) popisonine $R_3 = R_4 = Me,\ R_1 = R_2 = H$
(74) dauricoline $R_3 = Me,\ R_1 = R_2 = R_4 = H$

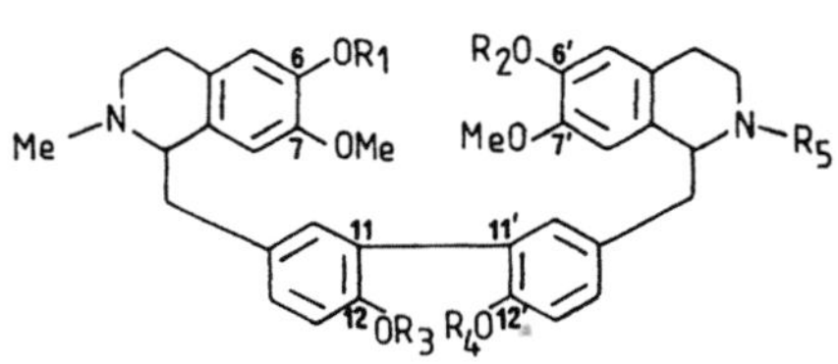

(75) pisopowine $R_1 = R_2 = R_3 = R_4 = R_5 = Me$
(76) pisopowidine $R_2 = R_3 = R_4 = R_5 = Me,\ R_1 = H$
(77) pisopowiarine $R_3 = R_4 = R_5 = Me,\ R_1 = R_2 = H$
(78) pisopowamine $R_1 = R_2 = Me,\ R_3 = R_4 = R_5 = H$
(79) pisopowetine $R_1 = R_2 = R_5 = Me,\ R_3 = R_4 = H$
(80) pisopowiaridine $R_3 = R_5 = Me,\ R_1 = R_2 = R_4 = H$

Fig. 15. Bisbenzylisoquinolines of *Popowia pisocarpa*

The structure of pisopowine (75) and the five related alkaloids (76) to (80) isolated from *Popowia pisocarpa* are interesting from the biogenetic view point. There is agreement that the biogenesis of bisbenzylisoquinolines results from the oxidative coupling of coclaurine-type monomers. Carbon-oxygen diaryl ether bonds are usually observed, but carbon-carbon diphenyl bonds can also exist. It is interesting to note that more than 25 years ago, it had been suggested (Barton and Cohen 1957) that coclaurine-type units might lead, by way of oxidative dimerization, to bisbenzylisoquinolines with the very same skeleton as we have found for the *Popowia pisocarpa* alkaloids. These substances might be intermediates, not yet isolated from other species, on the route leading to biphenyl alkaloids with a diaryl ether bond between the isoquinoline rings.

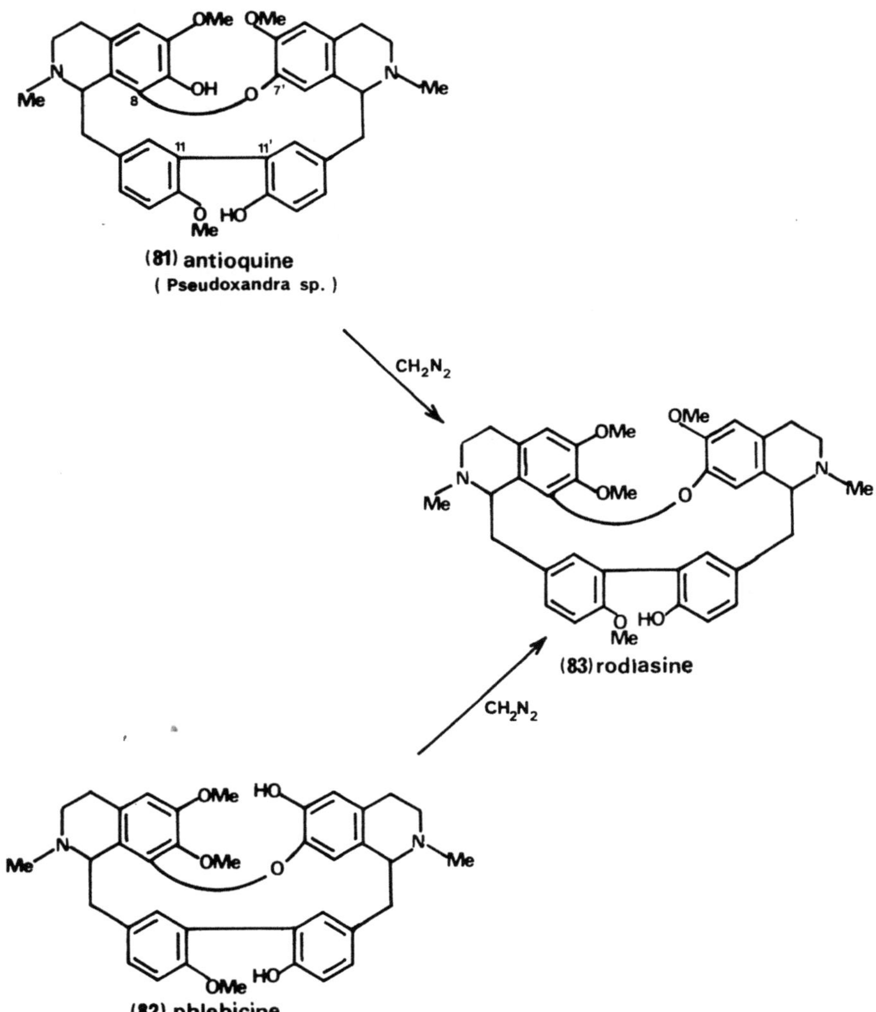

Fig. 16

3.2 *Pseudoxandra* Alkaloids

This last type of alkaloid has been found recently in a Columbian annonaceous tree belonging to the genus *Pseudoxandra* (Cortes et al. 1984c). The bark of this tree contains 1.2% of alkaloids, the main component of which is new and has been named antioquine (81). Antioquine (81) is a bisbenzylisoquinoline with two types of linkage between the monomers: a biphenyl bond between carbon atoms 11 and 11′, as in the case of pisopowine (75), and an ether linkage between carbon atoms 8 and 7′,

Antioquine in itself is not a particularly unique alkaloid because an isomer, phlebicine (82), had already been isolated from another annonaceous species, *Crematosperma polyphlebum* (Leboeuf et al. 1982a). Both antioquine (81) and phlebicine (82) have been converted into a common derivative, rodiasine (83) (Fig. 16). The interesting point in the case of *Pseudoxandra* is that we have found that it contains several antioquine derivatives with different degrees of oxidation.

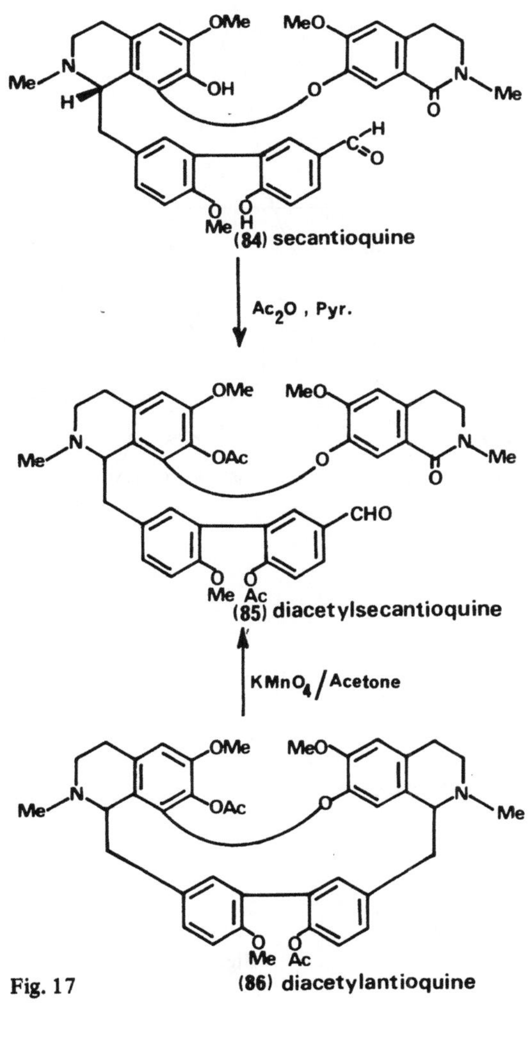

Fig. 17 (86) diacetylantioquine

In the first place, secantioquine (84) is a minor alkaloid characterized by the presence of an aldehyde carbonyl group and a conjugated lactam function, recognizable in its IR and ^{1}H NMR spectra. Its mass spectrum is characteristic of secobisbenzylisoquinolines and its UV spectrum, which exhibits a bathochromic and hyperchromic effect in the presence of alkali, suggests the presence of a hydroxy group para to the aldehyde function. The available spectral data lead to the proposal for secantioquine of a structure (84) which can be derived from antioquine (81) by oxidation. This has been confirmed by correlation with one of the oxidation products (85) of diacetylantioquine (86) (Cortes et al. 1984d) (Fig. 17).

The configuration of secantioquine at carbon-1 has been established by examination of its circular dichroism spectrum (Fig. 18). According to Shamma et al.

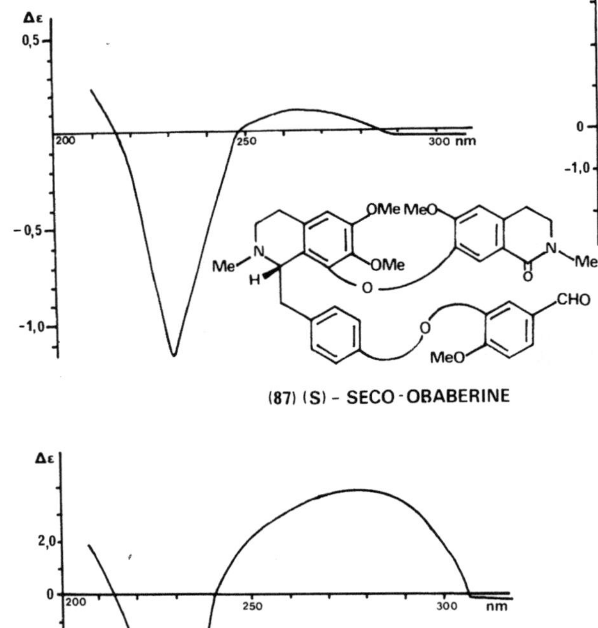

(87) (S) - SECO-OBABERINE

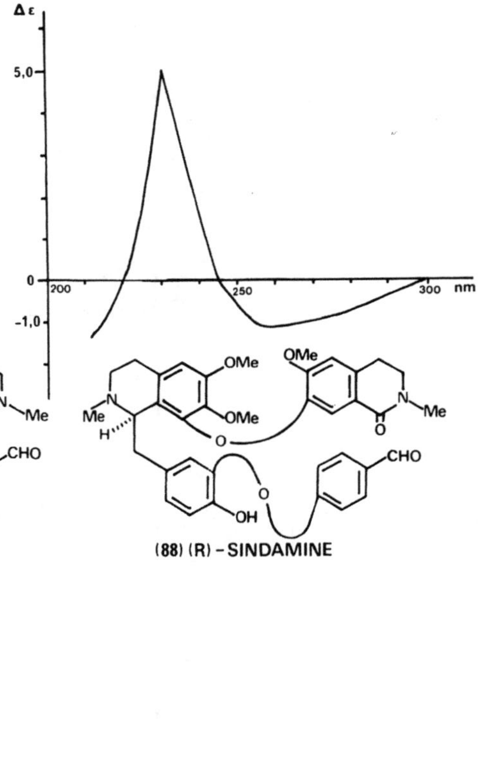

(88) (R) - SINDAMINE

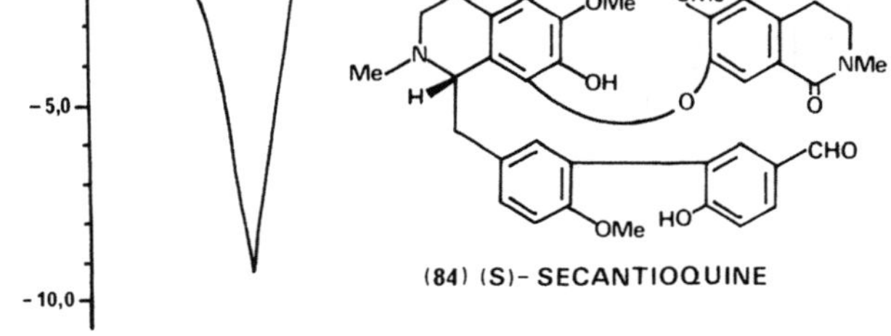

(84) (S)- SECANTIOQUINE

Fig. 18

(Leet et al. 1983), the configuration of secobisbenzylisoquinolines can be determined by the shape of their circular dichroism curve in the terminal region, between 210 and 220 nm. If the tail of the curve is positive, as in secoobaberine (87), configuration is 1(S), and if the tail of the curve is negative, as in sindamine (88), the configuration is 1(R).

All the cases described so far concern secobisbenzylisoquinolines joined by an ether bridge, but Shamma's rule can be safely extrapolated to the case of secantioquine (84), the first example of a secobisbenzylisoquinoline with a biphenyl linkage, as in every case the seco compound can be envisaged merely as a N-methylated benzylisoquinoline with a large, practically unconjugated substituent on the benzyl chromophore. Therefore we can conclude that secantioquine has the 1(S) configuration.

Two other alkaloids deserve to be singled out because of their originlality. They represent intermediate oxidation stages, either at carbon α as in pseudoxandrine (89), or at carbon α' as in oxandrine (90), which is the real intermediate between antioquine (81) and secantioquine (84) (Cortes et al. 1984e) (Fig. 19).

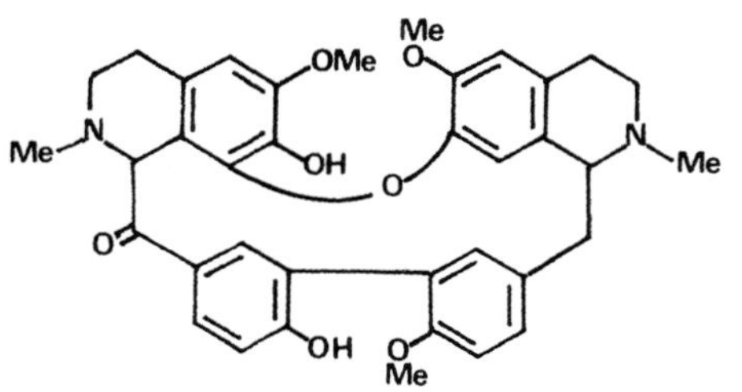

(89) pseudoxandrine

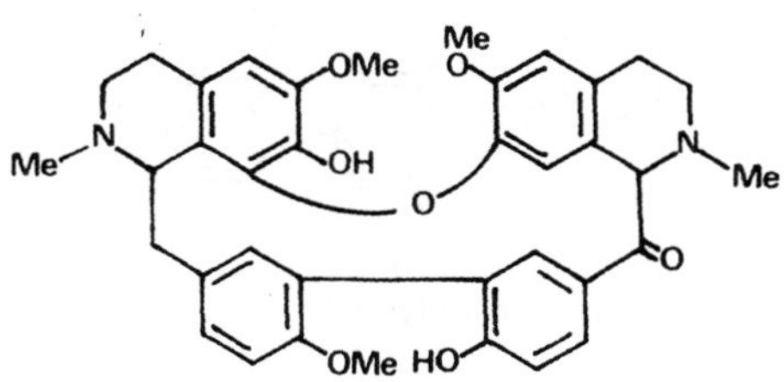

(90) oxandrine

Fig. 19

3.3 *Polyalthia* Alkaloids

Other types of dimers have been found recently in another Annonaceous species. Four bisaporphinoids have been isolated from *Polyalthia cauliflora* var. *beccarii* (Jossang et al. 1982). Three of these are beccapolinium (91), beccapoline (92), and beccapolydione (93). These alkaloids are characterized by a carbon-carbon bond between carbon atoms 4 and 7 of two aporphinoid moieties (Fig. 20).

It is interesting to consider how these alkaloids may be formed. It seems reasonable to postulate that they are formed from stephanine (94) in the course of its oxidation.

According to Shamma, Guinaudeau, and collaborators, the quaternary oxoaporphine thailandine (96) is probably an intermediate in the oxidation of stephanine (94) to oxostephanine (97) (Guinaudeau et al. 1981) (Fig. 21).

It has also been shown (Castedo et al. 1981) that oxoaporphines are obtained in better yield from aporphines than from noraporphines when Fremy's salt is used as the oxidising agent.

One can therefore assume that in *Polyalthia cauliflora* var. *beccarii,* in which both thailandine (96) and oxostephanine (97) have been found, tetradehydrostephanine (98) is formed on the route to these oxoaporphines. Tetradehydrostephanine (98) is very reactive, and its electron density at C-7 should be particularly high. This

(Polyalthia cauliflora var. beccarii)

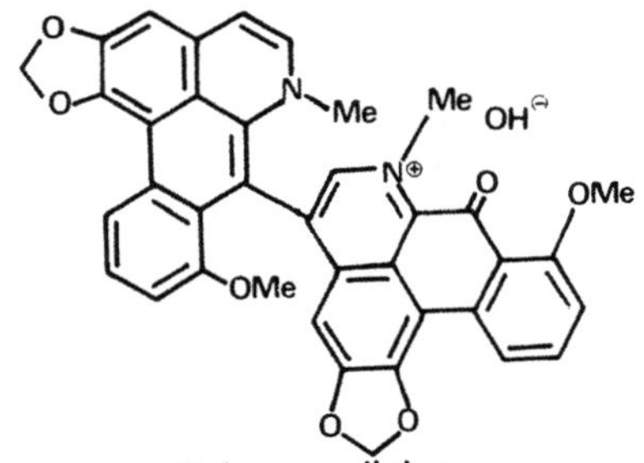

(91) beccapolinium

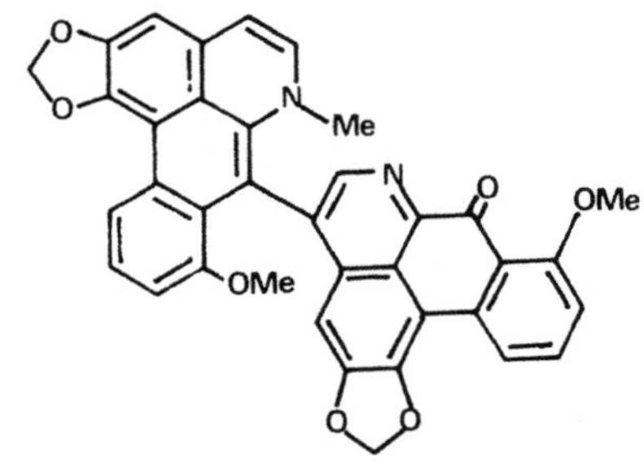

(92) beccapoline

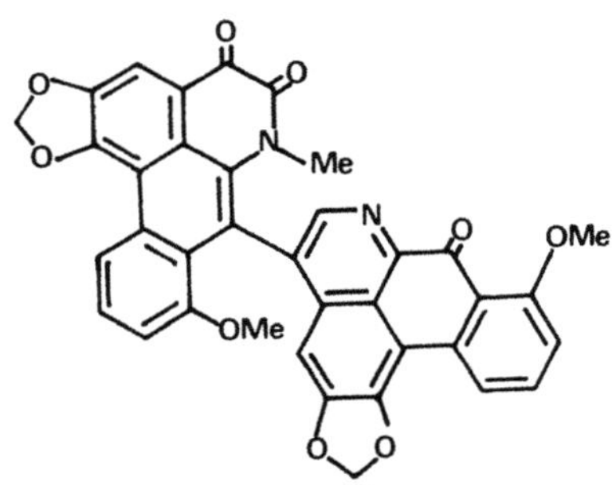

(93) beccapolydione

Fig. 20

alkaloid may thus react with preformed thailandine (96), in which C-4 is electron deficient (Jossang 1983a) (Fig. 22). Beccapolinium (91) would then result from a dehydrogenation of the C-7/C-4′ adduct of tetradehydrostephanine (98) and thailandine (96), the driving force for which would of course be the rearomatization of both aporphinoid systems.

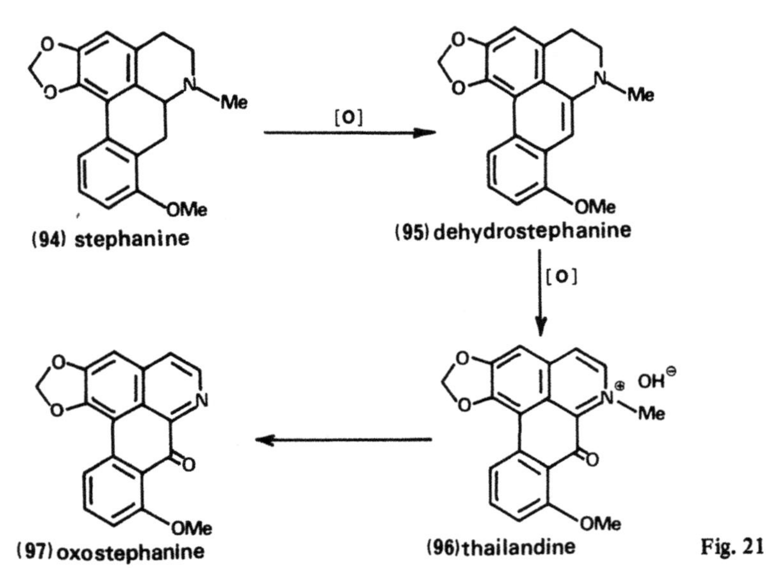

Fig. 21

Beccapoline (92) should result from the facile N-demethylation of beccapolinium (91) and beccapolydione (93) by further oxidation of the tetradehydrostephanine moiety.

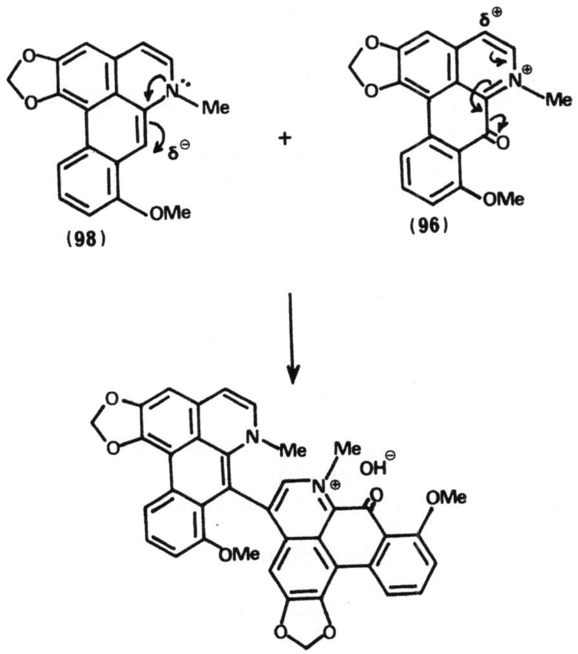

Fig. 22 (97)

4 Conclusions

In conclusion, these few examples illustrate the following facts:

– First, the alkaloid composition of the Annonaceae is much more complex than was recognized a decade ago, when to an alkaloid chemist the name Annonaceae only meant aporphinoids and berberines.

– Secondly, certain annonaceous genera or species are capable of carrying out apparently unique biosynthetic processes.

– Finally, it is now certain that some genera or species belonging to this family can perform relatively elaborate biosynthetic sequences, especially those based on intermolecular oxidative coupling.

In the case of the formation of intermolecular carbon-carbon bonds, this has been thought to be the province of phylogenetically more advanced families.

The demonstration of such capabilities in the Annonaceae may be interpreted either as a sign of specialization in a primitive family or, conversely, as an ancestral process which has been preserved in a broader range of taxa than had been recognized until now.

References

Achenbach H, Renner C, Addae-Mensah I (1982) Investigation of the constituents of *Hexalobus crispiflorus*. Liebigs Ann Chem 1623–1633

Ammar HA, Schiff PL Jr, Slatkin DJ (1983) Synthesis of 7,7-dimethylaporphine alkaloids. Heterocycles 20:451–454

Barton DHR, Cohen T (1957) Festschrift A Stoll. Birkhäuser, Basel, p 117

Castedo L, Puga A, Saa JM, Suau R (1981) Fremy's salt oxidation of some isoquinoline alkaloids. Biogenetic considerations. Tetrahedron Lett 22:2233–2236

Cavé A, Roblot F, Hocquemiller R, Moretti C, Jacquemin H (1981) Colloque International du C.N.R.S. n° 298: Substances Naturelles d'intérêt biologique du Pacifique (Nouméa, Août 1979), Paris

Chiaroni A, Riche C, Hocquemiller R, Rasamizafy S, Cavé A (1983) Structure cristalline de la guattescine; révision de structure, Tetrahedron 39:2163–2167

Cortes D, Ramahatra A, Dadoun H, Cavé A (1984a) La bélémine, intermédiaire biogénétique potentiel vers la biosynthèse des méthyl-7 aporphines. CR Acad Sci Ser II, 299:311–314

Cortes D, Ramahatra A, Dadoun H, Cavé A (1984b) Alcaloïdes du *Guatteria schomburgkiana*. J Nat Prod in press

Cortes D, Saez J, Hocquemiller R, Cavé A (1984c) Etude de l'antioquine et de ses dérivés, alcaloïde du *Pseudoxandra* af. *lucida* L. J Nat Prod in press

Cortes D, Saez J, Hocquemiller R, Cavé A (1984d) La sécantioquine, nouveau type de sécobisbenzyltétrahydroisoquinoléine. C R Acad Sci, Ser II 298:591–594

Cortes D, Saez J, Hocquemiller R, Cavé A (1984e) Oxandrine et pseudoxandrine, nouveau type d'alcaloïdes bisbenzylisoquinoléiques. (to be published)

Debourges D, Hocquemiller R, Cavé A, Lévy J (1984) La duguespixine, un intermédiaire vers la biogènese des alkyl-7-aporphines; hémisynthèse de la duguénaïne. J Nat Prod in press

Guinaudeau H, Shamma M, Tantisewie B, Pharadai K (1981) 4,5,6,6a-Tetradehydro-N-methyl-7-oxoaporphinium salts. J Chem Soc Chem Commun:1118–1119

Hassan C, Healey T, Waterman PG, Schwalbe CH (1982) Indolosesquiterpene and aporphine alkaloids from *Greenwayodendron (Polyalthia) suaveolens*. X-ray crystal structure of greenwayodendrin-3-one. J Chem Soc Perkin Trans I:2807–2812

Hocquemiller R, Rasamizafy S, Cavé A (1982a) Guattescine et guattescidine, nouveaux alcaloïdes aporphiniques de *Guatteria scandens*, Annonacées. Tetrahedron 38:911–915

Hocquemiller R, Debitus C, Roblot F, Cavé A (1982b) Guadiscine et guadiscoline, alcaloïdes aporphiniques originaux de *Guatteria discolor*, Annonacées. Tetrahedron Lett 23:4247–4250

Hocquemiller R, Rasamizafy S, Cavé A, Moretti C (1983) Alcaloïdes du *Guatteria scandens*. J Nat Prod 46:335–341

Hocquemiller R, Debitus C, Roblot F, Cavé A , Jacquemin H (1984a) Alcaloïdes des écorces de *Guatteria discolor* RE Fries, J Nat Prod 47:353–362

Hocquemiller R, Rasamizafy S, Cavé A (1984b) Alcaloïdes du *Guatteria trichostemon* (to be published)

Jossang A, Leboeuf M, Cavé A (1982) Un nouveau type d'alcaloïdes isoquinolinéiques, les bisaporphines. Tetrahedron Lett 23:5147–5150

Jossang A (1983a) Thèse Doct Etat Sci Pharm, Paris-Sud, Chatenay-Malabry

Jossang A, Leboeuf M, Cavé A (1983b) Un nouveau type d'alcaloïdes bisbenzylisoquinoléiques isolés de *Popowia pisocarpa* (Bl.) Endl. Annonacées. C R Acad Sci Ser II, 297:853–857

Lebœuf M, Cavé A, Bhaumik PK, Mukherjee B, Mukherjee R (1982a) The phytochemistry of the Annonaceae. Phytochemistry 21:2783–2813

Lebœuf M, Cortes D, Hocquemiller R, Cavé A (1982b) Guattouregidine, guattouregine et dihydromelosmine, alcaloïdes aporphiniques nouveaux de *Guatteria ouregou* Dun, Annonacées. C R Acad Sci Ser II, 295:191–194

Lebœuf M, Cortes D, Hocquemiller R, Cavé A (1982c) Structure de la gouregine, alcaloïde original apparenté aux cularines. Tetrahedron 38:2889–2896

Lebœuf M, Cortes D, Hocquemiller R, Cavé A (1983) Alcaloïdes de *Guatteria ouregou* Planta Med 48:234–245

Leet JE, Elango V, Hussain SF, Shamma M (1983) Chenabine and Jhelumine: secobisbenzylisoquinolines or simple isoquinoline-benzylisoquinoline dimers? Heterocycles 20:425–429
Mollov N,Philipov S (1979) Reactivity of aporphine alkaloids against formaldehyde. Chem Ber 112:3737–3739
Phoebe CH Jr (1980) Ph D thesis, Univ Pittsburgh
Roblot F, Hocquemiller R, Jacquemin H, Cavé A (1978) Alcaloïdes du *Duguetia calycina* Benoist. Plant Med Phytother 12:259–266
Roblot F, Hocquemiller R, Cavé A (1981) Duguecalyne et duguenaïne, alcaloïdes aporphiniques orginaux du *Duguetia calycina* Benoist, Annonacées. CR Acad Sci Ser II 293:373–377
Roblot F, Hocquemiller R, Moretti C, Cavé A (1983) Alcaloïdes du *Duguetia obovata*. J Nat Prod 46:862–873
Schumacher HM, Rüffer M, Nagakura N, Zenk MH (1983) Partial purification and properties of (S)-Norlaudanosoline synthase from *Eschscholtzia tenuifolia* cell cultures. Planta Med 48:212–220
Zabel V, Watson WH, Phoebe CH Jr, Knapp JE, Schiff PL Jr, Slatkin DJ (1982) Melosmine and melosmidine, 7,7-dimethyltetradehydroaporphine alkaloids from *Guatteria melosma*. J Nat Prod 45:94–101

The Chemistry and Pharmacology of Cularine Alkaloids

L. CASTEDO [1]

1 Introduction

The cularines are a group of isoquinoline alkaloids with the general structure (1), which incorporates an oxepine system in its molecular skeleton. If the nitrogen is secondary the alkaloid can be called a norcularine. The occurrence of cularine alkaloids in nature seems to be associated mainly with the Fumariaceae family (Shamma 1972, Shamma and Moniot 1978).

It was Manske who in 1938 first isolated the "parent" alkaloid (+)-cularine (1a) from plants belonging to the genera *Dicentra* and *Corydalis*, but its structure was not established until 1950 (Manske 1950). The molecule was later deduced from NMR studies to be V-shaped with ring C having a twist-boat conformation, the oxygen atom being the bow (2) (Bhacca et al. 1966). This was confirmed by an X-ray crystallographic analysis of cularine methiodide (Kametani et al. 1972, Shimanouchi et al. 1973, which likewise confirmed the S-configuration of its asymmetric centre, previously assigned by chemical correlation (Kunitomo et al. 1971). (+)-Cularimine (1b), (+)-cularidine (1c) and (+)-cularicine (1d) are the other three members of this group which were fully characterized before 1965 (Manske 1968, Shamma 1972). All the above alkaloids possess oxygenated functions at C-7, C-3′ and C-4′.

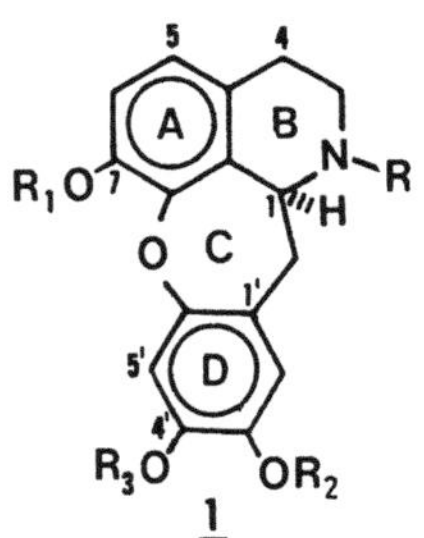

1

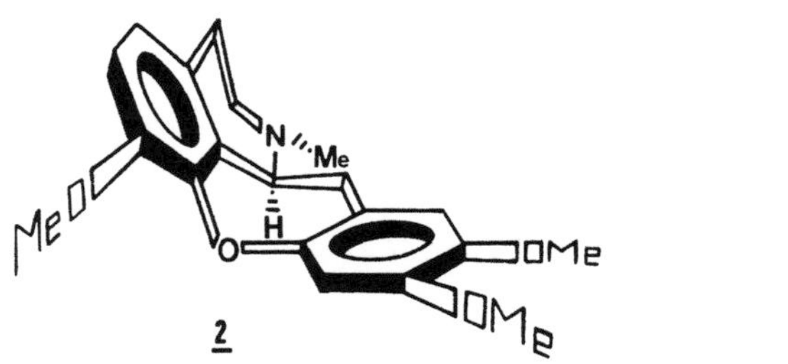

2

a: (+)-cularine: $R = R_1 = R_2 = R_3 = Me$
b: (+)-cularimine: $R_1 = R_2 = R_3 = Me$; $R = H$
c: (+)-cularidine: $R = R_2 = R_3 = Me$; $R_1 = H$
d: (+)-cularicine: $R = Me$; $R_1 = H$; $R_2 + R_3 = CH_2$

Fig. 1

1 Departamento de Quimica Orgánica de la Facultad de Química e Instituto de Productos Naturales Orgánicos (Sección de Alcaloides) del C.S.I.C. Santiago de Compostela, Spain

The Chemistry and Biology of Isoquinoline Alkaloids, ed. by Phillipson et al.

Only the cularine moiety of the cularinemorphine dimers (cancentrine and dehydroderivatives) have substituents at C-7, C-4′ and C-5′. In 1932 cancentrine was isolated by Manske from *Dicentra canadensis* (Goldie)Walp, but nearly 40 years passed before its structure was deduced (Clark et al. 1970).

Since 1982 substantial advances in the field of cularines have occurred with the isolation of a large number of new alkaloids with novel structures making research in this area of considerable current interest.

In our studies of *Sarcocapnos enneaphylla* (L.)DC., (+)-sarcocapnine (3a) was isolated and found to be the first cancentrine-type cularine obtained from a natural source (Campello et al. 1982). Monophenolic (+)-sarcocapnidine (3b) and diphenolic (+)-claviculine (3c) are further examples of this type of cularines substituted at C-7, C-4′ and C-5′, which are sometimes called isocularines and have been isolated by our group from *Sarcocapnos crassifolia* (Desf.) DC. (Boente et al. 1983a) and *C. claviculata* (L.)DC. (Boente unpublished). (+)-Sarcocapnidine has also been obtained by Blaschke's group (Blaschke and Scriba, 1983b) and (+)-claviculine by Guinaudeau's group (Guinaudeau and Allais 1984) from the latter source.

The number of C-7, C-3′ and C-4′ substituted cularines, also referred to as "classical" cularines, has been increased by a recent isolation of several new bases. In 1983, Guinaudeau and Allais reported obtaining (+)-culacorine (1e), (+)-norcularidine (1f) and (+)-norcularicine (1g) from *C. claviculata* (Allais and Guinaudeau 1983b, Guinaudeau and Allais 1984) and Blaschke et al., (+)-O-methylcularicine (1h), from the same source (Blaschke and Scriba 1983b). Further examples are (+)-celtine (li), (+)-celtisine (lj) and (+)-breoganine (le) which we have isolated very recently, the first two from *S. enneaphylla* (L.)DC. and the last from *S. crassifolia* and *C. claviculata* (Boente et al. 1984b). (+)-Breoganine must be identical to (+)-culacorine and it and (+)-celtisine are the first diphenolic cularines to have been characterized.

Quite recently we have isolated secocularine (4a) and secocularidine (4b) (Boente et al. 1984a) and later, norsecocularine (4c) (Castedo et al. 1984), which are the only well-characterized examples of a new group of isoquinoline alkaloids which we have named secocularines, as they are most probably derived from cularines by an

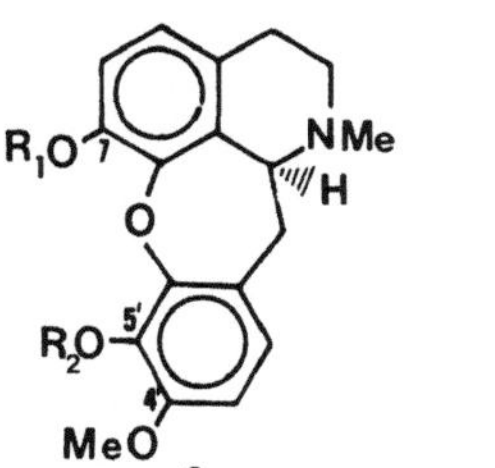

3

a: (+)-sarcocapnine: $R_1 = R_2 = Me$

b: (+)-sarcocapnidine: $R_1 = Me$; $R_2 = H$

c (+)-claviculine: $R_1 = R_2 = H$

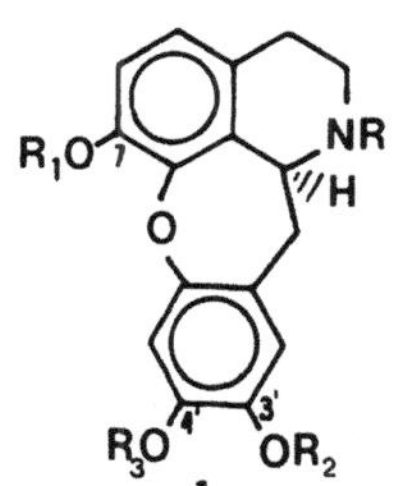

1

e: (+)-culacorine: $R = R_3 = Me$; $R_1 = R_2 = H$ ((+)-breoganine)

f: (+)-norcularidine: $R_2 = R_3 = Me$; $R = R_1 = H$

g: (+)-norcularicine: $R = R_1 = H$; $R_2 + R_3 = CH_2$

h: (+)-o-methylcularicine: $R = R_1 = Me$; $R_2 + R_3 = CH_2$

i: (+)-celtine: $R = R_1 = R_2 = Me$; $R_3 = H$

j: (+)-celtisine: $R = R_2 = Me$; $R_1 = R_3 = H$

Fig. 2

in vivo Hofmann degradation. The first was obtained from *S. crassifolia* and the other two from *C. claviculata.* From the same plant, we have also obtained noyaine (5) (Castedo et al. 1984), which is the only known example of a new group of secocularines, named C-secocularines as they are probably derived from cularines by cleavage of their ring C.

When in 1982 we isolated oxosarcocapnine (6a) (Campello et al. 1982) from *S. enneaphylla,* it was the first member of a new group of cularine bases, the oxoisocularines. This group has very recently been increased with the isolation of monophenolic oxosarcophylline (6c) (Castedo et al. 1984) and monophenolic oxosarcocapnidine (6b) (Boente et al. 1983a) from the same source and from *S. crassifolia,* respectively. In 1983 we also isolated the oxocularine alkaloids oxocompostelline (7a) from *S. enneaphylla* and oxocularine (7b) from *C. claviculata* (Boente et al. 1983b) which contain the substitution pattern of "classical" cularines. Oxocularine had also been isolated by Guinaudeau's group in 1983 from *C. claviculata* (Allais and Guinaudeau 1983b).

In 1983, Guineaudeau et al. reported (Allais and Guinaudeau 1983a) the isolation of (+)-limousamine (8a) from *C. claviculata,* which is the only known example of a 4-hydroxycularine, and very recently a 4-hydroxyisocularine, (+)-4-hydroxysarcocapnine (8b), has been isolated by our group from *S. enneaphylla,* together with yagonine (9) the first dioxocularine, and aristoyagonine (10), the first cularine alkaloid that incorporates a five-member lactam in ring B (Castedo et al. 1984). They can be considered the cularine counterparts of the 4-hydroxyaporphines, dioxoaporphines and aristolactams of the aporphine series (Shamma 1972, Shamma and Moniot 1978).

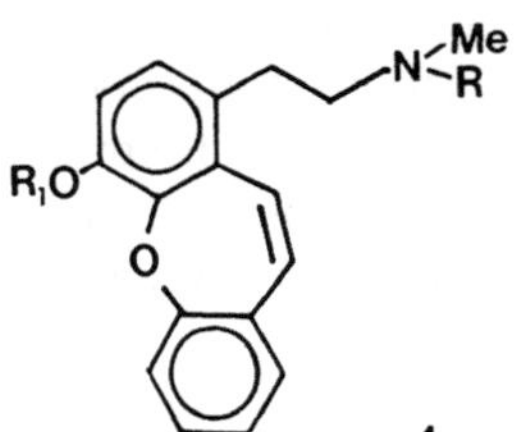

4

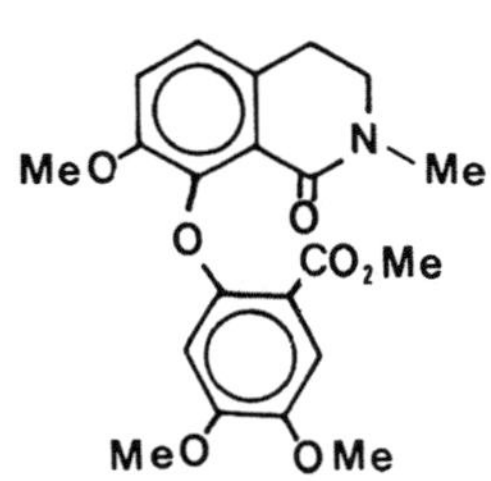

5

a) secocularine: $R = R_1 = Me$
b) secocularidine: $R = Me$; $R_1 = H$
c) norsecocularine: $R = H$; $R_1 = Me$

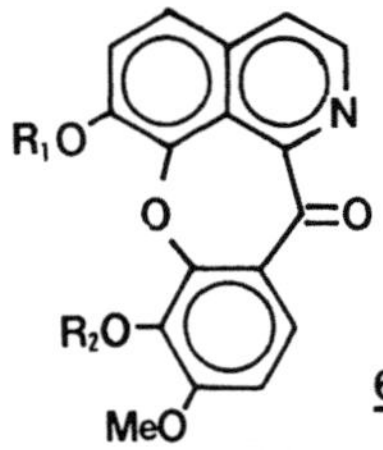

6

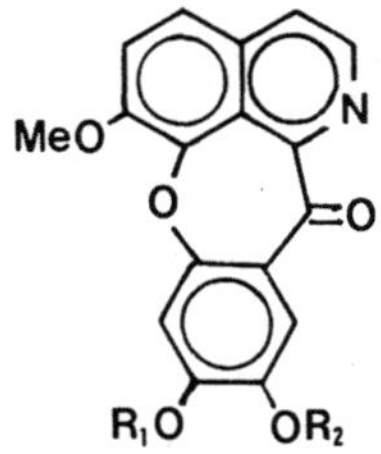

7

a) oxosarcocapnine: $R_1 = R_2 = Me$
b) oxosarcocapnidine: $R_1 = Me$; $R_2 = H$
c) oxosarcophylline: $R_2 = Me$; $R_1 = H$

a) oxocompostelline: $R_1 + R_2 = CH_2$
b) oxocularine: $R_1 = R_2 = Me$

Fig. 3

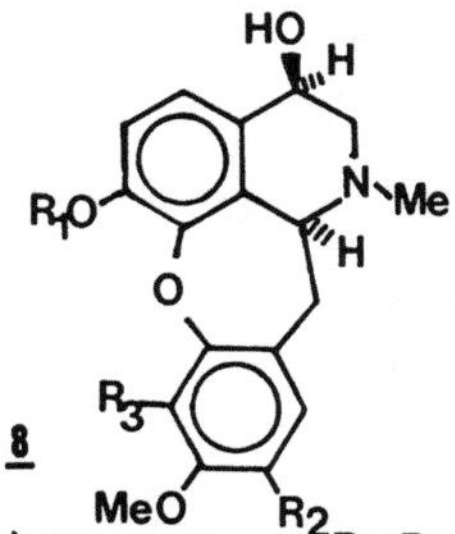

8

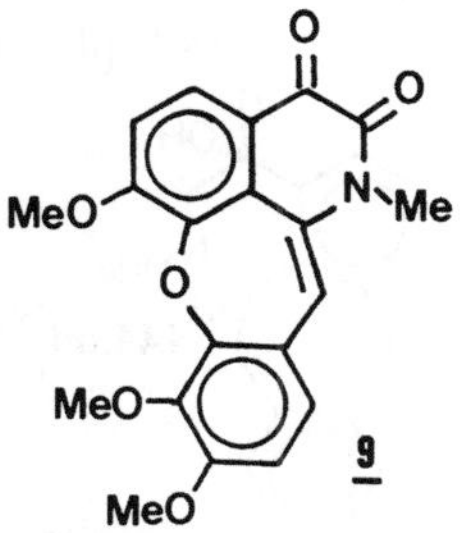

9

a) (+)- limousamine: $R_1=R_3=H; R_2=OMe$
b) (+)-4-hydroxysarcocapnine: $R_1=Me; R_2=H; R_3=OMe$

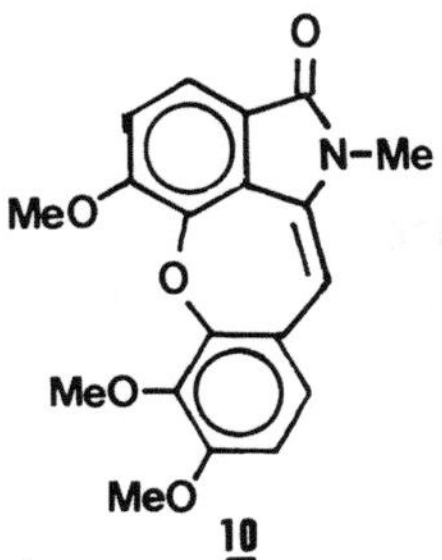

10

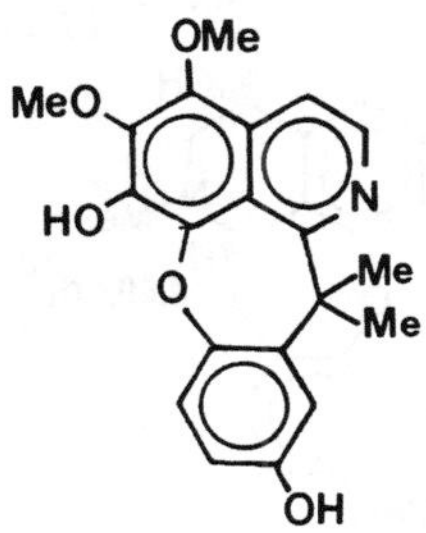

11

Fig. 4

Finally, to the above list of compounds can be added the recently characterized gouregine (11) (Leboeuf et al. 1982), which has an unusual oxygenation pattern and bears two methyl substituents at C-α and has been isolated from *Guatteria ouregou* Dun. (Annonaceae).

The configuration of 4-hydroxycularines cannot be assigned (Allais and Guinaudeau 1983a) on the basis of a simple comparison of the H-4 NMR chemical shift with that of similar epimeric pairs in the aporphine and berberine series, where a clear cut difference between the carbinol chemical shifts of the two epimers is observed, because we have found that this phenomenon does not occur in the cularine series. Comparison of the NMR spectra of synthetic limousamine (8a), 4-hydroxysarcocapnine (8b) and their respective epimers (8c) and (8d) with those of sarcocapnine (3a), cularidine (1c) and cularine (1a) reveals consistent chemical shifts for C-α protons as well as virtually the same coupling constants for those signals. Hence, the same twist-boat conformation (2) established for ring C in cularine (1a) can be assigned to the other cularine compounds. However, in all the above 4-hydroxycularines H-4 exhibits basically the same signal at the same position, with coupling constants of about 3Hz (Fig. 5). These values are in agreement with dihedral angles shown in Newman projections (8b′ and 8d′) of 4-hydroxysarcocapnine (8b) and its epimer (8d). Dreiding model analysis of 4-hydroxycularines reveals that such arrangements are only possible when the hydroxyl group is pseudoaxial in both epimers. Therefore, ring B must have a different conformation in each epimer, and stereomodels show that it can have either a half-chair conformation

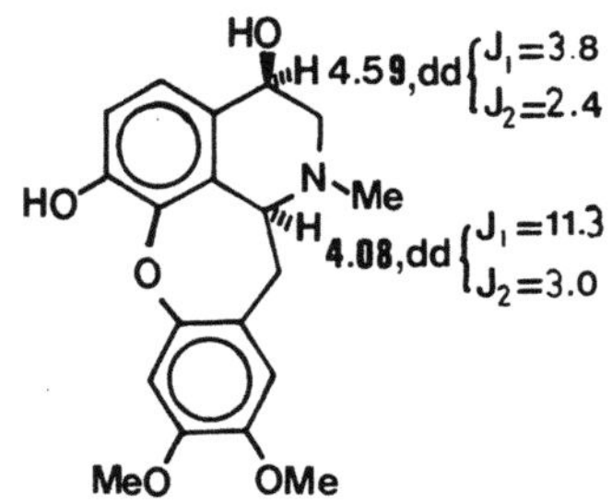

8a

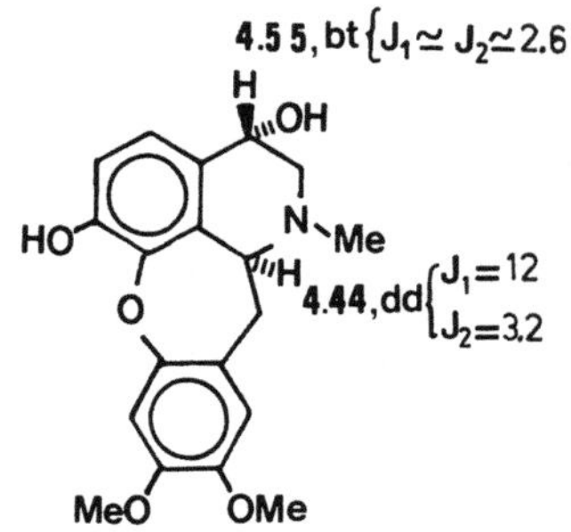

8c

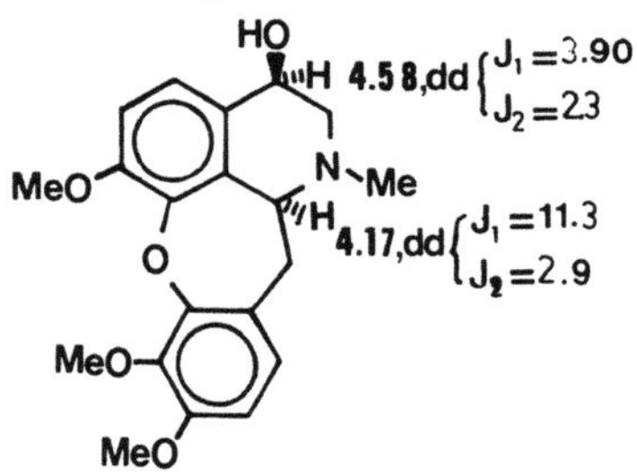

8b

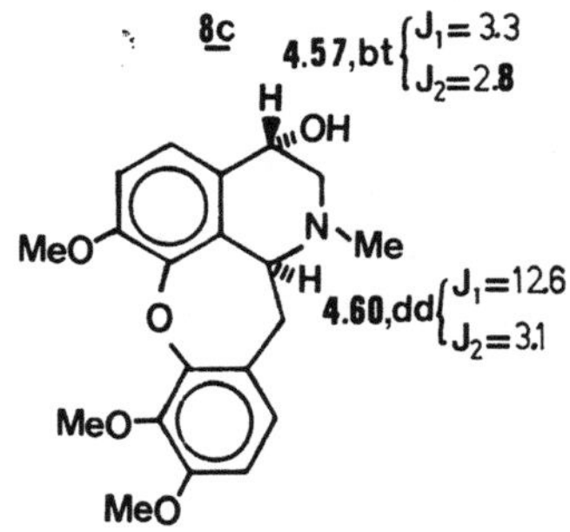

8d

Fig. 5

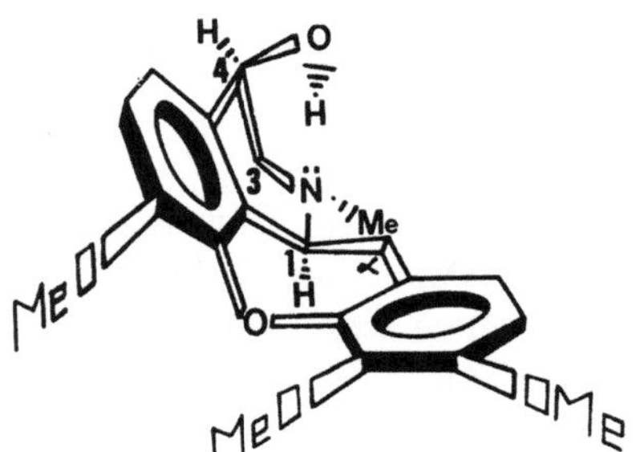

8b
SYN

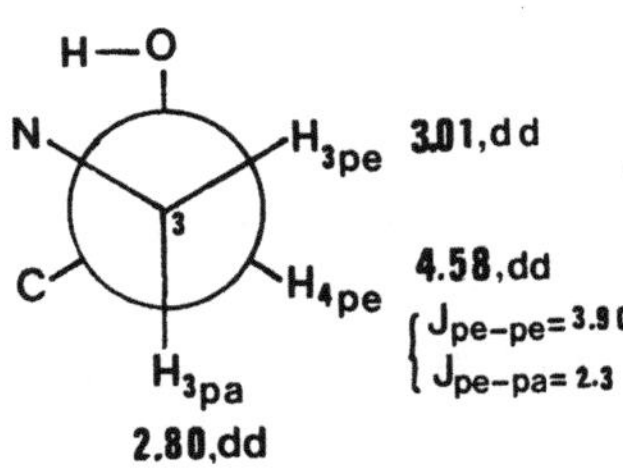

8b'

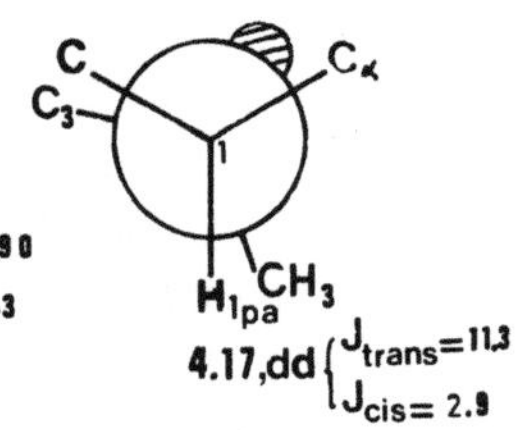

8b''

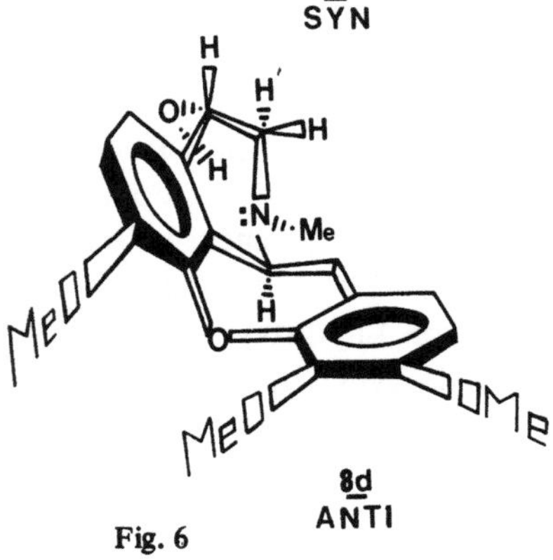

8d
ANTI

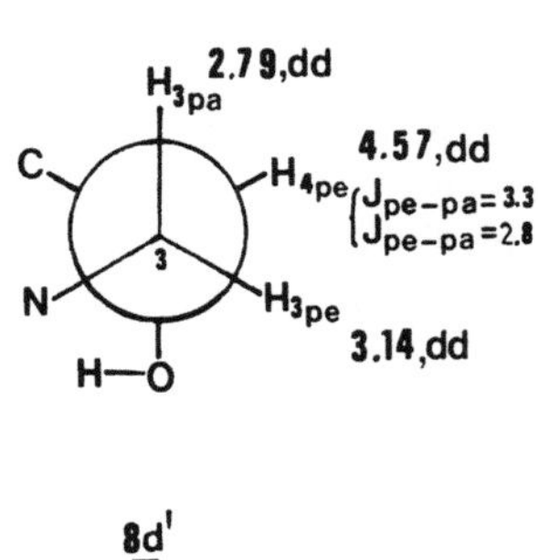

8d'

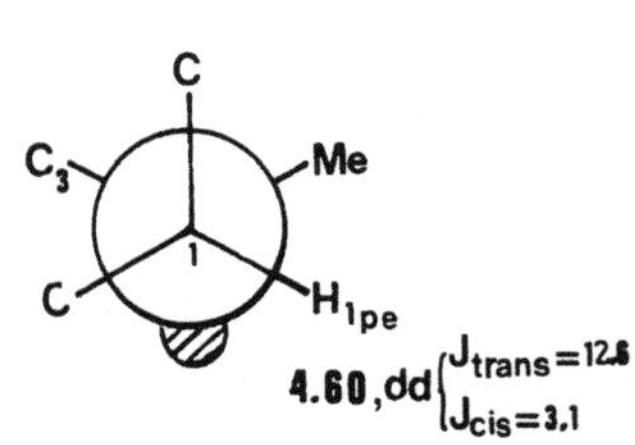

8d''

Fig. 6

or a distorted form, the two being easily interchanged by simply rotating the N-C_3-C_4 bonds, without ring C being involved.

This is further supported by the NMR studies which we have carried out to ascertain the configuration of 4-hydroxycularines (Castedo et al. 1984) in which a lanthanide shift reagent forming a complex with a C-4 hydroxyl means that the chemical shift of H-1 can only be affected in the epimer in which H-4 and H-1 are on different sides (anti) and if the hydroxyl is pseudoaxial. When the NMR spectrum of both 4-hydroxysarcocapnine (8b) and its C-4 epimer (8d) were taken in the presence of Eu $(fod)_3$ a strong shift of H-1 (slope: m=11) was only observed for 4-epihydroxysarcocapnine (8d) thus proving it has anti stereochemistry. 4-Hydroxysarcocapnine (8b) for which only a small shift of H-1 (slope: m=1.7) due to coordination of the europium atom to the nitrogen lone pair was observed, must therefore have syn configuration.

It is interesting that comparison of the NMR data of the two epimers shows that H-1 exhibits a dd at δ 4.17 ppm in 4-hydroxysarcocapnine (8b) and further down (4.6 ppm) in its epimer (8d). A similar phenomenon is also observed in limousamine (8a) and its epimer (8c), where H-1 appears as a dd at 4.08 and 4.44 ppm, respectively (Fig. 5). Hence, limousamine must indeed have syn stereochemistry as previously thought. The configuration of 4-hydroxycularines can thus be easily obtained from a simple NMR analysis of the H-1 chemical shift. In the compounds with syn configuration it should appear as a dd around 4.1 ppm and further down (around 4.5 ppm) in the anti compounds. This difference can be attributed to H-1's tending to be pseudoaxial in the former and pseudoequatorial in the latter, resulting in different environments for H-1 as is shown in Newmann projections (8b″) and (8d″) (Fig. 6).

Both a small repulsive peri effect between a pseudo-equatorial OH at C-4 and H-5 and a stabilizing intramolecular hydrogen bond between the hydroxyl and the nitrogen lone pair, together with greater flexibility of the skeleton, might be the likely causes of the hydroxyl group becoming pseudoaxial in both epimers of the cularine series. Interestingly, in the aporphine series the corresponding hydroxyl is pseudoaxial in the natural epimer such as cataline and pseudoequatorial in the purely synthetic epimer (Castedo et al. 1982).

2. Biogenesis and Chemistry

Three routes have been proposed for the biogenesis of cularine alkaloids in nature (Kametani et al. 1971a, b, Jackson et al. 1974). The least likely one starts from a diphenolic bisphenethylamine (12) and involves a sequence which includes phenolic oxidative coupling to an 11-member ring (13) followed by formation of the tetrahydroisoquinoline system through an iminium (14) as shown in Fig. 7. To date, such precursors are unknown in nature.

The other two routes begin with a diphenolic tetrahydrobenzylisoquinoline. In one, an intermediate, procularine, undergoes a dienone-phenol rearrangement to furnish the cularine or isocularine skeletons as outlined in Fig. 8. To date, no pro-

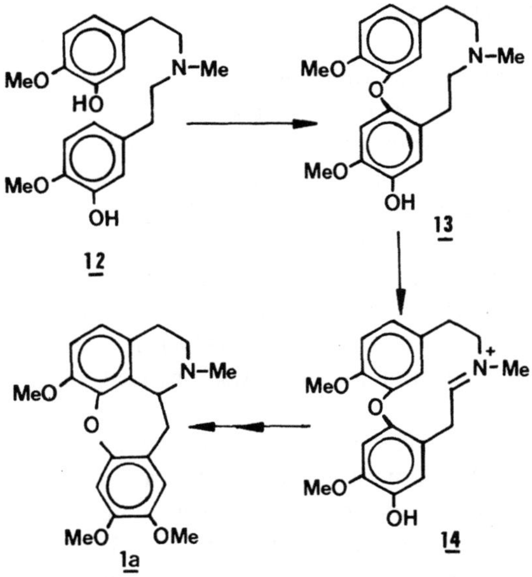

Fig. 7

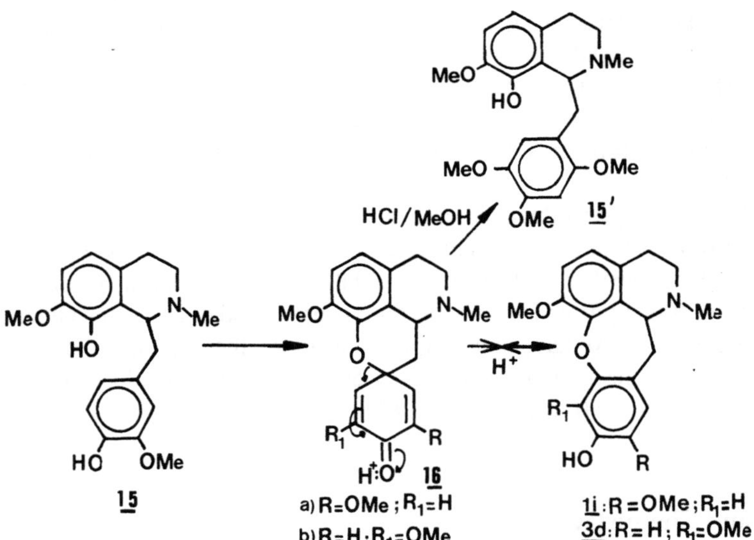

Fig. 8

cularines (16) or cularines monooxygenated in ring D are known to occur in nature. Furthermore, neither of the synthetic diastereomeric procularines (16a and 16b) obtained by intramolecular C-O oxidative coupling of diphenolic tetrahydroisoquinolines (15) with $K_3Fe(CN)_6$ has been found to rearrange in acid to cularine derivatives (1i and 3d) (Kametani et al. 1971a,b, Jackson et al. 1974). Phenolic benzylisoquinolines like (15′) were mainly obtained resulting from reaction of the spirodienone system with solvent (Birch et al. 1975, Choudhury 1974).

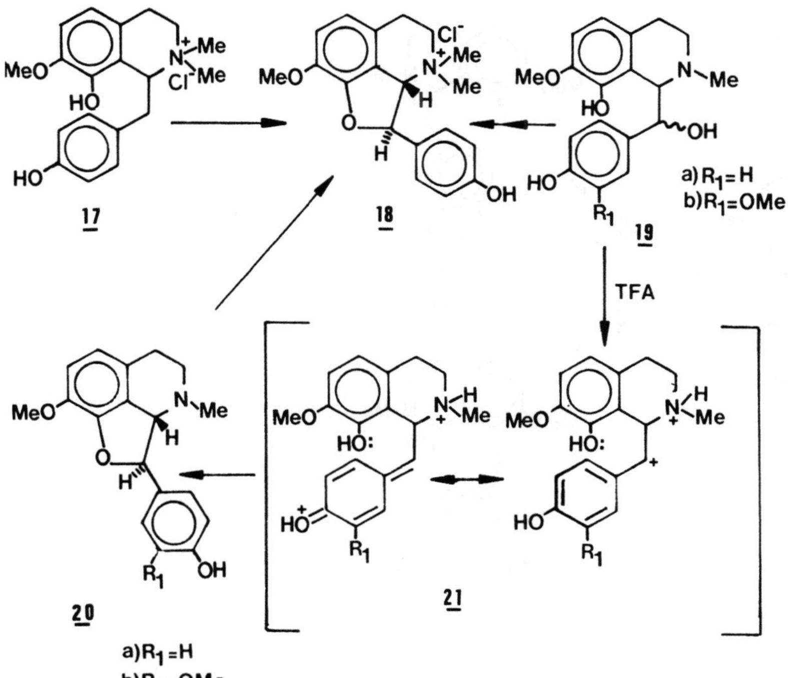

Fig. 9

On the other hand, tetrahydrobenzylisoquinolines such as (17) which posses the same type of diphenolic substitution as (15) have been proposed by Zarga et al. (1981) as the biogenetic precursors for quettamines (18), a new type of isoquinoline alkaloid recently characterized. This cyclization has been accomplished in 35% yield via hydroxybenzyltetrahydroisoquinoline (19a) by acid treatment followed by N-methylation as outlined in Fig. 9 (Chattopadhyay and Shamma 1982). In this process the probable intermediate is the benzyl cation (21a) which is structurally close to the corresponding benzylic free radical or quinone methide which is the probable intermediate that undergoes cyclization in the natural process. We have also observed the same cyclization when the hydroxybenzyl compound (19b) was treated with TFA which afforded the quettamine analogue (20b) in 30% yield (Rodríguez de Lera 1983). Interestingly enough, no cularine compound was observed which would have resulted from an alternative 1-4 addition of C-8 phenol to the dienone system.

In the second route postulated from tetrahydrobenzylisoquinolines, the biogenesis of cularines would proceed by direct phenolic oxidative coupling of another type of diphenolic compound having the phenol groups at C-8 and C-3′ as in (22). Para-coupling to C-3′ would afford the cularine skeleton (1), while ortho-coupling, the cancentrine type (isocularine) (3). The main drawback of this hypothesis hitherto has been that no natural benzylisoquinoline with the 7,8,3′,4′-substitution pattern was known.

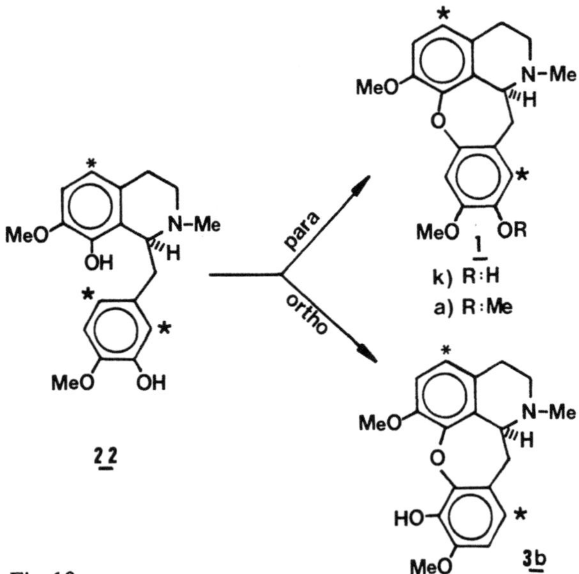

Fig. 10

In our studies of *S. crassifolia* (Boente et al. 1983a) and *C. claviculata* (Boente unpublished), however, we have isolated now (+)-crassifoline (22) together with (+)-sarcocapnidine (3b) and other cularine alkaloids. Moreover, phenolic oxidative coupling of crassifoline has been found to afford a mixture of sarcocapnidine (3b) and O-demethylcularine (1k) (Boente et al. 1983a, Kametani et al. 1971b). Very recently (+)-crassifoline (3b) has also been obtained from *C. claviculata* by Blaschke's group (Blaschke and Scriba 1983b) and later, independently, by Guinaudeau's group (Guinaudeau and Allais 1984). Finally, Blaschke and Scriba (1983a) have also found that when crassifoline labelled with tritium as in (22) is fed to the plant, radioactive cularine (1a) is obtained, which conclusively proves the role of crassifoline as the precursor of cularine alkaloids (Fig. 10).

On the other hand, it is possible to conceive other biogenetic mechanisms from crassifoline (22). That involving a C-C oxidative coupling would give a neoproaporphine (23) which might then rearrange to a dibenzazonine (25) via immonium (24) (Fig. 11). This would have a different substitution pattern from those dibenzazonines known to date and found mainly in the Menispermaceae and Leguminosae families. Support for this new biogenetic pathway from crassifoline (22) has just been obtained in our laboratory from a further study of *C. claviculata* which led us to the isolation of (+)-crassifolazonine (25) (Castedo et al. 1984), the first dibenzazonine with a new type of substitution pattern, which might be derived from crassifoline. Other alternative biogenetic pathways from crassifoline can be further envisioned and therefore new types of isoquinoline alkaloids are likely to be isolated in the near future.

As an example of the biogenetic origin of a 4-hydroxycularine alkaloid, (+)-4-hydroxysarcocapnine (8b) might derive either from (+)-sarcocapnine (3a) by further

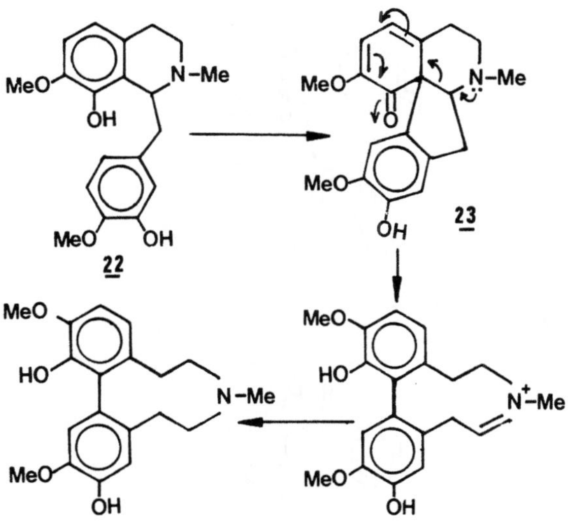

25 (+)–CRASSIFOLAZONINE 24

Fig. 11

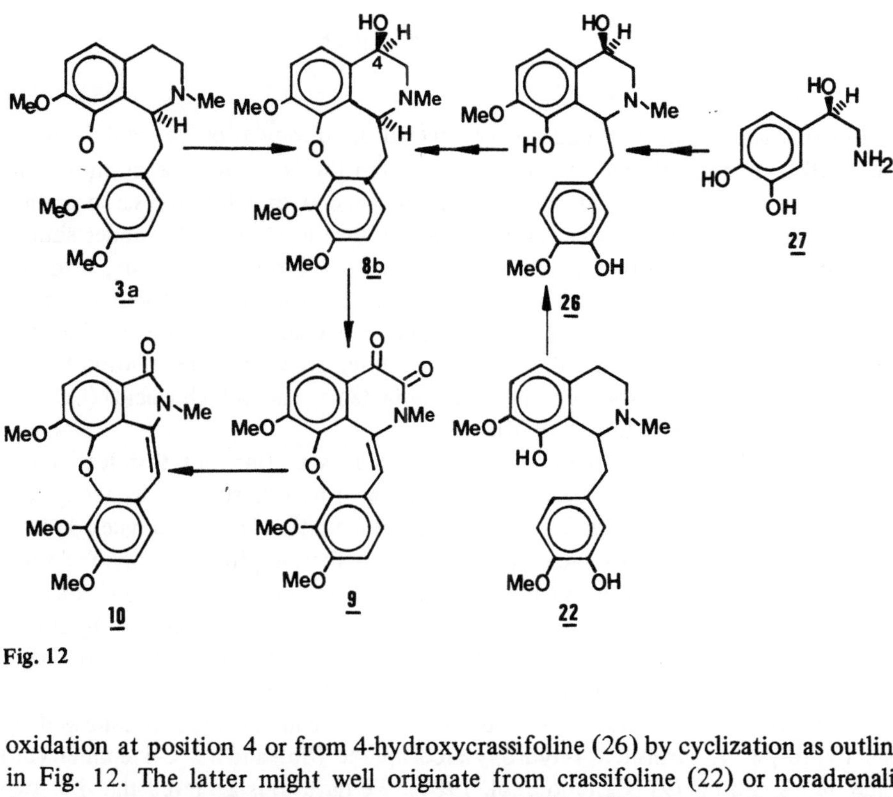

Fig. 12

oxidation at position 4 or from 4-hydroxycrassifoline (26) by cyclization as outlined in Fig. 12. The latter might well originate from crassifoline (22) or noradrenaline (27). Further oxidation of 4-hydrosarcocapnine (8b) would give the 4,5-dioxocularine,

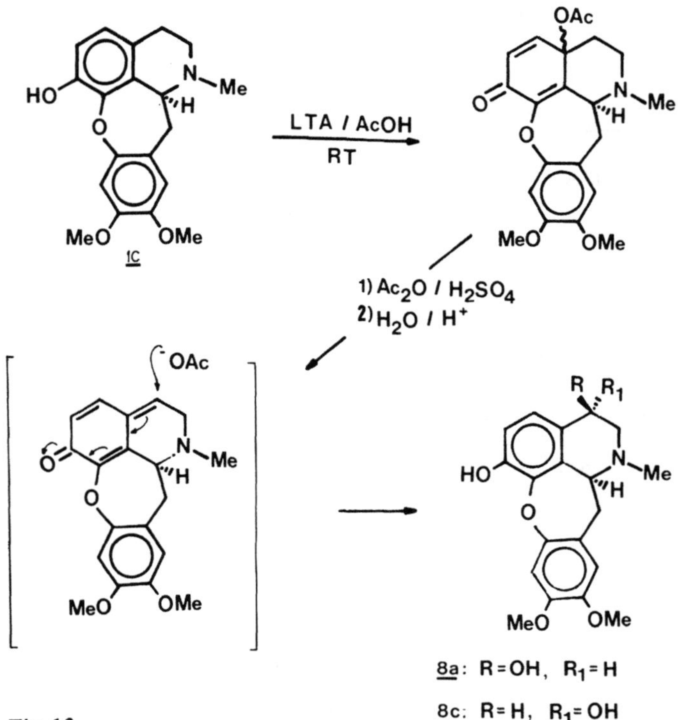

Fig. 13

yagonine (9), which can then undergo net overall decarbonylation to yield aristoyagonine (10). Significantly we have recently isolated the last three compounds from the same source and have just achieved the conversion of 4-hydroxysarcocapnine (8b) into yagonine (9) by oxidation with DDQ (Castedo et al. 1984) under similar conditions to those reported to convert cataline, a 4-hydroxyaporphine, into the corresponding dioxoaporphine, pontevedrine (Castedo et al. 1976). The above type of biogenetic sequence, 4-hydroxysarcocapnine → yagonine → aristoyagonine, has previously been postulated in the aporphine series regarding the relationship 4-hydroxyaporphines → dioxoaporphines → aristolactams → aristolochic acids (Castedo et al. 1976, Shamma and Moniot 1978).

Further studies on 4-hydroxycularines have led us to find out that lead tetraacetate oxidation of cularidine (1c) followed by acid hydrolysis gives in good yield both limousamine (8a) and its C-4 epimer (8c) in the ratio (1/1) (Rodríguez de Lera unpublished) (Fig. 13). This type of oxidation on thaliporphine, a phenolic aporphine, has been reported to yield stereospecifically and quantitatively the corresponding 4-hydroxyaporphine (Hoshino et al. 1975), which has C-4 hydroxyl pseudoaxial. On the other hand, the same oxidation of phenolic tetrahydrobenzylisoquinoline (27) yielded both C-4 acetoxyl epimers (28 and 29) in the ratio (1/9) (Fig. 14). When this mixture was subjected to Ullmann cyclization followed by basic hydrolysis it furnished 4-hydroxysarcocapnine (8b) and its C-4 epimer (8d) in the above ratio (1/9) (Castedo et al. 1984). We have thus achieved the first two syntheses of 4-hydroxycularines. Both limousamine (8a) and 4-hydroxysarcocapnine (8b) have the same configuration at C-4 as has been shown before. The above results

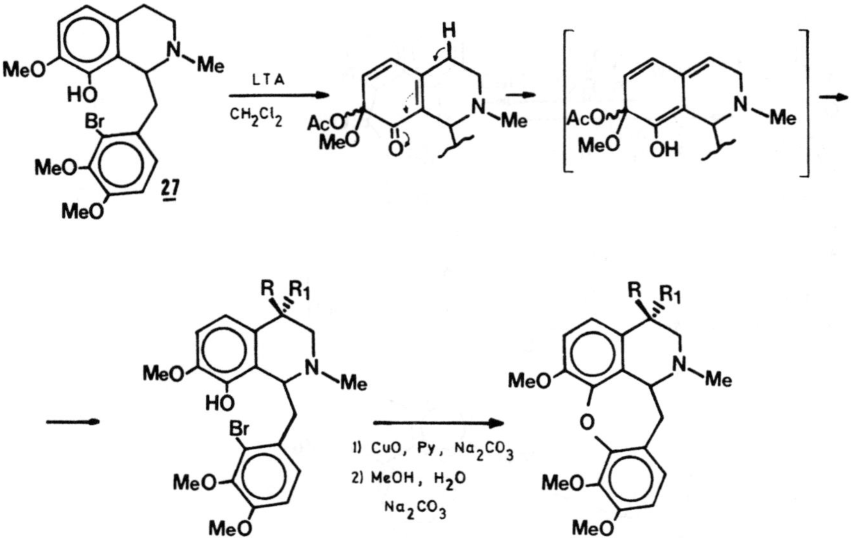

28 R_1 = H, R = OAc — **8b**: R = OH, R_1 = H

29 R_1 = OAc, R = H — **8d**: R = H, R_1 = OH

Fig. 14

also reveal that lead tetraacetate oxidation is less stereoselective in the cularine than in the aporphine series and therefore natural 4-hydroxycularines would best be obtained by oxidation of suitable phenolic cularines. If this synthesis were carried out starting from a phenolic tetrahydrobenzylisoquinoline, the unnatural epimer would be the main product.

The oxocularines are most probably derived from the corresponding cularines by further oxidation of dehydrocularines (Fig. 15). To date, no dehydrocularines have been isolated from natural sources. However, dehydrocularine (30) has just been obtained in 30% yield in our laboratory by photochemical oxidation of cularine in the presence of benzophenone. Dehydrocularine (30) was found to be much more unstable than the analogues in the aporphine series. Attempts to oxidize dehydrocularine (30) under various conditions to furnish C-secocularine noyaine (5) have given a complex mixture of products (Boente unpublished). Such C-seco-cularines are probably derived from the cularines through the dehydrocularines by in vivo cleavage of ring C. Significantly, cularine and noyaine have been isolated from the same source as, respectively, the major alkaloid and a very minor component (Boente unpublished).

However, conversion of cularines into the corresponding oxocularines has been achieved by Fremy's salt and lead tetraacetate in acetic acid, the former reagent giving much better yields than the latter. Thus treatment of sarcocapnine (3a) with Fremy's salt in pyridine-water has furnished oxosarcocapnine (6a) in 50% yield (Campello et al. 1982). Oxidation of phenolic cularines as (3b) to the

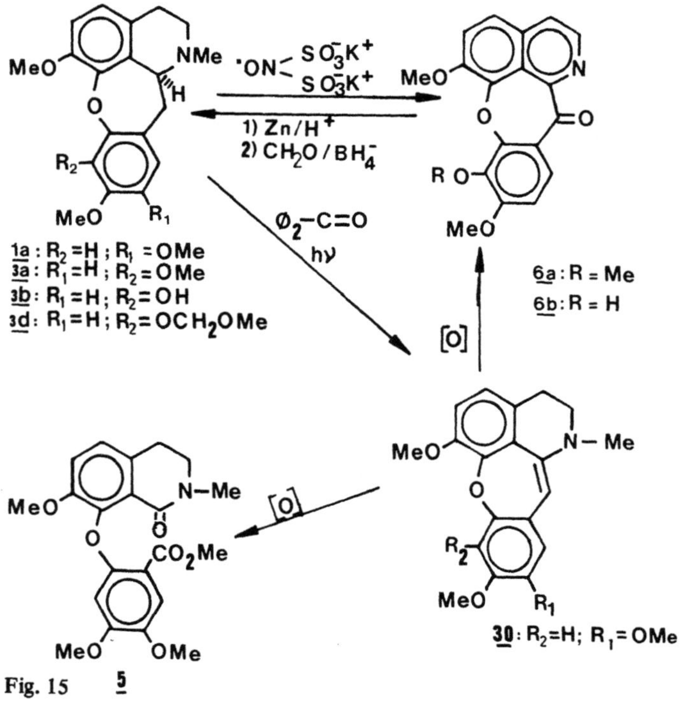

MeO
NMe
H
O
R2
MeO
R1
·ON< SO3-K+ SO3-K+
1) Zn/H+
2) CH2O/BH4-
MeO
N
O
O
R O
MeO
Ø2-C=O
hν
1a : R2 = H ; R1 = OMe
3a : R1 = H ; R2 = OMe
3b : R1 = H ; R2 = OH
3d : R1 = H ; R2 = OCH2OMe
6a : R = Me
6b : R = H
[O]
MeO
N–Me
O
R2
MeO
R1
30 : R2=H; R1=OMe
MeO
N–Me
O
O
CO2Me
MeO
OMe
[O]
5

Fig. 15

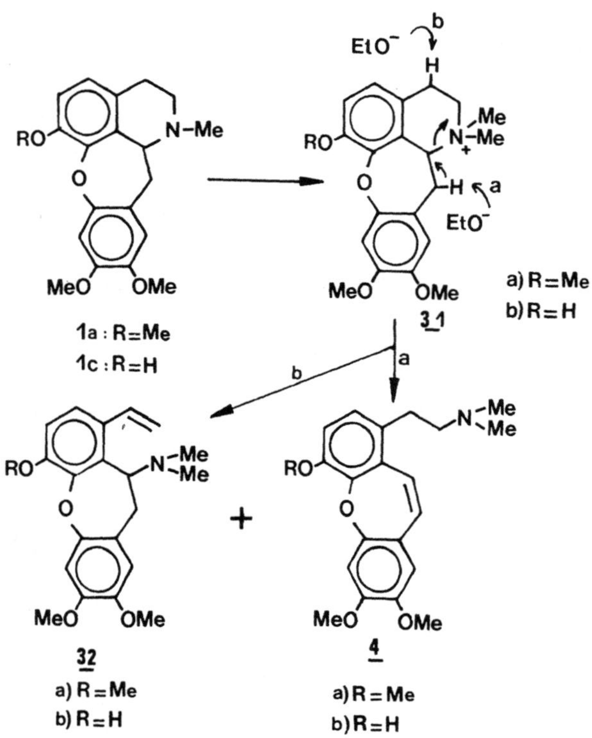

RO
N–Me
O
MeO
OMe
1a : R=Me
1c : R=H
EtO-
b
H
RO
Me
N+-Me
H
a
EtO-
MeO
OMe
31
a) R=Me
b) R=H
b
a
RO
N-Me
-Me
O
MeO
OMe
32
a) R=Me
b) R=H
+
RO
N-Me
-Me
O
MeO
OMe
4
a) R=Me
b) R=H

Fig. 16

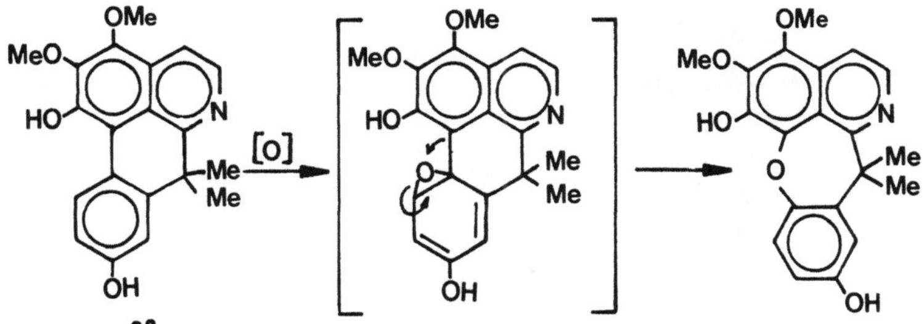

33

Fig. 17

11

corresponding oxocularines requires previous protection of phenolic groups. Thus treatment of O-methoxy-methyl protected sarcocapnidine (3d) with Fremy's reagent followed by acid deprotection afforded oxosarcocapnidine (6b) in 58% yield (Boente et al. 1983a) (Fig. 15). Only a 10% yield of oxocularine (7b) has been obtained by lead tetraacetate oxidation of cularine (1a) (Allais and Guinaudeau 1983b). The reverse process, oxocularine → cularine, has been achieved by Zn reduction and subsequent reductive N-methylation of the resulting norcularine (Campello et al. 1982) (Fig. 15).

The secocularines are probably derived by in vivo Hofmann degradation of quaternary cularine salts. A similar relationship between the phenanthrene alkaloids and the aporphines has been suggested (Shamma 1972). Hofmann degradation of cularine methiodide (31a) using sodium ethoxide in refluxing ethanol has given in 83% yield a mixture of secocularine (4a) and styrene (32a) in the ratio (6:1). Similar results have been obtained from monophenolic cularidine methiodide (31b) (Boente et al. 1984a) (Fig. 16).

Finally, the interesting and recently characterized alkaloid gouregine (11) incorporates the cularine skeleton in its structure, but bears an unusual oxygenated substitution pattern with two methyl groups at C-α. It has been suggested that it is biogenetically derived from the 1,2,3,9-tetraoxygenated aporphine melosmine (33) found in the same source, by further oxidation, a process that has been achieved in 90% yield in the laboratory using Fenton's reagent (Leboeuf et al. 1982). (Fig. 17). Therefore gouregine should not be considered a real cularine alkaloid.

3 Synthesis

There are two general approaches for the synthesis of cularines, both involving the formation of the diaryl ether linkage as the key step (Fig. 18). One route involves the formation of this bond by an Ullmann reaction in the initial stages of synthesis and the subsequent construction of rings C and B. The other approach begins with the synthesis of an appropriately substituted benzylisoquinoline and leaves to the last stage of the sequence the introduction of the diarylether function via intramolecular Ullmann condensation or phenolic oxidative coupling.

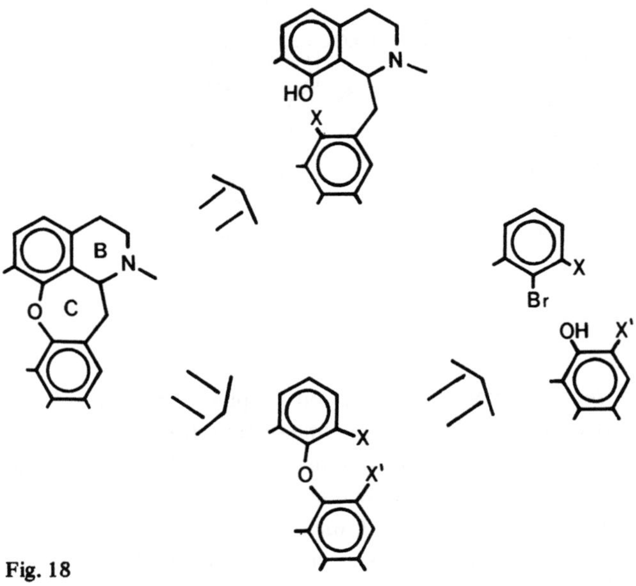
HO
X
B
N
O
C
X
Br
OH
X'
X
O
X'

Fig. 18

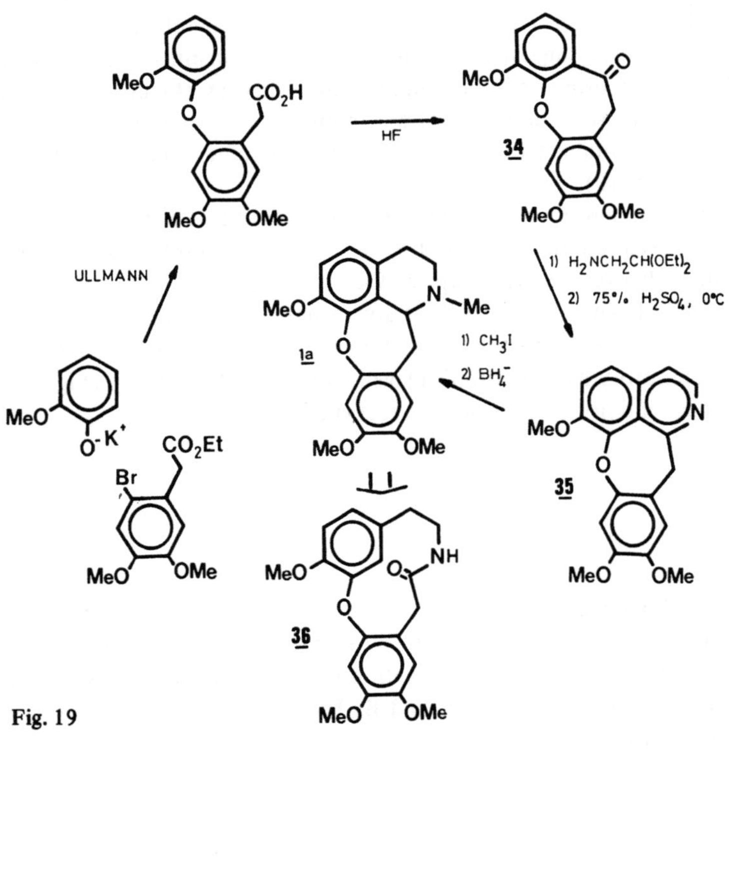
MeO
CO2H
O
MeO
OMe
HF
MeO
O
34
MeO
OMe
1) H2NCH2CH(OEt)2
2) 75% H2SO4, 0°C
ULLMANN
MeO
N
Me
1a
O
1) CH3I
2) BH4-
MeO
OMe
MeO
O-K+
CO2Et
Br
MeO
OMe
MeO
NH
O
36
MeO
OMe
MeO
N
O
35
MeO
OMe

Fig. 19

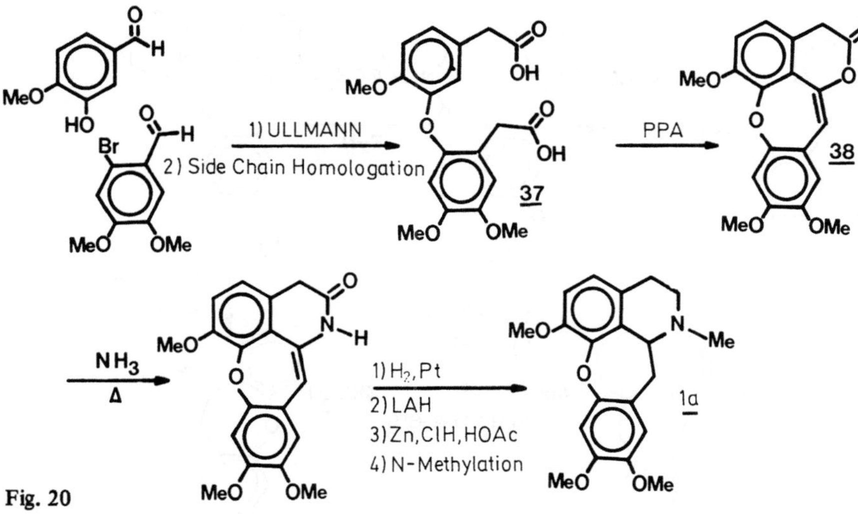

Fig. 20

As an example of the former strategy the first synthesis of (±)-cularine (1a) via keto-oxepine (34) (outlined in Fig. 19) furnished ring B of (35) in only 8.35% yield by the Schlittler-Müller method (Kametani and Fukumoto 1963). Other attempts at ring B formation introduced variations in this scheme such as the Bischler-Napieralski reaction on an intermediate (36). However, a low yield of cyclized product also resulted (Kametani et al. 1963).

An alternative route to (±)-cularine (1a) which has been shown to the possible begins with an Ullmann condensation and proceeds via cyclisation of dicarboxylic acid (37) (Fig. 20). However, this sequence is also too long and some of the stages occurred with low yield. Attempts to synthesize phenolic cularines by this approach have led only to O-alkyl derivatives that were not convertible into the phenolic bases (Kametani et al. 1970). However, variations in the original synthesis of cularine have been introduced to accommodate the sensitivity of the required phenol protecting group to Lewis acids. The diphenyl ether (39) was converted to the morpholinamide (40) which under Bischler-Napieralski conditions cyclized to the enamine (41), which upon hydrolysis furnished the strategic ketone (42). The Bobbitt modification of the Pomeranz-Fritsch cyclization was applied to add the two carbon atoms and the nitrogen atom required to complete the carbon skeleton of cularicine (1d). The 4-ethoxy group of the norcularicine derived from this reaction was removed by hydrogenolysis over Pt under acid conditions and then reductive N-methylation completed the synthesis of (+)-cularicine (1d) (Fig. 21) (Noguchi and McLean 1975).

In the second general approach two routes have been employed for the formation of the biphenylether linkage in the final step. The best results were obtained by the Ullmann reaction, which has been achieved, usually in good yield, via two different types of intermediates, as is shown antithetically in Fig. 22. The best yields in the cyclization step were obtained when (43) and (44) contained a tertiary nitrogen. The synthesis of these precursors was the main difficulty of this approach, as 7,8-disubstituted isoquinolines could not be prepared by the Bischler-Napieralski re-

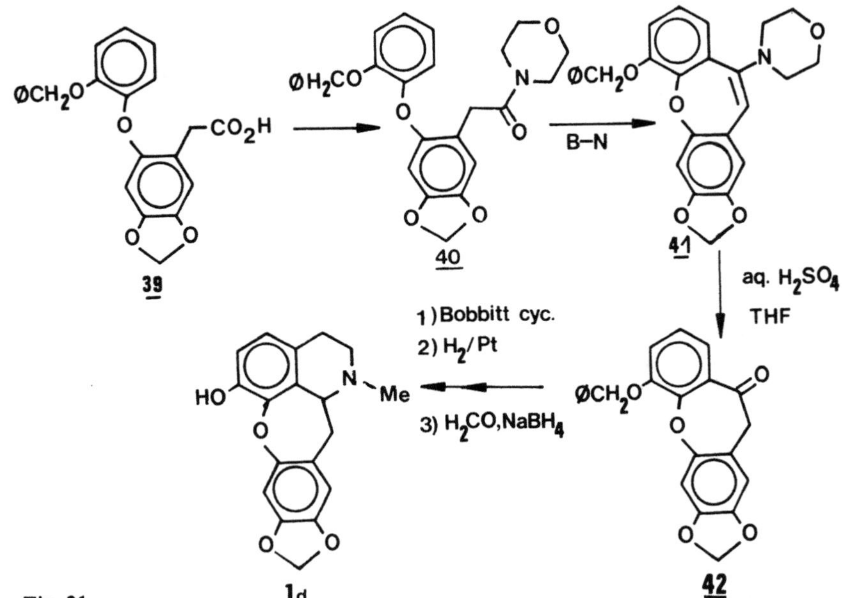

Fig. 21

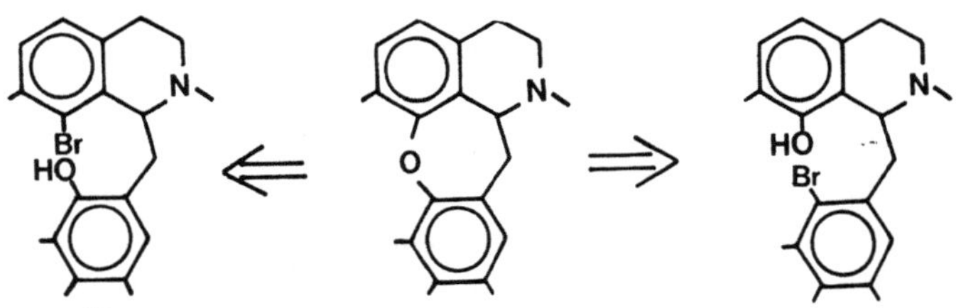

Fig. 22

action. Similar limitations applied to the related Pictet-Spengler synthesis. This problem was circumvented in a synthesis of (±)-cularine (1a) by using an ethoxy-carbonyl-amino-β-phenethylamide (45) in order to activate the para-position and thus effect the required Bischler-Napieralski cyclization (45) → (46) (Fig. 23). Conventional steps then led to the phenol (47) which gave (±)-cularine (1a) via the Ullmann reaction (Ishiwata et al. 1970).

In an alternative approach, preparation of the key intermediate (44) by the Bischler-Napieralski reaction or the Pictet-Spengler method was made feasible by introducing a bromine at C-5 as a blocking substituent to force cyclization to occur in the desired fashion. This route is exemplified by the synthesis of (±)-cularine (1a) outlined in Fig. 23 (Hsu et al. 1972). (±)-Cularimine, (±)-cularicine and (±)-isocularine have also been synthesized in this way (Shamma and Moniot 1978).

A remarkable improvement in the synthesis of such intermediates (44) is now available which involves the akylation of Reissert compounds (53) at C-1. Moreover,

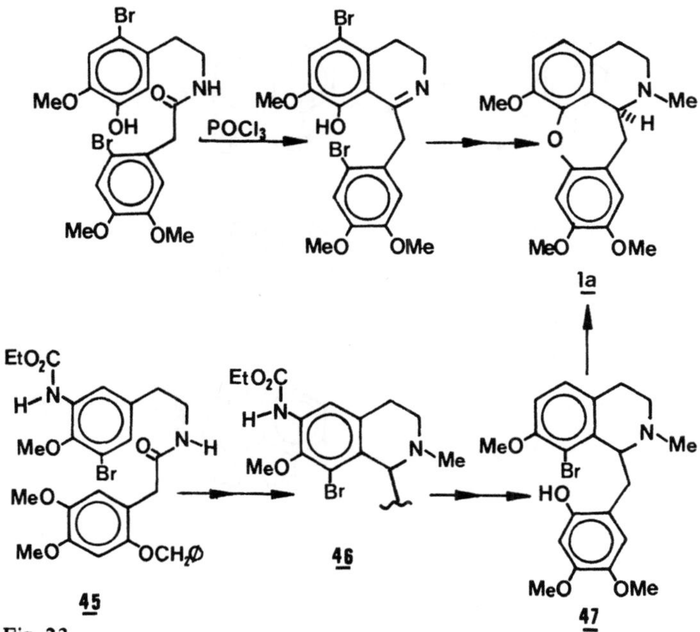
Fig. 23

these can now be efficiently prepared via the slight variant of the Pomeranz-Fritsch cyclization developed by Jackson et al. (1974), which is particularly suitable for the required 7,8-oxygenated isoquinolines (52) as outlined in Fig. 24. An improvement of the Jackson modification is the new and simple methodology developed recently by Boger et al. (1981) which does not depend on catalytic reduction for the preparation of the required N-tosyl benzylaminoacetal (48). It is based on a simple treatment of benzylic mesylates or halides like (50) with the sodium anion of N-tosyl aminoacetaldehyde-dimethylacetal (49) (Fig. 24). Subsequent cyclization of (48) by acid treatment affords N-tosyl dihydroisoquinoline (51), which by base catalyzed elimination of p-toluenesulfonic acid gives the required isoquinoline (52). These variants also provide a direct route to fully aromatic isoquinolines like (52), in contrast to the Bobbitt modification, which leads mainly to tetrahydroisoquinolines.

We have successfully applied these modifications to a new synthesis of (±)-cularine (1a), (±)-O-methylcularicine (1h), (Rodríguez de Lera 1983), (±)-sarcocapnine (3a) and (±)-4-hydroxysarcocapnine (8b) (Rodríguez de Lera, unpublished), using an Ullmann cyclization in the final step as outlined for (±)-sarcocapnine (3a) in Fig. 25. Interestingly enough, attempts to prepare oxocularine (7b) by this convergent approach via benzoylisoquinoline (54) resulted in a complex mixture of products. This intermediate (54) was easily obtained by the Reissert approach using the suitable bromobenzaldehyde derivative followed by hydrolysis and PDC oxidation of the resulting benzyl alcohol. An alternative synthesis of (±)-cularine was achieved recently by Shono et al. (1981), also using Ullmann cyclization in the final step. In this synthesis, the formation of the required phenolic bromobenzylisoquinoline (57b)

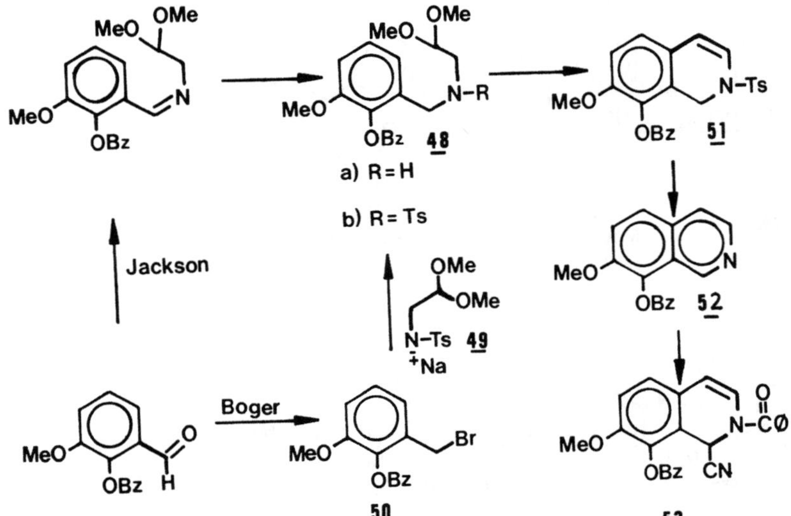

Fig. 24

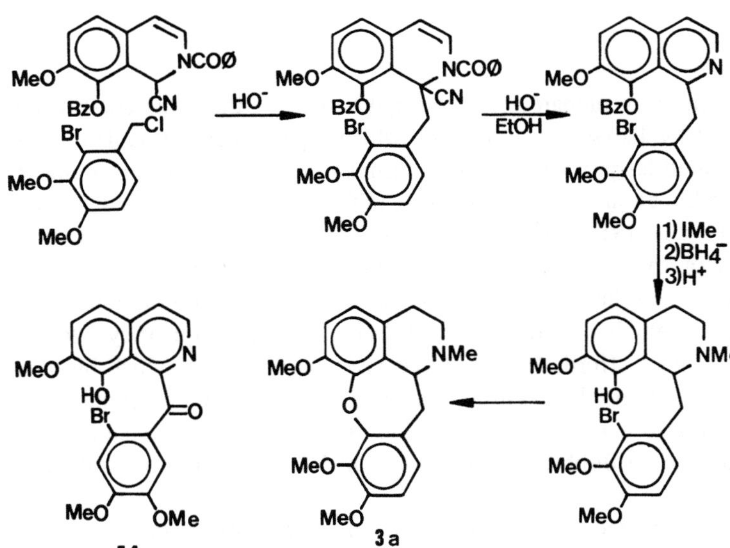

Fig. 25

was achieved in moderate yield by alkylation of the isoquinoline nucleus via an electrochemical reduction of an isoquinolinium salt (55) in the presence of the appropriate bromobenzylbromide derivative (56) (Fig. 26).

The second general approach to cularines has been a biogenetic one involving phenolic oxidative coupling of a diphenolic benzylisoquinoline like crassifoline

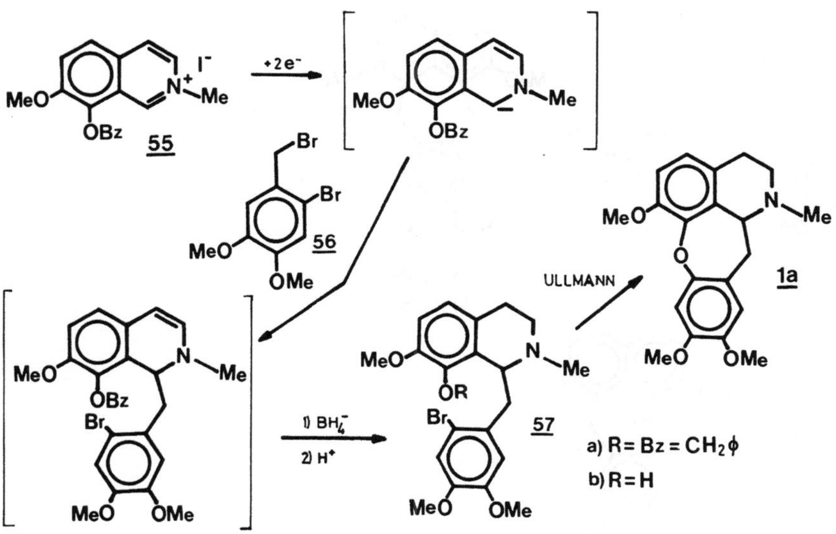

Fig. 26

(22) which has been found to give a very low yield of a mixture of sarcocapnidine (3b) and O-demethylcularine (1k) by treatment with $K_3Fe(CN)_6$ as mentioned above (Fig. 10). By using VOF_3 in TFA as the oxidation reagent we have found that crassifoline-borane complex (58a) gives a 40% yield of O-demethylcularine (1k) (Fig. 27) (Rodríguez de Lera 1983). Furthermore, we have also achieved this oxidative coupling on the borane complex (58b) of a monophenolic tetrahydrobenzylisoquinoline obtaining cularine (1a) in 70% yield. However, this oxidative cyclization could not be achieved on the corresponding benzylisoquinoline (59). We have also unsuccessfully attempted the above C-O strategic coupling via catechol intermediates by using diphenylselenoxide; through "iminium benzenium" ions prepared in situ; and by intramolecular aromatic substitution induced by a base or photochemically on suitable intermediates.

Finally, we have synthesized cularines by a new route based on a novel cyclization in the key step to form the strategic C-O bond. This cyclization consists of a phenoxide ion attack upon an intermediate benzyne (61) generated by the action of dimsyl sodium on an aromatic 6′-bromo-benzylisoquinoline (60) (Fig. 28), which is readily prepared via the Reissert method. In this synthesis, the intermediate benzyne (61b) was attacked intramolecularly by nitrogen and phenoxy nucleophiles to give a 20% yield of the cularine compound (62b) via O-attack, together with 65% yield of pyrrocoline (63b) (N-attack). Attempts to reduce the N-attack via quaternization or through a more rigid intermediate such as benzoylisoquinoline (54) were not successful as other competing processes now operated which hindered the cyclization. When the Ullmann reaction was also attempted on 6′-bromobenzylisoquinoline (60) a complex mixture of products resulted. Oxidation of the cularine compound (62b) is effected quickly and efficiently by several oxidation agents e.g. air, Fremy's salt, giving rise to the oxocularine oxocompostelline (7a) (Fig. 28)

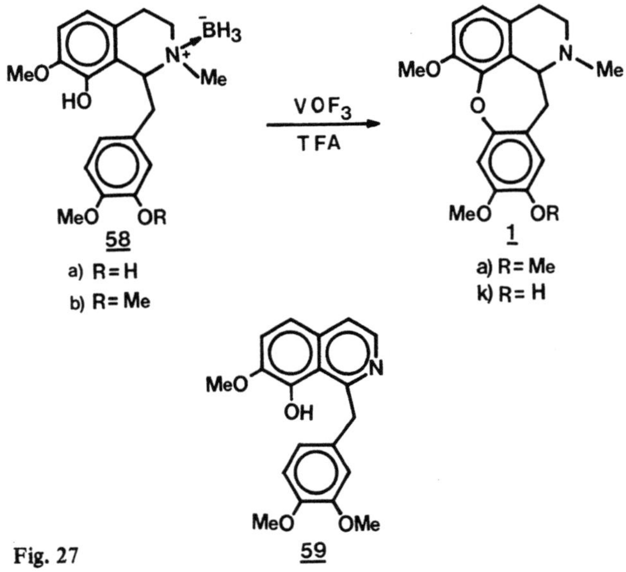
BH3
N+
Me
MeO
HO
MeO
OR
58
a) R=H
b) R=Me
VOF3
TFA
N
Me
MeO
O
MeO
OR
1
a) R=Me
k) R=H
MeO
N
OH
MeO
OMe
59

Fig. 27

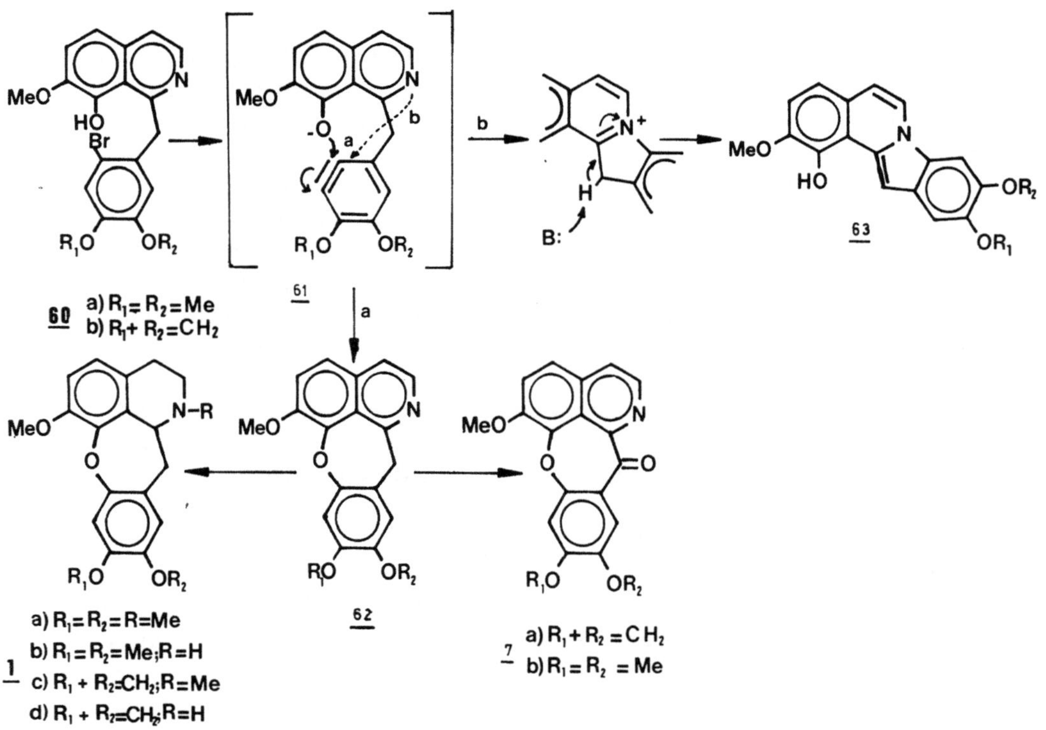
MeO
N
HO
Br
R1O
OR2
60 a) R1=R2=Me
b) R1+R2=CH2
MeO
N
-O
a
b
R1O
OR2
61
b
N+
H
B:
MeO
N
HO
OR2
OR1
63
a
MeO
N-R
O
R1O
OR2
a) R1=R2=R=Me
b) R1=R2=Me;R=H
1 c) R1+R2=CH2;R=Me
d) R1+R2=CH2;R=H
MeO
N
O
RO
OR2
62
MeO
N
O
O
R1O
OR2
a) R1+R2=CH2
7 b) R1=R2=Me

Fig. 28

(Boente et al. 1983b). The same approach was utilized to synthesize oxocularine (7b), 50% of pyrrocoline (63a) and 25% of the cularine compound (62a) being obtained in the cyclization step.

The new synthetic procedure offers the advantage over other methods of allowing the preparation of N-norcularines, cularines and oxocularines from a common intermediate, tetradehydronorcularines like (62) (Fig. 28).

4 Pharmacology

The therapeutic value of cularine alkaloids is virtually unknown due chiefly to the small number of cularine alkaloids that were known until quite recently. A few pharmacological properties of cularine were reported in 1940 when it was compared with hydrastine and papaverine (Reynolds 1940). Recently, in a study to establish the structural features responsible for the important biological activities of the alkaloids (–)-emetine, (–)-cryptopleurine and (–)-tylocrebrine, which belong to different families, a large number of related compounds have been examined. Cularine was found to exhibit no cross-resistance to emetine-resistant mutants of Chinese hamster ovary cells (Gupta et al. 1980).

In spite of the present lack of significant pharmacological data, it is by no means improbable that in the future cularine alkaloids may be found to possess some therapeutic properties. On the basis of the N-O-O triangle hypothesis put forward by Cheng and Zee-Cheng (1981), the cularines should be tested for possible antineoplastic activity.

Acknowledgements. I am greatly indebted to a number of graduate and postdoctoral students and several colleagues who have been referenced throughout this article for their contributions and valuable suggestions. Financial support was provided by the Comisión Asesora (Spain).

References

Allais DP, Guinaudeau H (1983a) Limousamine: a 4-hydroxylated cularine alkaloid. Heterocycles 20(10):2055–2057

Allais DP, Guinaudeau H (1983b) Three new cularine alkaloids: culacorine, norcularicine and oxocularine. J Nat Prod 46:881–883

Bhacca NS, Craig JC, Manske RHF, Roy SK, Shamma M, Slusarchyk WA (1966) The configuration and conformation of cularine, Tetrahedron 22:1467–1475

Birch AJ, Jackson AH, Shanon PVR, Stewart GW (1975) Phenol oxidation. Part IV. Synthesis and novel ring opening of spirocyclic dienones related to the benzylisoquinoline alkaloid cularine, J. Chem Soc Perkin I:2492–2501

Blaschke G, Scriba G (1983a) The biosynthesis of cularine. Tetrahedron Lett 24:5855–5856

Blaschke G, Scriba G (1983b) New papaveraceae alkaloids including an 7,8-oxygenated isoquinoline, the postulated precursor of the cularine alkaloids. Z Naturforsch 38c:670–671

Boente JM, Castedo L, Cuadros R, Rodríguez de Lera A, Saá JM, Suau R, Vidal MC (1983a) New phenolic isocularine and benzylisoquinoline alkaloids and their biogenetic relationship. Tetrahedron Lett 24:2303–2306

Boente JM, Castedo L, Rodríguez de Lera A, Saá JM, Suau R, Vidal MC (1983b) Oxocompostelline and oxocularine, structure and synthesis. Tetrahedron Lett 24:2295–2298

Boente JM, Castedo L, Domínguez D, Fariña A, Rodríguez de Lera A, Villaverde MC (1984a) Secocularines: a new group of isoquinoline related alkaloids present in the Fumariaceae, Tetrahedron Lett 25:889–892

Boente JM, Castedo L, Rodríguez de Lera A, Saá JM, Suau R, Vidal MC (1984b) Tetrahedron Lett 25:1829–1830

Boger DL, Brotherton CE, Kelley MD (1981) A simplified isoquinoline synthesis. Tetrahedron 3977–3980

Campello MJ, Castedo L, Saá JM, Suau R, Vidal MC (1982) Cancentrine-type cularine alkaloids: sarcocapnine and oxosarcocapnine. Tetrahedron Lett 23:239–240

Castedo L, Suau R, Mouriño A (1976) A revised structure for pontevedrine. Tetrahedron Lett 501–502

Castedo L, Riguera R, Sardina J, García-Blanco S, Fonseca I (1982) On the structure of cataline. Heterocycles 19:1591–1593

Castedo L et al. (1984) Int Symp Chem Biol Isoquinoline Alkaloids, Phytochem Soc Eur, Londdon

Chattopadhyay S, Shamma M (1982) A convergent synthetic route to the quettamines and the 7,8-dioxygenated tetrahydrobenzylisoquinoline alkaloids. Heterocycles 19:697–701

Cheng CC, Zee-Cheng RKY (1981) The N-O-O triangulation hypothesis - an assessment after one decade. Heterocycles 15:1275–1283

Choudhury AM (1974) Acid-catalysed rearrangement of a spirocyclohexadienone involving an external nucleophile. J Chem Soc Perkin I:132–134

Clark GR, Manske RHF, Palenik GJ, Rodrigo R, MacLean DB, Baczynskyj L,Gracey DEF, Saunders JK(1970) The structure of cancentrine: A novel dimeric benzylisoquinoline alkaloid. J Am Chem Soc 92:4998–4999

Guinaudeau H, Allais P (1984) New isoquinoline alkaloids from *Corydalis claviculata.* Heterocycles 22(1):107–111

Gupta RS, Krepinsky JJ, Siminovitch L (1980) Structural determinants responsible for the biological activity of (–)-emetine, (–)-cryptopleurine and (–)-tylocrebrine: structure-activity relationship among related compounds. Mol Pharmacol 18:136–143

Hoshino O, Hara H, Ogawa M, Umezawa B (1975) A stereospecific synthesis of (±)-cataline. J Chem Soc Chem Commun 306–307

Hsu HC, Kikuchi T, Aoyagi S, Iida H (1972) Studies on the synthesis of 7,8-disubstituted 1-benzylisoquinolines and related compounds. Synthesis of dl-cularimine and dl-cularine J Pharm Soc Jpn 92:1030–1033

Ishiwata S, Fujii T, Miyaji N, Satoh Y, Itakura K (1970) Synthesis of aminoisoquinolines and related compounds. X. A modified synthesis of (±)-cularine. Chem Pharm Bull (Japan) 18: 1850–1855

Jackson AH, Stewart GW, Charnock GA, Martin JA (1974) Phenol oxidation, part III. Synthesis of the benzylisoquinoline alkaloid cularine. J Chem Soc Perkin I:1911–1920

Kametani T, Fukumoto K (1963) Cularine and related compounds Part VI. A total synthesis of (±)-cularine. J Chem Soc:4289–4296

Kametani T, Fukumoto K, Shibuya S, Nakano T (1963) Cularine and related compounds. V. Investigation of the Ullmann reaction products of methyl (2-bromo-4,5-dimethoxyphenyl) acetate with phenolic compounds, and an approach to the total synthesis of cularine. Chem Pharm Bull 11:1299–1305

Kametani T, Iida H, Kibayashi C (1969) Synthesis of heterocyclic compounds. CCLXXIX. Synthesis of 0-isobutylcularidine. J Heterocycl Chem 6(1)61–71

Kametani T, Iida H, Kibayashi C (1970) J Hetereocyclic Chem 7:339

Kametani T, Fukumoto K, Fujihara M (1971a) Biosynthesis of cularine, J Chem Soc Chem Commun:352–353

Kametani T, Fukumoto K, Fujihara M (1971b) Biogenetic-type synthesis of cularine. Bio-org Chem 1:40–50

Kametani T, Honda T, Shimanouchi H, Sasada Y (1972) Absolute configuration of the alkaloid cularine: an X-ray structure determination. Chem Commun:1072–1073

Kunitomo J, Morimoto K, Yamamoto K, Yoshikawa Y, Azuma K, Fujitani K (1971). The absolute configuration of cularine: A chemical correlation to L(S)-laudanosine. Chem Pharm Bull (Japan) 19:2197–2199

Leboeuf M, Cortes D, Hoquemiller R, Cavé A, Chiaroni A, Riche C (1982) Structure de la gouregine, alcaloïde original apparente aux cularines. Tetrahedron 38:2889–2896

Manske RHF (1950) The alkaloids of fumariaceous plants. XLIII. The structures of cularine and of cularimine. J Am Chem Soc 72:55–59

Manske RHF (1968) In: Manske RHF (ed) The alkaloids vol X. Academic Press, London New York, pp 463–465

Noguchi I, McLean DB (1975) Synthesis of (±)-cularicine. Can J Chem 53(1):125–130

Reynolds AK (1940) The pharmacological actions of cularine. J Pharmacol Exp Ther 69: 112–116

Rodríguez de Lera A (1983) PhD thesis, Univ Santiago

Shamma M (1972) The isoquinoline alkaloids, Academic Press, London New York

Shamma M, Moniot JL (1978) Isoquinoline alkaloids research 1972–1977. Plenum Press, New York

Shimanouchi H, Sasada Y, Honda T, Kametani T (1973) Crystal and molecular structure and absolute stereochemistry of cularine methiodide. J Chem Soc Perkin II:1226–1230

Shono T, Miyamoto T, Mizukami M, Hamaguchi H (1981) Electroreductive synthesis of 1-(bromobenzyl)-isoquinoline derivatives and its application to cularine synthesis. Tetrahedron Lett 25:2385–2388

Zarga MHA, Miana GA, Shamma M (1981) The Quettamines: a new class of isoquinoline alkaloids. Tetrahedron Lett 22:541–544

Bisbenzylisoquinoline Alkaloids

P. L. SCHIFF, Jr.[1]

1 Introduction

The bisbenzylisoquinoline alkaloids are represented by over 225 different compounds which have been isolated from plants in 14 different plant families (Guha et al. 1979, Schiff Jr. 1983). It is the purpose of this presentation to review the distribution, chemical classification and major reactions of importance in structure elucidation of these alkaloids. In addition, a tabular summary of new alkaloids isolated and identified in the last several years will be presented.

2 Distribution

The bisbenzylisoquinoline alkaloids are particularly abundant in plants of the families Berberidaceae, Menispermaceae, Monimiaceae and Ranunculaceae as shown in Table 1 (Guha et al. 1979, Schiff Jr. 1983). The most prodigious source of these alkaloids are plants of the family Menispermaceae where species representing 21 genera have produced no less than 112 different alkaloids. In Table 2 (Guha et al. 1979, Schiff Jr. 1983), it can be seen that plants in 11 genera of the Menispermaceae produce from 1–5 different alkaloids, while those in 8 genera produce from 11–20 different alkaloids. The genus *Cyclea* is known to be the richest source of these alkaloids in the Menispermaceae, with 23 different alkaloids having been isolated from 7 species. The second richest source of these alkaloids by family are species of the genus *Thalictrum* in the plant family Ranunculaceae (Tables 1 and 2). This genus is the source of no less than 66 different bisbenzylisoquinoline dimers and serves as the single richest generic source of these alkaloids. Six different genera in the family Monimiaceae have been a source of 28 different bisbenzylisoquinoline alkaloids with the genus *Daphnandra* being a source of 27 of these alkaloids (Tables 1 and 2). Finally, species of the genera *Berberis* and *Mahonia* of the family Berberidaceae have produced 20 different alkaloids with 19 of them being found among numerous species of the genus *Berberis* (Tables 1 and 2). In summary, plants of the family Menispermaceae seem to have the greatest likelihood of producing bisbenzylisoquinoline alkaloids while plants of the genera *Berberis* (Berberidaceae), *Daphnandra* (Monimaceae) and *Thalictrum* (Ranunculaceae) are the richest generic sources of these alkaloids.

1 Department of Pharmacognosy, School of Pharmacy, University of Pittsburgh, Pittsburgh, Pennsylvania 15261, USA

The Chemistry and Biology of Isoquinoline Alkaloids, ed. by Phillipson et al.

Table 1. Distribution of the bisbenzylisoquinoline alkaloids by family

Family	Genera	Plants	Alkaloids
Aristolochiaceae	1	1	1
Annonaceae	5	5	7
Berberidaceae	2	42	20
Buxaceae	1	1	1
Euphorbiaceae	1	1	2
Hernandiaceae	2	2	5
Lauraceae	3	3	11
Magnoliaceae	1	2	3
Menispermaceae	21	63	112
Monimiaceae	6	11	28
Nymphaceae	1	1	3
Ranunculaceae	3	51	66
Rhamnaceae	1	2	3
Umbelliferae	1	1	1

Table 2. Distribution of the bisbenzylisoquinoline alkaloids by genus

Family	Genus	Alkaloids
Aristolochiaceae	*Aristolochia*	1
Annonaceae	*Crematosperma*	1
	Guatteria	3
	Isolona	2
	Phaeanthus	2
	Uvaria	1
Berberidaceae	*Berberis*	19
	Mahonia	6
Buxaceae	*Buxus*	1
Euphorbiaceae	*Andrachne*	2
Hernandiaceae	*Gyrocarpus*	2
	Hernandia	3
Lauraceae	*Dehassia*	1
	Lindera	1
	Nectandra	9
Magnoliaceae	*Magnolia*	3
Menispermaceae	*Abuta*	14
	Anisocyclea	4

Table 2 (cont.)

Family	Genus	Alkaloids
	Anomospermum	1
	Arcangelisia	2
	Chondodendron	11
	Cissampelos	15
	Cocculus	12
	Cyclea	23
	Epinetrum	3
	Limacia	3
	Limaciopsis	2
	Menispermum	6
	Pachygone	4
	Paracyclea	4
	Pleogyne	2
	Pycnarrhena	12
	Sciadotenia	3
	Stephania	17
	Synclisia	5
	Tiliacora	15
	Triclisia	12
Monimiaceae	*Atherosperma*	3
	Daphnandra	27
	Doryphora	6
	Dryadodaphne	2
	Laurelia	1
	Nemuaron	1
Nymphaceae	*Nelumbo*	3
Ranunculaceae	*Isopyrum*	3
	Thalictrum	66
	Xanthorhiza	2
Rhamnaceae	*Colubrina*	3
Umbelliferae	*Heracleum*	2

3 Classification

3.1 Structural

Bisbenzylisoquinoline alkaloids are structurally constructed of two monomeric benzylisoquinoline units (usually benzyltetrahydroisoquinolines) joined by one or more ether bridges. In addition, direct carbon-to-carbon biphenyl linkages and methyleneoxy linkages are also found. If one considers the "head" of the monomer to be the isoquinoline portion and the "tail" to be the benzyl portion, permutations of joining these monomeric units together include tail-to-tail, head-to-head, head-

Table 3. Mode of linkage of bisbenzylisoquinoline alkaloids

Class	Type	Number of alkaloids
Tail-to-tail	Head Head | | Tail—O—Tail	31
Head-to-head	Head—O—Head | | Tail Tail	0
Head-to-tail	Head Tail | | Tail—O—Head	4
	Head—O—Tail | | Tail—O—Head	28
	Head—CH_2O—Tail | | Tail—O—Head	6
	Head—CH_2O—Tail \ O / | | Tail—O—Head	2
Head-to-head and tail-to-tail	Head—O—Head | | Tail—O—Tail	120
	Head—O—Head \ O / | | Tail—O—Tail	22
	Head—O—Head | | Tail——Tail	10
	Head—O—Head \ O / | | Tail——Tail	10
	Head—CH_2O—Head \ O / | | Tail—O—Tail	2

Head = Isoquinoline-portion
Tail = Benzyl-portion

to-tail, and head-to-head plus tail-to-tail. An inspection of the data found in recent reviews (Guha et al. 1979, Schiff Jr, 1983) plus a consideration of information found in *Chemical Abstracts* through 1983 can be summarized in Table 3. This table shows that the most common type of dimeric linkage to be head-to-head plus tail-to-tail (164 alkaloids), with single ether bridges between both heads and both tails to be the most frequent (120 alkaloids) of that group. Tail-to-tail linked dimers (31 alkaloids) and head-to-tail linked dimers (40 alkaloids) are almost equal in occurrence with head-to-head linked dimers being unknown to date. Biphenyl linkage appears restricted to the benzyl portions of the molecule (tail-to-tail) (20 alkaloids). Multiple oxygen bridged isoquinoline portions are invariably observed as dibenzo-p-dioxin compounds (32 alkaloids) with the tail-to-tail linkage being either a single ether group (22 alkaloids) or a biphenyl attachment (10 alkaloids). Finally, alkaloids with methyleneoxy bridges are most commonly observed with this linkage being head-to-tail (8 alkaloids) than head-to-head (2 alkaloids).

3.2 Numerical-Symbolic

A consideration of the oxidation state of the isoquinoline portion of these alkaloids reveals that the great majority exist as bisbenzyltetrahydroisoquinolines containing two tertiary nitrogen atoms (NCH_3 groups) (135 alkaloids). The second most common group are those alkaloids containing one secondary nitrogen and one tertiary nitrogen (NCH_3) (41 alkaloids) while there are only 3 alkaloids which are bis-secondary amines. Thirteen alkaloids have been isolated as salts with 5 being mixed tertiary-quaternary, 5 being monoquaternary and 3 being bisquaternary. The most commonly occurring alkaloid N-oxides are monotertiary-monotertiary-N-oxide (5 alkaloids) while only 1 alkaloid is monosecondary-monotertiary-N-oxide. Some 13 dimeric alkaloids are monoimines-monotertiary bases with 2 of them having imine groups conjugated with carbonyl functions while only one alkaloid contains bisimine functionality. Only 4 alkaloids have been isolated which are benzylisoquinoline-benzyltetrahydroisoquinoline (tertiary) bases. Finally, 2 alkaloids are known to be benzylisoquinoline (quaternary)-benzyltetrahydroisoquinoline (tertiary) and only 1 base is a true bisbenzylisoquinoline bisquaternary compound.

As ever-increasing numbers of these alkaloids were isolated, the need for a systematic classification became apparent. In 1976, a type of classification using numbers and symbols was devised (Shamma and Moniot 1976) which systematically described each half of the dimer with respect to its oxygenation pattern and both type(s) and position(s) of coupling. At that time, some 26 types of dimers were described and at least 3 have been added in the meantime (Guha et al. 1979, Schiff Jr. 1983).

4 Major Reactions of Importance in Structure Elucidation

4.1 Sodium-Liquid Ammonia Cleavage

Almost 50 years ago, the cleavage of diaryl ethers with Na in liquid NH_3 was reported (Sartoretto and Sowa 1937). It is believed that Na in liquid NH_3 is a rich source of electrons and that addition of an electron to the system forms an ion radical which is subsequently protonated by the ammonia prior to cleavage. It was not until 1951, however, that this reaction was applied to the bisbenzylisoquinoline alkaloid isotetrandrine in determination of the steric configuration of the two asymmetric centres in the alkaloid (Tomita et al. 1951). Since that time, this reaction has been utilized on at least 104 different bisbenzylisoquinoline alkaloids and remains one of the most useful methods of structure elucidation of these dimers (Guha et al. 1979, Schiff Jr. 1983). In general, phenolic alkaloids are usually first methylated or ethylated and the resulting nonphenolic dimer cleaved into two predominant monomers, although complex reaction mixtures are not uncommon. The cleavage products may be separated into nonphenols and phenols in a conventional manner and mixtures subsequently chromatographically separated and identified. Several examples from the recent literature serve to illustrate the versatility of this reaction.

Cleavage of O-ethylthalistine (1) afforded the monomers (2) and (3) demonstrating both the nature of the cleavage (which usually occurs in a manner such that the diaryl ether oxygen atom remains on the aromatic ring which is substituted para to the ether oxygen (Sartoretto and Sowa 1937, Guha et al. 1979, Schiff Jr. 1983) and the lability of the 2-position in 1,2,3-trioxygenated benzylisoquinolines (W.-N. Wu et al. 1980) (Fig. 1).

Cleavage of the monoquaternary alkaloid O-ethylthalirabine (4) yielded the monomers (5), (6) and (7) illustrating the same features as with O-ethylthalistine and additionally showing the Emde-like reaction on the quaternary centre (Liao et al. 1978) Fig. 1).

Sciadoferine (8), a dimeric alkaloid containing an imine function, was first reduced then ethylated prior to cleavage into (9) and (10) (Galeffi et al. 1978) (Fig. 2).

More recent examples of the utility of this cleavage include the structure elucidation of the alkaloid osornine which contains two ether bridges (Leet et al. 1983) and the alkaloids vanuatine, vateamine and malekulatine, each of which contains a single ether bridge (Bruneton et al. 1983). Alkaloids containing a dibenzodioxin group require a two stage reductive cleavage, utilizing NaH along with Na/NH_3 in the first stage followed by Na/NH_3 in the second stage. This dual cleavage will produce monomeric products and was of use in the elucidation of structure of isotrilobine (Inubushi et al. 1963) and cocsulinine (Joshi et al. 1974).

Deuterium labelling has been used in several ways to aid in the identification of cleavage products. Heating O-methyloxyacanthine with DCl and D_2O in a sealed tube prior to cleavage results in the deuteration of only those protons ortho to aromatic methoxy groups (Inubushi et al. 1972). Cleavage of O-methylbelarine with deuterated liquid ammonia afforded two monomeric products with deuterum labelling only in the positions of cleavage of the two aryl ethers (Falco et al. 1971).

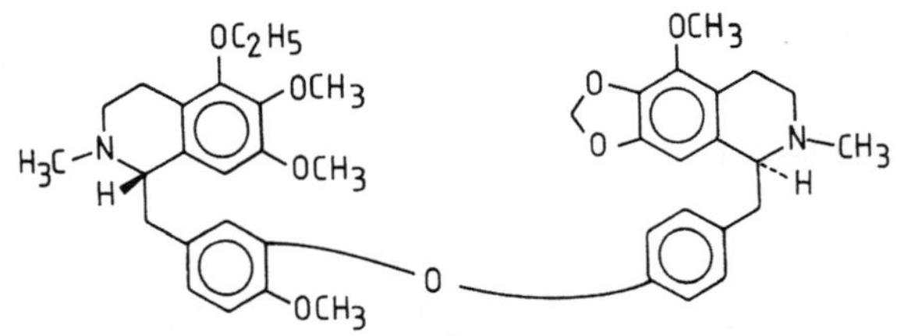

1

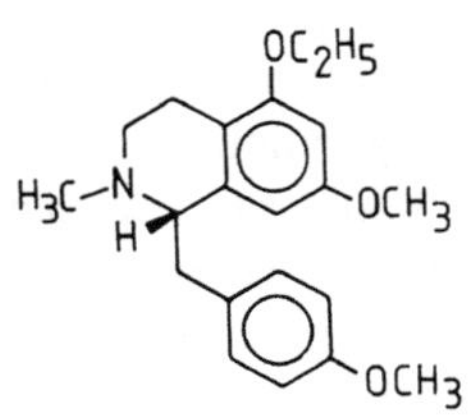

2

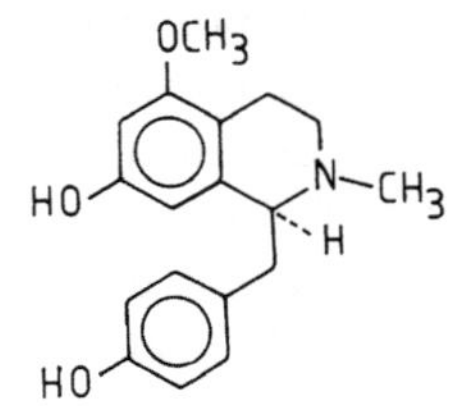

3

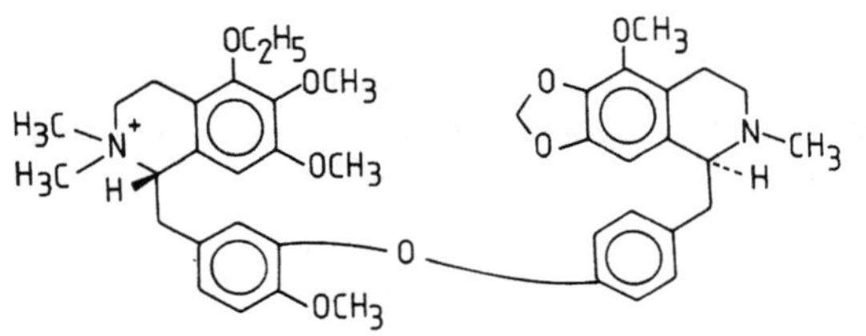

4

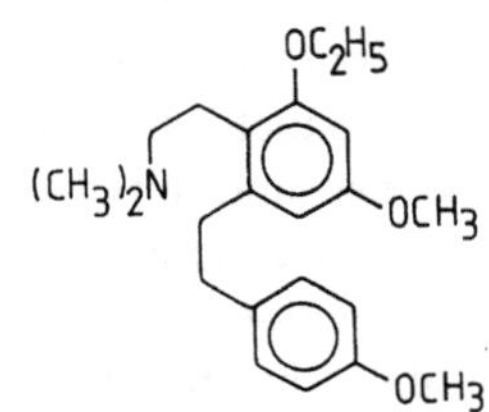

5

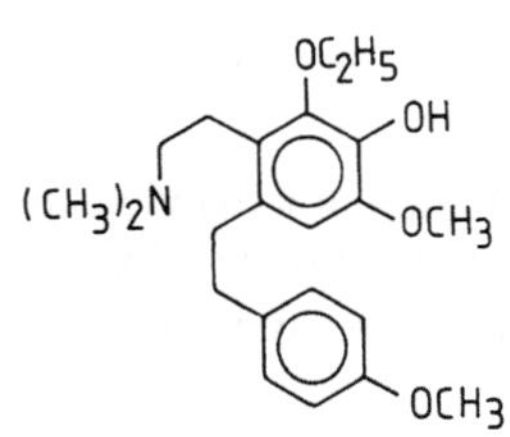

6

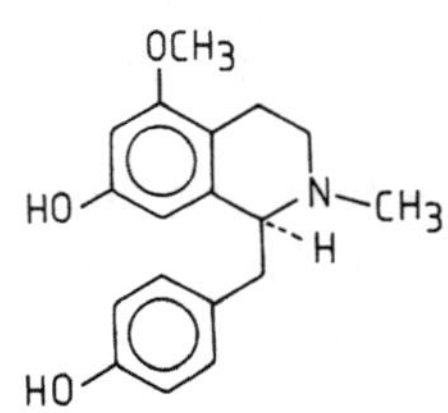

7

Fig. 1

Treatment of krukovine with deuterated diazomethane prior to cleavage gave the appropriately labelled deuteromethylethers in one of the cleavage products (Saá et al. 1976). Deuteriomethylation prior to cleavage was also useful in the structure determination of grisabine and grisabutine (Ahmad and Cava 1977).

4.2 Potassium Permanganate Oxidation

Oxidation with potassium permanganate in aqueous acidic solution has been of utility in determining the nature and substitution of the lower half (benzyl half) of various bisbenzylisoquinoline alkaloids. In general, the major oxidative product is either a biphenyletherdicarboxylic acid or a biphenyldicarboxylic acid. It is

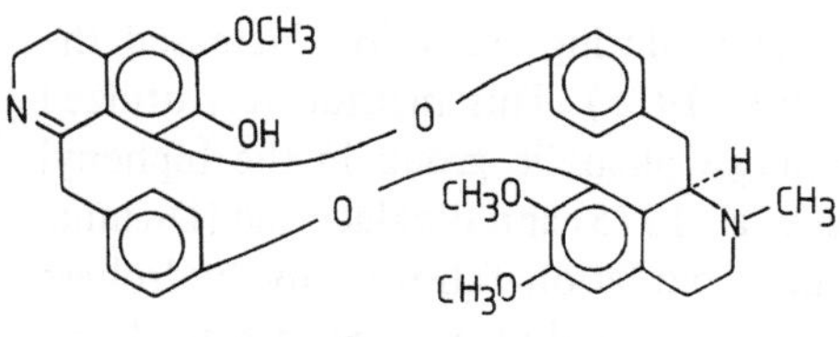

8

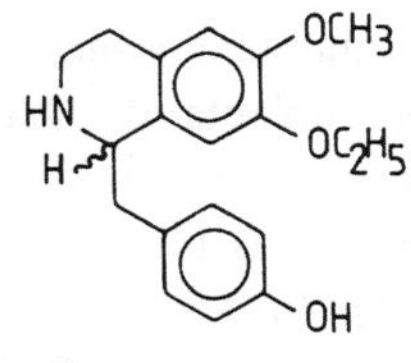

9

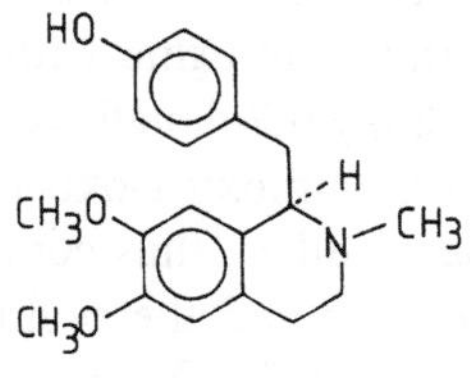

Fig. 2

10

necessary to alkylate phenolic groups in this portion of the molecule prior to oxidation so that the dimeric nature of the product is not lost via phenolic oxidation. Oxidation of thalidasine in this manner afforded 2-methoxy-4′, 5-dicarboxybiphenylether (Kupchan et al. 1969) while similar oxidation of O-ethylfuniferine gave 2-methoxy-2′-ethoxy-5,5′-dicarboxybiphenyl (Tackie et al. 1973) and confirmed the biphenyl nature of the benzyl portion of funiferine. Earlier studies frequently used potassium permanganate oxidation of Hofmann degradation products to afford the appropriate dicarboxylic acids. Examples of this include the alkaloids dryadine and dryadodaphnine (Bick et al. 1969) and tiliacorine (Anjaneyulu et al. 1969).

Potassium permanganate in acetone is a milder form of oxidation which was used in the structure elucidation of baluchistanamine (11), a secobisbenzylisoquinoline alkaloid that occurred along with oxyacanthine (12) in *Berberis baluchistanica*. It was observed that oxyacanthine could be converted to baluchistanamine by treatment with potassium permanganate in acetone (Shamma et al. 1974). In addition, other structurally similar bisbenzylisoquinoline alkaloids underwent oxidation in an analogous manner to afford an aldehydolactam (5–35% yield) which was formed by cleavage at the benzylic bond of the isoquinoline moiety unsubstituted at C-8 or C-8′; Later research with lactamic secobisbenzylisoquinoline alkaloids demonstrated that both in vivo and in vitro oxidations occur at the less sterically hindered benzylic site (C-8 or C-8′) (Leet et al. 1982).

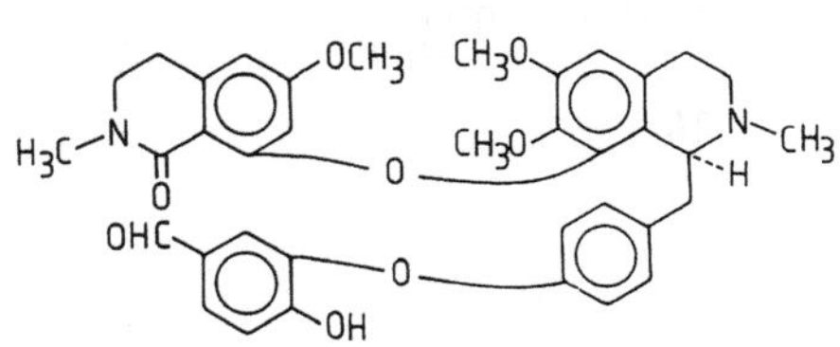

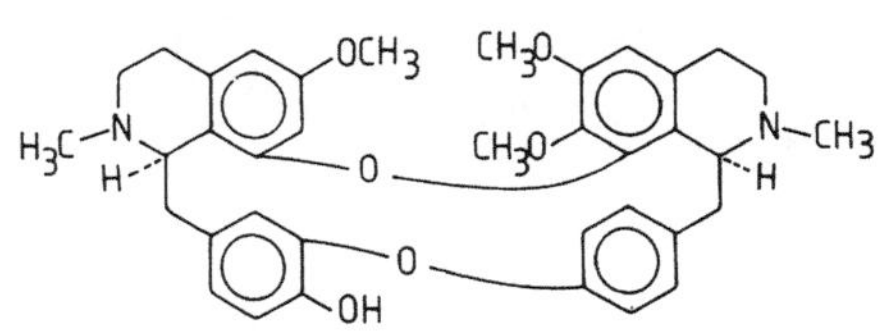

11

12

Fig. 3

Protection of phenolic groups as acetyl esters during oxidation increased the yield of the aldehydolactams (Shamma and Foy 1975). This method was utilized in the establishment of the position of the single phenolic group in the biphenyl portions of the alkaloids tiliacorine (Shamma et al. 1975) and dinklacorine (Dwuma-Badu et al. 1979). In the case of the former, oxidation followed by hydrolysis generated a para-hydroxybenzaldehyde function which showed a pronounced and characteristic bathochromic shift in strong base. This was not true in the latter case even though both alkaloids could be converted to a common methyl ether. In another problem, oxidation of O,O-diethylthalirugine with potassium permanganate in acetone afforded 2-methyl-5-ethoxy-6,7-dimethoxy-1-oxo-1,2,3,4-tetrahydroisoquinoline + 2-methyl-6-methoxy-7-ethoxy-1-oxo-1,2,3,4-tetrahydroisoquinoline + 2-methoxy-4′,5-dicarboxybiphenyl ether, thus confirming the positions of the phenolic groups in the isoquinoline rings and the points of attachment for the benzyl groups to the top rings (Wu et al. 1978). Other recent examples of the utility of this oxidation in the elucidation of structure of new alkaloids in this class include thalirugidine (W.-N. Wu et al. 1978), thaliracebine (Liao et al. 1978), thalibrunine (J. Wu et al. 1980), thalibrine (W.-N. Wu et al. 1980), thalmirabine (W.-N. Wu et al. 1980) and thalistine (W.-N. Wu et al. 1980).

4.3 Photo-oxidation (Oxidative Photolysis)

Photo-oxidative cleavage was first successfully used on the alkaloid isotetrandrine (Bick et al. 1971). In this study, ultraviolet irradiation of isotetrandine (13) in the presence of oxygen gave the aminolactam (14) (30%) (after reduction with sodium borohydride) and the dialdehyde (15) (50%) (Fig. 4).

The reaction was of great utility in the determination of structure of the dibenzo-p-dioxin bisbenzylisoquinoline alkaloids micranthine and telobine (Bick et al. 1972) and the alkaloid phlebicine (Cava et al. 1974). In each case, the products resembled those derived from fragmentation via electron impact and the top half of the alkaloid was characterized as an aminolactam (up to 30% yield) with the carbonyl function being on that side of the alkaloid unsubstituted at C-8 or C-8′. The lower half of the alkaloid was consistently identified as a biphenyletherdialdehyde or a biphenyldialdehyde (less than 50% yield) and characterized as the reduced diol. The initial

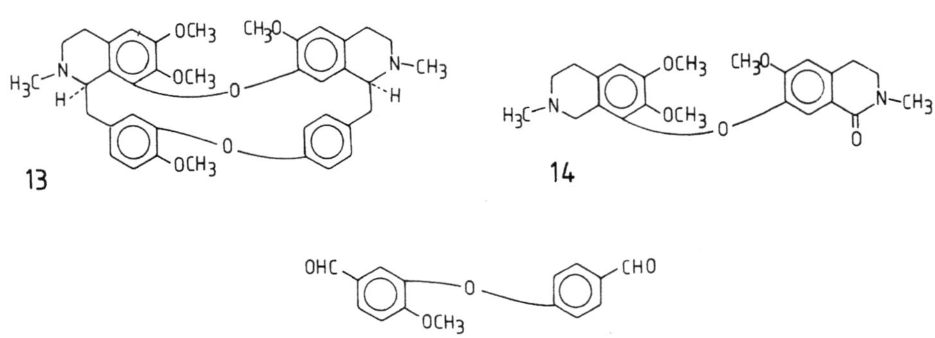

Fig. 4

step of the reaction apparently involved the formation of amine radical cations followed by homolytic fission of the doubly benzylic bonds. Hence, oxidative photolysis emerged as still another alternate to Na/NH_3 cleavage in the elucidation of structure of bisbenzylisoquinoline alkaloids by conversion to simpler products.

4.4 Cerium Oxidation

The use of ceric ammonium nitrate (Ce^{4+}) in the elucidation of structure of bisbenzylisoquinoline alkaloids appeared some 7 years after the introduction of photolytic oxidation. In a similar manner to photo-oxidation, oxidation of tetrandrine (16) with ceric ammonium nitrate afforded, after workup with sodium borohydride, the diamine (17) (95%) and the diol (18) (90%) (Bick et al. 1978) (Fig. 5).

The initial step in the reaction is apparently the transfer of electrons from nitrogen atoms to ceric ions, as Ce^{4+} is a well-established one-electron oxidant. Radical cations are generated which probably undergo fragmentation via homolytic fission at the doubly benzylic bonds. In general, 8 mol of ceric ammonium nitrate are used to oxidize 1 mol of bisbenzylisoquinoline alkaloid with yields of around 90% and products which do not require chromatography for purification. The reaction was used on hernadezine, O-methylmicranthine, berbamine (monophenolic), thalibrunine (monophenolic) and tenuipine (methylenedioxy group present) to afford slightly lower yields of the nonphenolic diamine top portions (85–90%) and complicated mixtures of the phenolic or methylenedioxy containing lower portions (Bick et al. 1978). The reaction was of great utility in the structural revision of thalibrunine (J. Wu et al. 1980) and the structure determination of pachygonamine (Abd El Kawi 1983).

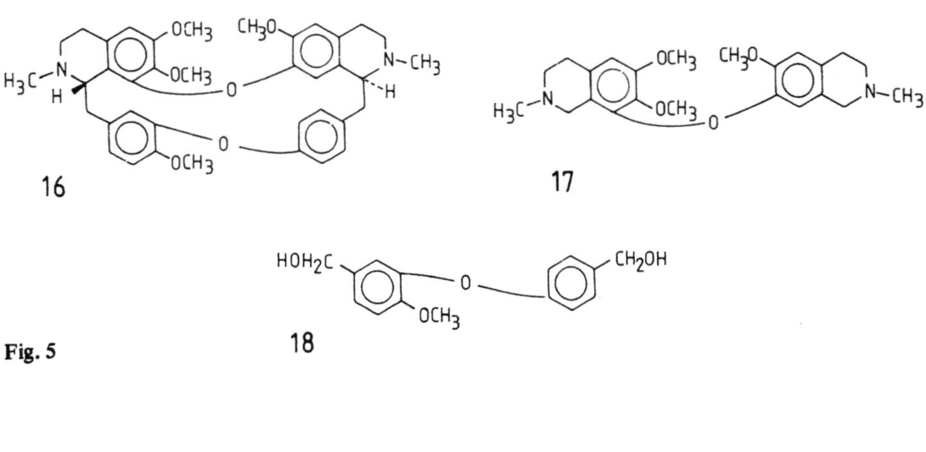

Fig. 5

5 Nuclar Overhauser Enhancement NMR Studies

The most useful technique to be recently utilized in the elucidation of structure of bisbenzylisoquinoline alkaloids has been the application of NMR nuclear Overhauser enhancement (NOE). In this technique, the spin-decoupled high resolution NMR spectrum is first obtained. Then, alternate irradiation of selected proton resonance regions with high and low r.f. power produces altered absorptions of interacting protons, either in terms of resonance line-widths corresponding to spin-spin couplings and/or signal intensities representing NOE. The use of FT-NOE difference spectrometry allows enhancements even as low as 0.5% to be observed. This technique was apparently first used in the bisbenzylisoquinoline alkaloid series in the structure elucidation of the biphenyl alkaloid tiliamosine (Guha et al. 1976). Since then, the technique has been used on numerous occasions in this series including the structure elucidation of dihydrodaphnine diacetate (Neuhaus et al. 1981); colorflammine (Van Beek et al. 1982); temuconine (Guinaudeau et al. 1982); sindamine, punjabine and gilgitine (Leet et al. 1982); osornine and chillanamine (Leet et al. 1983); repanduline (Neuhaus et al. 1983); vanuatine, vateamine and malekulatine (Bruneton et al. 1983). This technique may prove to be a superior alternative to Na/NH_3 cleavage, particularly when only extremely small quantities of alkaloid have been isolated (Guinaudeau et al. 1982).

6 New Alkaloids

Table 4 is a summary of the sources, methods of identification and references for most new alkaloids isolated from 1981 through 1983.

Table 4. New Bisbenzylisoquinoline alkaloids (1981 through 1983)

Alkaloid	Plant	Method	Reference
Berbacolorflamimine	*Pycnarrhena longifolia* (Menispermaceae)	FD-MS, ^{13}C-NMR, $NaBH_4$ reduction	Van Beek et al. (1982)
Calafatimine	*Berberis buxifolia* Lam. (Berberidaceae)	UV, ^{1}H-NMR, MS, conversion (CH_2O+$NaBH_4$) to calafatine	Fajardo et al. (1981)
Chillanamine	*Berberis buxifolia* Lam. (Berberidaceae)	UV, MS, ^{1}H-NMR, CD, comparison with osornine	Leet et al. (1983)
Colorflammine	*Pycnarrhena longifolia* (Menispermaceae)	FD-MS, $NaBD_4$ & $NaBH_4$ reduction, NOE, ^{13}C-NMR	Van Beek et al. (1982)
Cycleanine N-oxide	*Synclisia scabrida* Miers (Menispermaceae)	^{1}H-NMR, EI-MS, FD-MS, conversion to cycleanine (H_2SO_3) (TLC), preparation from cycleanine (H_2O_2) (TLC)	Ohiri et al. (1983)
Daphnine	*Daphnandra repandula* (Monimiaceae)	X-ray crystallography	Guilhem and Bick (1981)
N-Demethylcycleanine	*Stephania glabra* (Roxb.) Miers. (Menispermaceae)	Conversion to cycleanine (CH_2O + HCOOH)	Bhakuni and Gupta (1982)
Dihydrodaphnine diacetate (derivative of daphnine)	*Daphnandra repandula* (Monimiaceae)	^{1}H-NMR, NOE	Neuhaus et al. (1981)
N,N'-Dimethyllindoldhamine (Same as guattegaumerine)	*Polyalthia nitidissima* (Annonaceae)	UV, ^{1}H-NMR, MS, preparation via N-methylation (CH_2O + $NaBH_4$) of lindoldhamine	Jössang et al. (1983)
Gilgitine	*Berberis lycium* Royle (Berberidaceae)	UV, IR, ^{1}H-NMR, MS, CD	Leet et al. (1982)
Guattegaumerine	*Guatteria gaumeri* Greenman (Annonaceae)	UV, IR, ^{1}H-NMR, MS, CD, preparation via N-methylation (CH_2O + $NaBH_4$) of lindoldhamine	Dehaussy et al. (1983)

Table 4 (cont.)

Alkaloid	Plant	Method	Reference
Hernandezine-N-oxide	*Thalictrum sultanabadense* Stapf. (Ranuculaceae)	UV, ^{1}H-NMR, MS, conversion ($Zn + H_2SO_4$) to hernandezine	Mukhamedova et al. (1981)
Isodaurisoline	*Polyalthia nitidissima* (Annonaceae)	UV, ^{1}H-NMR, MS, optical rotation, comparison with daurisoline	Jössang et al. (1983)
Isogilletine-N-oxide	*Triclisia gilletii* (De Wild.) Staner (Menispermaceae)	UV, IR, ^{1}H-NMR, MS, conversion (H_2SO_3) to gilletine isomer	Owusu et al. (1981)
Magnolamine	*Michelia fuscata* Blume (Magnoliaceae)	Structure revision confirmation via reductive cleavage (Na/NH_3) of O,O,O-triethyl-magnolamine and synthesis	Tanaka et al. (1981)
Malekulatine	*Hernandia peltata* Meissner (Hernandiaceae)	UV, MS, ^{1}H-NMR, NOE, reductive cleavage (Na/NH_3)	Bruneton et al. (1983)
N-Methylpachygonamine	*Pachygone ovata* (Poir.) Miers ex Hook (Menispermaceae)	UV, MS, ^{1}H-NMR, Ce^{4+} oxidation, methylation ($CH_2O + HCOOH$)	Sultanbawa et al. (1983) Abd El Kawi (1983)
7-O-Methyllindoldhamine	*Polyalthia nitidissima* (Annonaceae)	UV, ^{1}H-NMR, MS, optical rotation, conversion ($CH_2O + NaBH_4$) to isodaurisoline	Jössang et al. (1983)
7′-O-Methyllindoldhamine	*Polyalthia nitidissima* (Annonaceae)	UV, ^{1}H-NMR, MS, optical rotation, conversion (CH_2O+$NaBH_4$) to daurisoline	Jössang et al. (1983)
Nortiliacorinine A	*Tiliacora racemosa* Colebr. (Menispermaceae)	Absolute configuration (S,S) and biosynthesis via incorporation of specific precursors	Bhakuni et al. (1981)

Table 4 (cont.)

Alkaloid	Plant	Method	Reference
Osornine	*Berberis buxifolia* Lam. (Berberidaceae)	UV, MS, ^{1}H-NMR, NOE, reductive cleavage (Na/NH_3) of O-methylosornine	Leet et al. (1983)
Pachygonamine	*Pachygone ovata* (Poir.) Miers ex Hook. (Menispermaceae)	UV, MS, ^{1}H-NMR, Ce^{4+} oxidation, methylation ($CH_2O + HCOOH$)	Sultanbawa et al. (1983) Abd El Kawi (1983)
Phaeantharine	*Phaeanthus ebracteolatus* (Presl) Merill. (Annonaceae)	Structure revision via UV, FD-MS, ^{1}H-NMR, ^{13}C-NMR	Van Beek et al. (1983)
Punjabine	*Berberis lycium* Royle (Berberidaceae)	UV, IR, ^{1}H-NMR, MS, CD	Leet et al. (1982)
Repanduline	*Daphnandra repandula* (Monimiaceae)	^{1}H-NMR, NOE established regioisomerism and relative stereochemistry	Neuhaus et al. (1983)
Sindamine	*Berberis lycium* Royle (Berberidaceae)	UV, IR, ^{1}H-NMR, MS, CD, preparation from berbamine	Leet et al. (1982)
Temuconine	*Berberis valdiviana* Phil. (Berberidaceae)	UV, ^{1}H-NMR, NOE, MS, CD, comparison with berbamunine	Guinaudeau et al. (1982)
Tiliacorinine-2′-N-oxid	*Tiliacora triandra* Diels (Menispermaceae)	Conversion to tiliacorinine (PCl_3), ^{1}H-NMR	Wiriyachitra and Phuriyakorn (1981)
Vanuatine	*Hernandia peltata* Meissner (Hernandiaceae)	UV, MS, ^{1}H-NMR, NOE, reductive cleavage (Na/NH_3) of O-O-dimethyl-vanuatine then CD	Bruneton et al. (1983)
Vateamine	*Hernandia peltata* Meissner (Hernandiaceae)	UV, MS, ^{1}H-NMR, NOE, reductive cleavage (Na/NH_3) O,O,O-triethylvateamine	Bruneton et al. (1983)

References

Abd El Kawi M (1983) Isolation and structure elucidation of alkaloids from Indian and Sri Lankan *Pachygone ovata* Miers (Menispermaceae). Ph D Dissertation, Univ Pittsburgh, 288 pp

Ahmad R, Cava MP (1977) Grisabine and grisabutine, new bisbenzylisoquinoline alkaloids from *Abuta grisebachii.* J Org Chem 42: 2271–2273

Anjaneyulu B, Govindachari TR, Sathe SS, Viswanathan N, Gopinath KW, Pai BR (1969) Alkaloids of *Tiliacora racemosa* Colebr. Tetrahedron 25: 3091–3105

Bhakuni DS, Gupta S (1982) The alkaloids of *Stephania glabra.* J Nat Prod 45: 407–411

Bhakuni DS, Singh AN, Jain S (1981) The structure, absolute configuration and biosynthesis of nortiliacorinine A. Tetrahedron 37: 2651–2655

Bick IRC, Douglas GK, Taylor WI (1969) Alkaloids of *Dryadodaphne novoguineensis.* J Chem Soc (C): 1627–1632

Bick IRC, Bremner JB, Wiriyachitra P (1971) Photo-oxidation of some benzylisoquinoline-derived alkaloids. Tetrahedron Lett: 4795–4797

Bick IRC, Bremner JB, Leow HM, Wiriyachitra P (1972) Alkaloids of *Daphnandra* species. Part XI. Some dibenzo-p-dioxin-type biscoclaurine alkaloids from an unnamed species. The structure of micranthine. J Chem Soc Perkin I: 2884–2889

Bick IRC, Bremner JB, Cava MP, Wiriyachitra P (1978) Reaction of bisbenzylisoquinoline alkaloids with ceric ammonium nitrate: a new degradative procedure. Aust J Chem 31:321–326

Bruneton J, Shamma M, Minard RD, Freyer AJ, Guinaudeau H (1983) Novel biogenetic pathways from (+)-reticuline. Three dimeric alkaloids: (+)-vanuatine, (+)vateamine and (+)-malekulatine. J Org Chem 48: 3957–3960

Cava MP, Wakisaka K, Noguchi I, Edie DL, daRocha AI (1974) Phlebicine, a new biphenylbisbenzylisoquinoline alkaloid from *Crematosperma polyphlebum.* J Org Chem 39:3588–3591

Dehaussy H, Tits M, Angenot L (1983) Guattegaumerine, a new bisbenzylisoquinoline alkaloid from *Guatteria gaumeri.* Planta Med 49:25–27

Dwuma-Badu D, Withers SF, Ampofo SA, EL Azizi MM, Slatkin DJ, Schiff PL Jr, Knapp JE (1979) Constituents of West African Medicinal Plants. XXIII. The position of the phenolic function in dinklacorine, a confirmation of structure. J Nat Prod 42:116–119

Fajardo V, Garrido M, Cassels B (1981) Calafatimine, a new bisbenzylisoquinoline alkaloid. Heterocycles 15:1137–1139

Falco MR, Devries JX, Macciò Z, Bick IRC (1971) Belarine, a new bisbenzylisoquinoline alkaloid. Chem Commun: 1056–1057

Galeffi C, La Bua R, Messana I, Zapata Alcázar R, Marini-Bettòlo GB (1978) The Alkaloids of *Sciadotenia toxifera* Krukoff and A.C. Smith. Gazz Chim Ital 108:97–100

Guha KP, Das PC, Mukherjee B, Mukherjee R, Juneau GP, Bhacca NS (1976) Structure of tiliamosine: a new diphenyl bisbenzylisoquinoline alkaloid from *Tiliacora racemosa.* Tetrahedron Lett: 4241–4244

Guha KP, Mukherjee B, Mukherjee R (1979) Bisbenzylisoquinoline Alkaloids-A Review. J Nat Prod 42:1–84

Guilhem J, Bick IRC (1981) Daphnine, a unique bisbenzylisoquinoline alkaloid: X-ray crystal structure of its dihydriodide. J Chem Soc Chem Commun: 1007–1009

Guinaudeau H, Cassels BK, Shamma M (1982) The use of nuclear magnetic resonance nuclear Overhauser enhancements in the structural elucidation of bisbenzylisoquinoline alkaloids. Heterocycles 19:1009–1012

Inubushi Y, Nomura K, Miyawaki M (1963) Alkaloids of Menispermaceous plants CXCVII. Structure of trilobine and isotrilobine. 13. Isotrilobine. Yakugaku Zasshi 83:282–288; Chem Abstr (1963) 59:5212

Inubushi Y, Kikuchi T, Ibuka T, Saji I (1972) A convenient method for the determination of the attached position of diphenyl ether linkage in bisbenzylisoquinoline alkaloids. Tetrahedron Lett: 423–425

Joshi PP, Bhakuni DS, Dhar MM (1974) Structure and stereochemistry of cocsulinine. Ind J Chem 12:517–518

Jössang A, Leboeuf M, Cabalion P, Cavé A (1983) Alkaloids from Annonaceae. XLV. Alkaloids of *Polyalthia nitidissima*. Planta Med 49:20–24

Kupchan SM, Yang T-H, Vasilikiotis GS, Barnes MH, King ML (1969) Tumor inhibitors. XLII. Thalidasine a novel bisbenzylisoquinoline alkaloid tumor inhibitor from *Thalictrum dasycarpum*. J Org Chem 34: 3884–3888

Leet JE, Hussain SF, Minard RD, Shamma M (1982) Sindamine, punjabine and gilgitine: three new secobisbenzylisoquinoline alkaloids. Heterocycles 19:2355–2360

Leet JE, Fajardo V, Freyer AJ, Shamma M (1983) Some dimeric benzylisoquinoline alkaloids with an unusual oxygenation pattern. J Nat Prod 46:908–912

Liao W-T, Beal JL, Wu W-N, Doskotch RW (1978) Alkaloids of *Thalictrum*. XXVI. New hypotensive and other alkaloids from *Thalictrum minus* race B. J Nat Prod 41:257–270

Mukhamedova S, Maekh SKh, Yunusov SYu (1981) Alkaloids of *Thalictrum sultanabadense*. Khim Prir Soedin: 250–251

Neuhaus D, Rzepa HS, Sheppard RN, Bick IRC (1981) Assignment of the structure of dihydrodaphnine diacetate by nuclear Overhauser effect difference spectroscopy. Tetrahedron Lett 22:2933–2936

Neuhaus D, Sheppard RN, Bick IRC (1983) Structural and conformational study of repanduline using long-range nuclear Overhauser effect difference spectroscopy. J Am Chem Soc 105: 5996–6002

Ohiri FC, Verpoorte R, Svendsen AB (1983) Cycleanine N-oxide, a new alkaloid from *Synclisia scabrida*, Planta Med 47:87–89

Owusu PD, Knapp JE, Slatkin DJ, Schiff PL Jr (1981) Constituents of West African Medicinal Plants. XXVIII. Additional alkaloids of *Triclisia gilletii*. J Nat Prod 44:61–65

Saá JM, Lakshmikantham MV, Mitchell MJ, Cava MP (1976) Krukovine, a new bisbenzylisoquinoline alkaloid from *Abuta splendida* Krukoff and Moldenke. J Org Chem 41:317–319

Sartoretto PA, Sowa FJ (1937) The cleavage of diphenyl ethers by sodium in liquid ammonia. I. Ortho and para substituted diphenyl ethers. J Am Chem Soc 59:603–606

Schiff PL Jr (1983) Bisbenzylisoquinoline Alkaloids. J Nat Prod 46:1–43

Shamma M, Foy JE (1975) A controlled oxidation of bisbenzylisoquinolines. Tetrahedron Lett: 2249–2252

Shamma M, Moniot JL (1976) The systematic classification of bisbenzylisoquinolines. Heterocycles 4:1817–1824

Shamma M, Foy JE, Miana GA (1974) Baluchistanamine, a novel type dimeric isoquinoline alkaloid. J Am Chem Soc 96:7809–7811

Sultanbawa MUS, Sotheeswaran S, Balasubramaniam S, Abd El-Kawi M, Slatkin DJ, Schiff PL Jr (1983) Pachygonamine and N-methylpachygonamine, new bisbenzylisoquinoline alkaloids from *Pachygone ovata*. Heterocycles 20:1927–1932

Tackie AN, Dwuma-Badu D, Knapp JE, Schiff PL Jr (1973) The structure of funiferine, a biphenyl alkaloid from *Tiliacora funifera*. Lloydia 36:66–71

Tanaka H, Harad A, Ichino K, Ito K (1981) Alkaloids of *Michelia fuscata* Blume: The structure and synthesis of magnolamine. Heterocycles 16:1275–1279

Tomita M, Fujita E, Murai F (1951) Alkaloids of Menispermaceae. LXXXI. Structure of biscoclaurine alkaloids. 2. Steric configuration of the two asymmetric centres in the isotetrandrine molecule. J Pharm Soc Jpn 71:1035–1039; Chem Abstr (1952) 46:5060

Van Beek TA, Verpoorte R, Svendsen AB (1982) Colorflammine and berbacolorflammine, two new orange-colored bisbenzylisoquinoline alkaloids from *Pycnarrhena longifolia*. J Org Chem 47:898–900

Van Beek TA, Verpoorte R, Svendsen AB, Santos AC, Olay LP (1983) Revised structure of phaeantharine. J Nat Prod 46:226–231

Wiriyachitra P, Phuriyakorn B (1981) Alkaloids of *Tiliacora triandra*. Aust J Chem 34:2001–2004

Wu J, Beal JL, Doskotch RW (1980) Alkaloids of *Thalictrum*. XXVIII. The structure of thalibrunine, a reinvestigation and revision. J Org Chem 45:208–212

Wu W-N, Beal JL, Fairchild EH, Doskotch RW (1978) Alkaloids of *Thalictrum*. XXV. Six new bisbenzylisoquinoline alkaloids from *Thalictrum rugosum*. J Org Chem 43:580–585

Wu W-N, Liao W-T, Mahmoud ZF, Beal JL, Doskotch RW (1980) Alkaloids of *Thalictrum minus* Race B. J Nat Prod 43:472–481

Natural Degradative Routes for the Aporphines

M. SHAMMA [1]

1 Degradation of Aporphines

While it is recognized that aporphines may be biosynthesized in different ways (direct coupling, rearrangement of proaporphines, etc.), it is also true that they are catabolized in plants by a variety of routes.

The most common degradative pathway is through quaternization of the basic nitrogen followed by Hofmann elimination to furnish a phenanthrene possessing a β-dimethylaminoethyl side chain. Such a phenanthrene alkaloid can then suffer

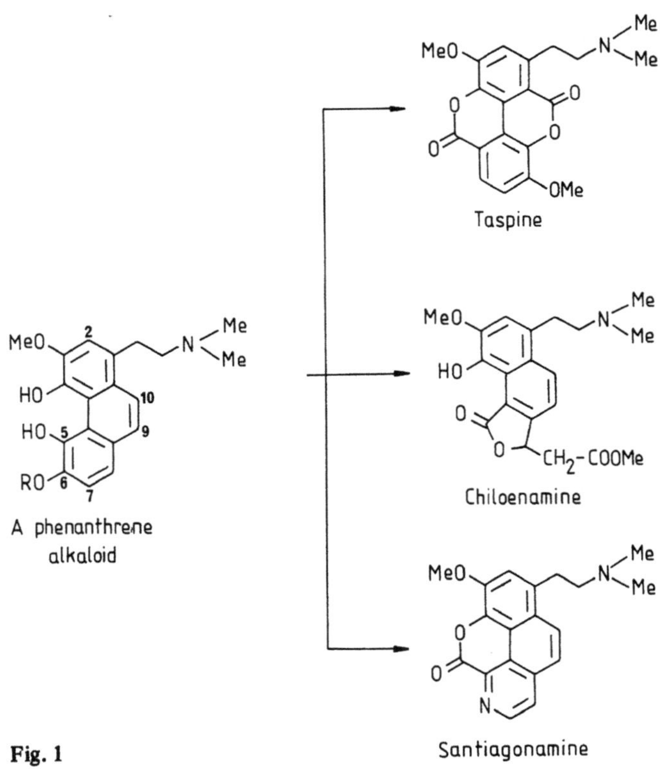

Fig. 1

1 Department of Chemistry, The Pennsylvania State University, University Park, Pennsylvania 16802, USA

The Chemistry and Biology of Isoquinoline Alkaloids, ed. by Phillipson et al.

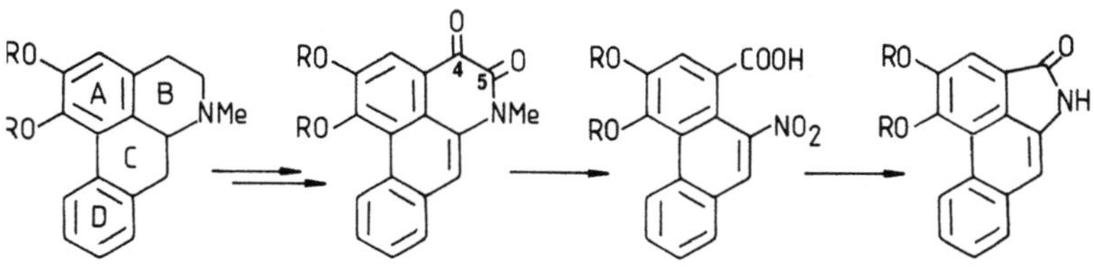

An aporphine A 4,5-dioxoaporphine An aristolochic acid An aristolactam

Fig. 2

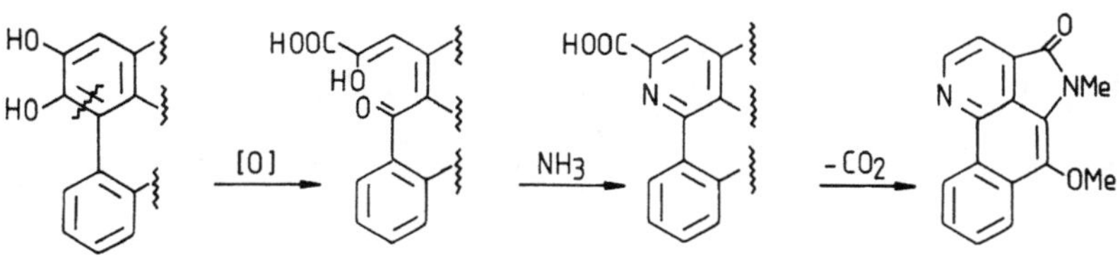

Eupoloramine

Fig. 3

oxidative cleavage of the C-9,10 double bond, a transformation which can eventually lead to taspine.

Alternatively, a phenanthrene alkaloid may also be oxidized by catechol dioxygenase enzymes at the C-5,6 or at the C-6,7 bonds, to produce such novel alkaloids as chiloenamine and santiagonamine (Shamma et al. 1983, Valencia et al. 1984) (Fig. 1).

Oxidation of ring B of an aporphine, through the probable intermediacy of a 4,5-dioxoaporphine, is also possible. The product is an aristolochic acid which may in turn be reduced and dehydrated to an aristolactam (Shamma and Guinaudeau 1984) (Fig. 2).

There is even a case known of oxidative fission of both rings A and B of an aporphine. In this instance, ring A is cleaved with the likely assistance of a catechol dioxygenase. The product, following ammonia capture and decarboxylation, is the unusual pyridinolactam eupolaramine (Bowden et al. 1976) (Fig. 3).

A curious situation is presented by the unusual aporphine alkaloid melosmine (Zabel et al. 1982). This compound is aromatic in ring B and dialkylated at C-7, so that it cannot readily be degraded by any of the above routes. Here, nature adopts a completely different course. The cularine-type alkaloid gouregine is the metabolic product, resulting from oxidation of ring D with subsequent expansion of ring C of melosmine as shown in Fig. 4 (Leboeuf et al. 1982).

Oxoaporphine formation is another natural means by which aporphines may be metabolized. The generation of oxoaporphines from aporphines is a purely oxidative process. However, oxoaporphines can lose the elements of carbon monoxide, possibly through a photochemical process, to form tetracyclic bases of the rufescine type (Silverton et al. 1977). Alternatively, it is conceivable that in other instances, the carbon monoxide unit is not lost, but is incorporated instead into the lower ring to form a tropolonoisoquinoline as represented by the alkaloid grandirubrine (Menachery and Cava 1980) (Fig. 5). It is worth pointing out in this respect that rufescine and gradirubrine co-occur in plants of the genus *Abuta*, family Menispermaceae (Buck 1984).

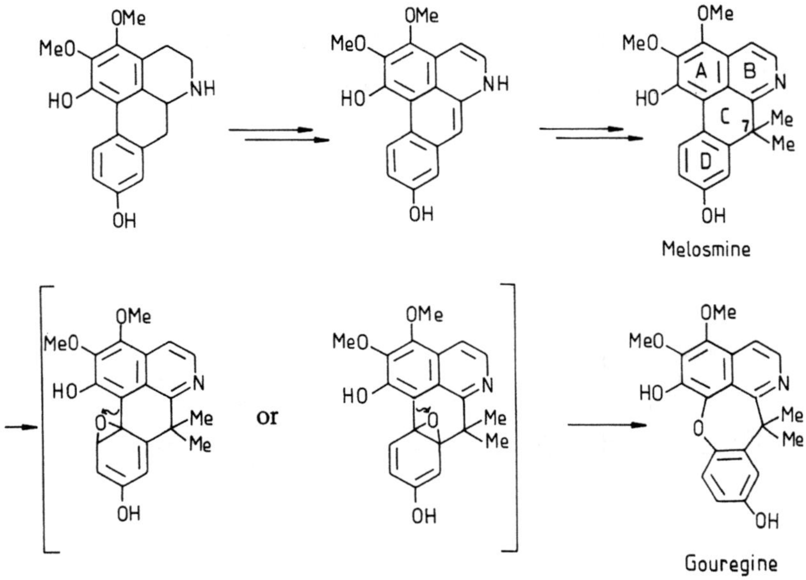
OMe
MeO
HO
NH
OH
A
B
C
D
N
7
Me
Me
Melosmine
or
Gouregine

Fig. 4

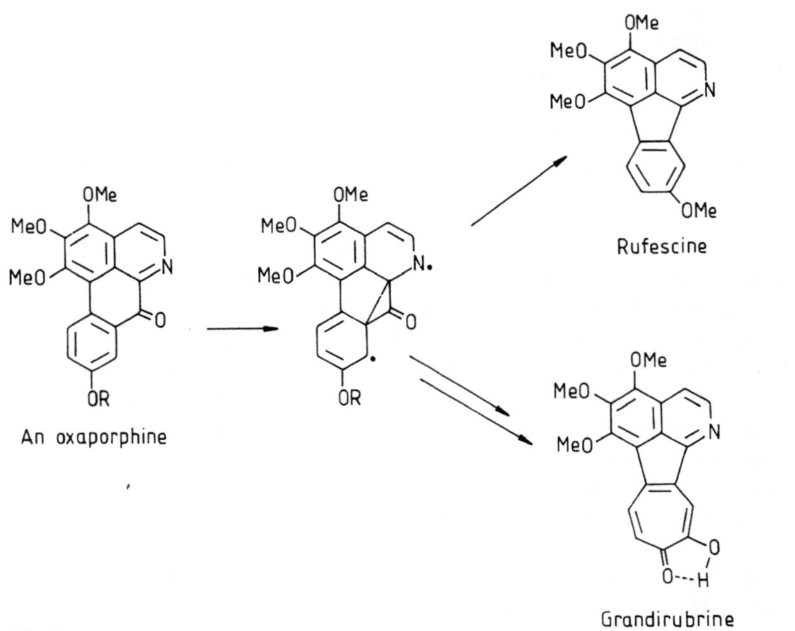
OMe
MeO
MeO
N
O
OR
An oxaporphine
N•
Rufescine
Grandirubrine

Fig. 5

References

Bowden BF, Freeman HC, Jones RDG (1976) Molecular and crystal structure of eupolauramine, a novel type of alkaloid from *Eupomatia laurina.* J Chem Soc Perkin II:565–662

Leboeuf M, Cortés D, Hocquemiller R, Cavé A, Chiaroni A, Riche C (1982) Structure de la geourégine, alcaloide original apparenté aux cularines. Tetrahedron 38:2889–2896

Menachery MD, Cava MP (1980) Grandirubrine, a new tropoloisoquinoline alkaloid. Heterocycles 14:1943–1945

Shamma M, Guinaudeau H (1984) Tetrahedron 40:4795–4822

Shamma M, Lan H-Y, Freyer AJ, Leet JE, Urzúa A, Fajardo V (1983) Chiloenamine and Chiloenine: Two unusual isoquinoline derived alkaloids. A new insight into the catabolism of aporphines. J Chem Soc Commun 799–800

Silverton JV, Kabuto C, Buck KT, Cava MP (1977) Structure of imerubrine, a novel condensed tropoloneisoquinoline alkaloid. J Am Chem Soc 99:6708–6712

Valencia E, Patra A, Freyer AJ, Shamma M, Fajardo V (1984) Santiagonamine: a new aporphinoid alkaloid incorporating a phenanthrene skeleton. Tetrahedron Lett 25:3163–3166

Zabel V, Watson WH, Phoebe CH Jr, Knapp PL, Schiff PL Jr and Slatkin DJ (1982) Melosmine and melosmidine, 7,7'-dimethyltetradehydroaporphine alkaloids from *Guatteria melosma.* J Nat Prod 45:94–101

For alternate suggestions for the biogenesis of rufescine and grandirubrine, see Buck KT (1984) in The Alkaloids, Vol XXIII, ed. by Brossi A, Academic Press, NY, pp 320–323

Synthesis and Structure-Activity Relationships of Aporphines as Dopamine Receptor Agonists and Antagonists

J. L. NEUMEYER [1]

1 Introduction

The aporphine alkaloids are one of the largest groups within the isoquinoline alkaloids (Shamma 1972, Fig. 1). In the naturally occurring aporphine alkaloids, positions 1 and 2 are always oxygenated and frequently other positions are also substituted with hydroxyl, methoxyl or methylenedioxy groups. In a few cases a hydroxyl function is located at C-7, while in the case of steporphine, oxygenation is at C-4. Glaucine and bulbocapnine were among the first naturally occurring aporphines to have their structures elucidated (Manske 1954) and usually the natural alkaloids have proved to be optically active, possessing either the R or S absolute configuration. For a recent review of aporphine biosynthesis see Shamma and Guinaudeau, 1984.

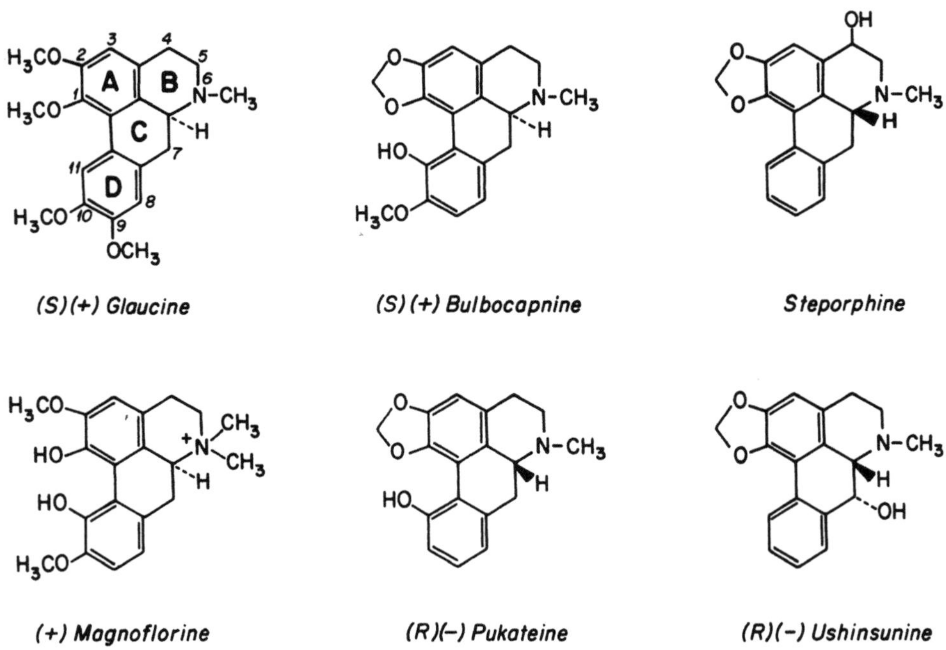

Fig. 1. Naturally occurring aporphine alkaloids

1 Section of Medicinal Chemistry, College of Pharmacy and Allied Health Professions, Northeastern University, Boston, Massachusetts 02115, USA

The Chemistry and Biology of Isoquinoline Alkaloids, ed. by Phillipson et al.

Bulbocapnine has been of pharmacological interest for many years since it exerts a cataleptic effect in animals and has recently been shown to act as dopamine (DA) antagonist. This alkaloid has the opposite configuration (6aS) to apomorphine (APO) at C-6a (Fig. 1).

As early as 1869 (Matthiessen and Wright), it was noted that acid catalyzed rearrangement of (–)-morphine provided a new base (–)-APO. Thus, APO was the first aporphine base whose structure was elucidated (Pschorr et al. 1902) and eventually confirmed by synthesis of APO-dimethyl ether. The chemistry of APO is thus closely linked to chemical studies involving morphine, although APO has little pharmacologic similarity to the narcotic analgesic from which its name was derived. The pharmacology and the clinical actions of APO were extensively evaluated between 1869 and 1969 (For review, see Neumeyer et al. 1981a). Many of its potentially useful actions in neuropsychiatric and other patients had been virtually forgotten or abandoned until the post-DOPA era and the explosive surge of interest in DA systems a century later. Apomorphine (4a), apocodeine (9a) and morphothebaine (11a) (Fig. 3), the three semi-synthetic aporphine alkaloids obtained from morphine, codeine and thebaine, respectively, all contain a methyl group attached to the nitrogen and have an R absolute configuration at the 6a-carbon.

Reports by Ernst and others (Ernst 1966, Ernst and Smelik 1967) drew attention to structural similarities between APO and DA and stimulated the systematic evaluation of APO and its congeners in relation to dopaminergic activity. However, the evaluation of the structure-activity relationship of systematically substituted aporphines, especially those unsubstituted in the A ring, and substituents other than methyl on the nitrogen atom, required the development of a practical method for the synthesis of the aporphine carbon skeleton. Such methods made possible the synthesis of several previously unavailable structural and stereoisomers of apomorphine (Neumeyer et al. 1967, 1969).

Our studies have focused on delineating the portions of the aporphine molecular structure responsible for dopaminergic properties and the interactions of such "rigid" analogues of DA with DA receptors. The process of drug design might thus be improved if DA receptors and their mode of interaction with active substances were known in more precise molecular detail. Such information can then be used to design conformationally defined structures in which pharmacophoric groups are oriented in the appropriate spatial arrangement for optimal receptor interaction. Since DA is an achiral and conformationally flexible molecule, little information concerning interactions with DA receptors can be obtained with this neurotransmitter.

2 Synthesis of Aporphines and Benzylisoquinolines

2.1 Synthesis of Monooxygenated Aporphines

In earlier studies, we evaluated (±)-11-hydroxy-(1), (±)10 hydroxy-(2) and (±)8-hydroxy-monohydroxy aporphines (3) (Fig. 2). The synthetic approach was the Reissert alkylation-Pschorr cyclization route developed earlier. We also wished to

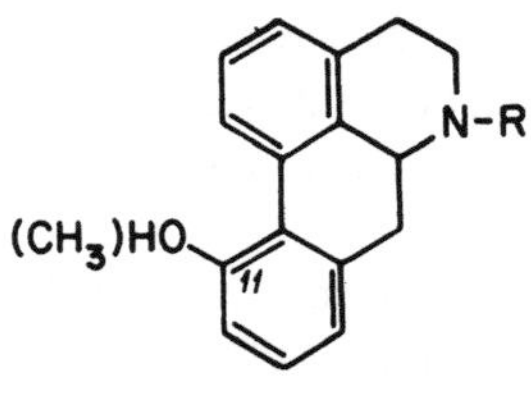

1 *11-Hydroxyaporphines*

a) (±)-$R = CH_3$; n-C_3H_7; CH_2[illegible]

b) (±) $R = CH_3$; n-C_3H_7

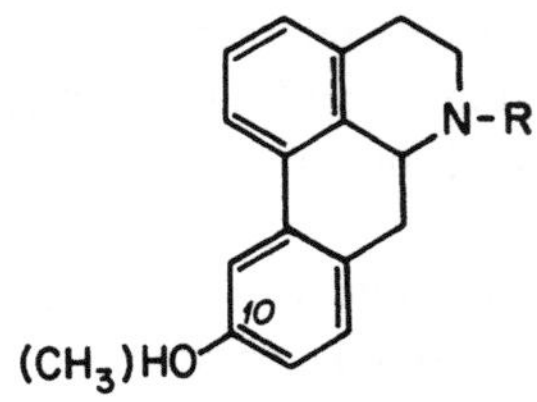

2 *10-Hydroxyaporphines*

a) (±) $R = CH_3$

b) $R = n$-C_3H_7

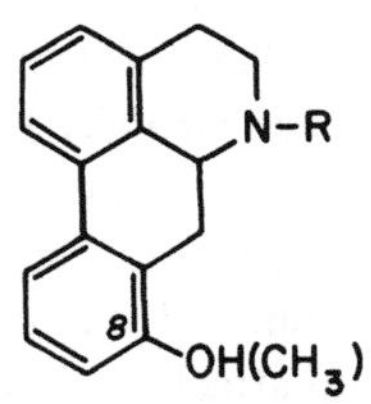

3 *8-Hydroxyaporphines*

a) $R = CH_3$

b) $R = n$-C_3H_7

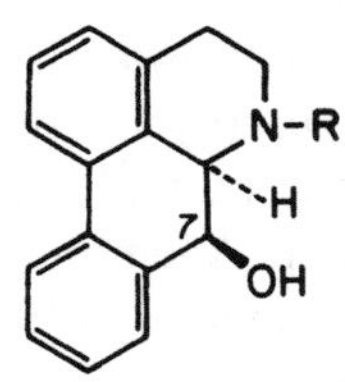

7-Hydroxyaporphines

a) $R = CH_3$

b) $R = H$

Fig. 2. Monooxygenated aporphines

assess the effect produced by the complete elimination of the 10- and 11- hydroxyl groups in APO and the insertion of an hydroxyl group at C-7 (Fig. 2). The 7-hydroxyaporphines have a structural relationship to such sympathomimetic amines as pseudoephedrine and epinephrine (Neumeyer et al. 1974, Neumeyer and Granchelli 1970). The DA agonist activity of such racemic monohydroxyaporphines as 11-hydroxyaporphines (1) and to a lesser extent 10-hydroxyaporphines (2) indicated that the catechol system is not a structural requirement for dopamine-like activity in such molecules. Similarly, comparison of (±)1b with (±)2b in the ^{3}H-APO binding assay (IC_{50}=36 vs. 700 nM, respectively) with bovine caudate tissue confirmed that an hydroxyl group at the 11-position plays an important role in receptor binding (Neumeyer et al. 1981c). These results led to the preparation of (–) 11-hydroxyaporphine, and its N-propyl analog (–)1a and b for further biological studies, since it had previously been established that DA agonist activity in APO resides principally in the 6aR (levorotatory) isomer (Saari et al. 1973). It is possible to obtain levorotatory 1a and 1b by the resolution of the (±)-11-methoxyaporphine (Neumeyer et al. 1974) followed by ether cleavage as carried out for the enantiomers of APO (Saari et al. 1973). We chose, however, to develop a more direct stereoselective synthesis of 1a and 1b from morphine. The strategy for this synthesis involved the selective removal of one of the hydroxyl groups on the aporphine ring via the phenyltetrazole ether (Ram and Neumeyer 1982), a procedure which we have applied to the conversion of morphothebaine to apomorphine (Ram and Neumeyer 1981).

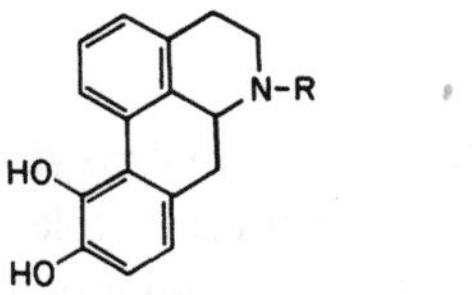

4 10,11-Dihydroxyaporphines

a) R = CH_3, APO
b) R = n-C_3H_7, NPA
c) R = CH_2CH_2Cl, NCA
d) R = H, Norapomorphine
e) R = CH_2CH_2OH

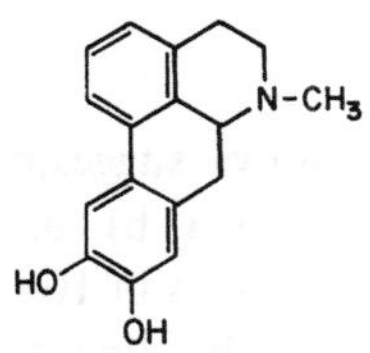

5 9,10-Dihydroxyaporphine (Isoapomorphine)

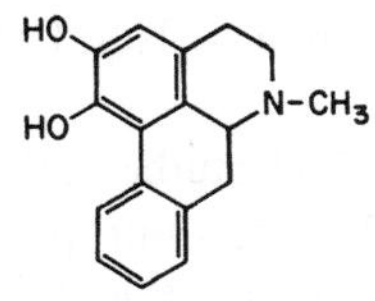

6 1,2-Dihydroxyaporphine

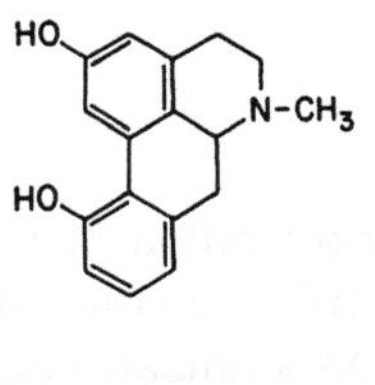

7 2,11-Dihydroxyaporphine

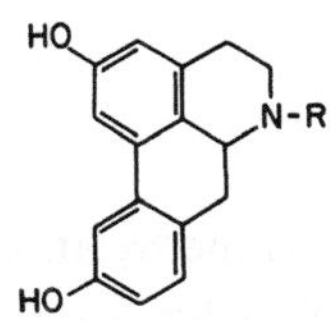

8 2,10-Dihydroxyaporphines

a) R = CH_3
b) R = n-C_3H_7

Fig. 3. Dioxygenated aporphines-catechols and non-catechols

2.2 Synthesis of Dioxygenated Aporphines (Fig. 3)

2.2.1 Catechols

The synthesis of (R)APO (−)-4a was first reported by the treatment of morphine (Matthiessen and Wright 1869) with concentrated hydrochloric acid. (R)-N-n-propylnorapomorphine (−)-4b, (R)NPA can be similarly prepared from N-n-propylnormorphine (Koch et al. 1968). (RS)APO (±)-4a (Neumeyer et al. 1970) and (RS)-NPA (±)-4b (Neumeyer et al. 1973a) were synthesized by a multi-step process from isoquinoline and vanillin. (R)-APO (−)-4a and (R)-NPA (−)-4b have also recently been prepared from the morphine alkaloid thebaine (Ram and Neumeyer 1981). The electrochemical synthesis of aporphines via cathodic cyclization of iodobenzylisoquinolinium salt was used to advantage for the synthesis of (RS)-APO-dimethylether and several other oxygenated and non-oxygenated aporphines (Gottlieb and Neumeyer 1976). The isomeric (RS) 9,10-dihydroaporphine ((±)-isoapomorphine) was also synthesized via the Reissert alkylation-Pschorr cyclization procedure (Neumeyer et al. 1973b). The (R) and (S) 1,2-dihydroxyaporphines (R)- and (S)-6 are obtained by the O-demethylation of (R) and (S) nuciferine as previously reported (Neumeyer et al. 1973b). (R)-NPA (−)-4b) was converted to (RS)-NPA-dimethylether by dehydrogenation of the 10,11-0,0′-dimethylether of (R)-NPA(−)-4b followed by reduction. (+)-Dibenzoyl-D-tartaric acid was used to resolve (RS)-NPA dimethylether. Ether cleavage gave (S)-NPA (Neumeyer et al. 1983b).

2.2.2 Dihydroxyaporphines – Non-Catechols

Previous pharmacological studies referred to above suggested that an 11-hydroxyl group as in 11-hydroxy-N-n-propylnoraporphine (1b) plays an important role in receptor binding. Although the 10-hydroxy group as in 10-hydroxy-N-n-propylnoraporphine (2b) was considerably less potent in binding to DA receptors, it was desirable to investigate the corresponding 2,11-dihydroxyaporphines and 2,10-dihydroxyaporphines. Thus, (–)2,11-dihydroxyaporphine (7) can be obtained by reduction of pukateine (15) with lithium and liquid ammonia and 2,10-dihydroxyaporphines 8a and 8b by selective removal of the 11-hydroxyl group on 2-O-methylmorphothebaine 11a via the phenyltetrazole ether (Ram and Neumeyer 1984).

2.2.3 "Protected" Catechols (Fig. 4)

We have repeatedly observed that O-methylation or esterification of one or more of the hydroxy groups, particularly at the 10- or 11- position, considerably reduced the potency of aporphines against ^{3}H-APO binding or DA stimulated production of cAMP. Thus morphothebaine [(–)-11a], or its 2-O-methylether [(–)-12a], and apocodeine [(–)-9a], contain methyl ether moieties in the 10-position. Similarly when the 10,11-position was masked by a methylenedioxy bridge as in 13a and 13b, IC_{50} values for inhibition of high affinity binding of ^{3}H-APO were even more drastically reduced (Table 1). This reduction in potency may be brought about by the bulk of the additional methyl or methylene group which may interfere with ligand-receptor interactions at one or more of the putative binding sites (see Fig. 8). These findings are consistent with previous in vivo observations of DA agonist inactivity of such O-methylated or 10,11-methylenedioxy aporphines as apocodeine and methylenedioxy apomorphine (MDO-APO, 13a). It has been recognized for many years that pharmacologic limitations of APO and NPA are, notably, their short duration of action and poor oral bioavailability. We have observed (Baldessarini et al. 1982) that (–)-10,11-methylenedioxy-N-n-propylnoraporphine (MDO-NPA, 13b) was an orally effective and long-acting apomorphine derivative that may act as a prodrug of NPA to exert activity at DA receptors in the brain. Thus, we prepared several methylene-substituted methylenedioxy analogs derived from NPA or APO 13b–13g that might serve as prodrugs of APO or NPA with potentially useful characteristics, such as prolonged or orally effective in vivo activity at DA receptors (Campbell et al. 1982). The most active of these was MDO-NPA (13b). When MDO-NPA was further evaluated after pretreatment with a large dose of the hepatic oxidase inhibitor, SKF-525 A, it was found that stereotyped behavioral responses to doses of MDO-NPA of 0.3 to 3.0 $mgkg^{-1}$ i.p. were prevented, as were the strong motorinhibitory effects of low doses of MDO-NPA (0.05 to 0.2 $mgkg^{-1}$, i.p.). In contrast, pretreatment with SKF-525A itself did not produce noticeable changes in behavior, and failed to prevent stereotyped behavioral responses to NPA (Campbell et al. 1982).

In order to study the metabolism of APO and NPA in vivo we have examined methods for the detection and quantitation of APO, NPA and MDO-NPA in biological fluids (Maksoud et al. 1983). For such studies, we required reference samples of

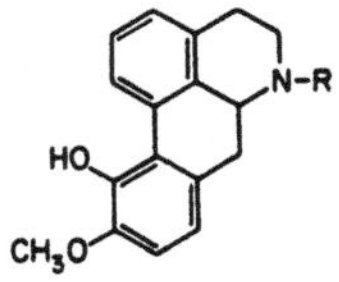

9 *11-Hydroxy-10-methoxy-aporphines*
a) R = CH_3, Apocodeine
b) R = n-C_3H_7

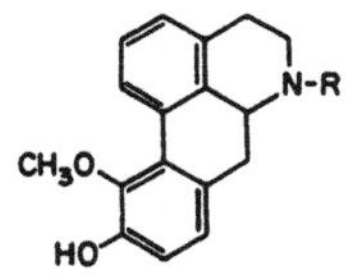

10 *11-Methoxy-10-hydroxy-aporphines*
a) R = CH_3, Isoapocodeine
b) R = n-C_3H_7

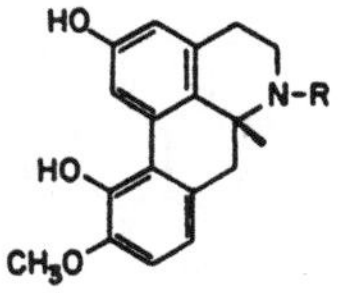

11 *2,11-Dihydroxy-10-methoxyaporphines*
a) R = CH_3, Morphothebaine
b) R = n-C_3H_7

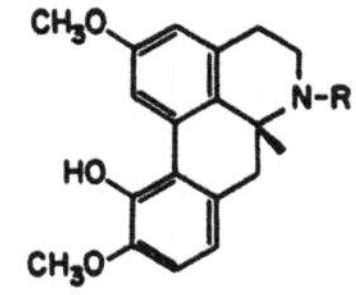

12 *2,10-Dimethoxy-11-hydroxyaporphines*
a) R = H_3, 2-O-Methyl-morphothebaine
b) R = n-C_3H_7

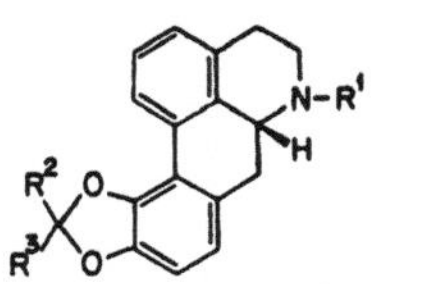

13 *10,11-Methylenedioxy-aporphines*
a) R^1 = CH_3, R^2, R^3 = H, MDO-APO
b) R^1 = n-C_3H_7, R^2,R^3 = H, MDO-NPA
c) R^1, R^2, R^3 = CH_3
d) R^1 = CH_3, R^2, R^3 = C_6H_5
e) R^1 = CH_3, R^2, R^3 = =O
f) R^1 = n-C_3H_7, R^2 = CH_3, R^3 = C_2H_5
g) R^1 = n-C_3H_7, R^2 = CH_3, R^3 = $CH_3(CH_2)_4$

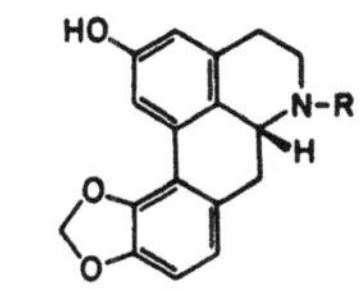

14 *10,11-Methylenedioxy-2-hydroxyaporphines*
a) R = CH_3
b) R = n-C_3H_7

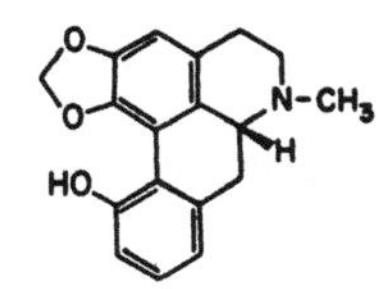

15 *Pukateine*

Fig. 4. Protected catechol-methyl ethers and methylenedioxy analogs

apocodeine (9a) and its n-propyl analog (9b) as well as isoapocodeine (10a) and its heretofore unknown N-propyl analog (10b). Apocodeine (9a) and its N-propyl analog (9b) were prepared by the acid-catalyzed rearrangement of codeine by a procedure previously described (Granchelli et al. 1980). The synthesis of (–)-11-methoxy-10-hydroxy-N-n-propylnorisoapocodeine, (10b) has recently been achieved (Davis et al. 1985) and also by an alternate method (Smith and Davis 1978) involving the selective microbial O-demethylation of apomorphine dimethyl ether to yield isoapocodeine (10a). We applied this latter procedure to the preparation of (–)-10b using *Cunninghamella elegans* and the product obtained was identical in all respects to that obtained via total synthesis from morphine (Davis et al. 1985).

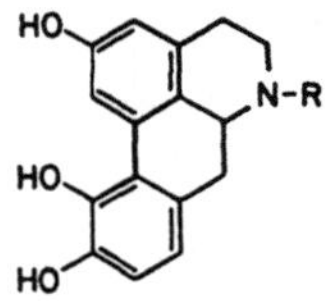

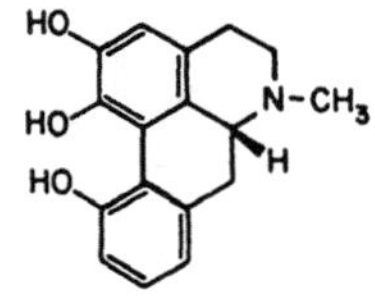

16 2,10,11-Trihydroxyaporphines **17** 1,2,11-Trihydroxyaporphine

a) R = CH_3, THA

b) R = n-C_3H_7, TNPA

c) R = H

d) R = C_2H_5, TNEA

e) R = $CH_2CH{=}CH_2$

Fig. 5. Trihydroxyaporphines

2.3 Synthesis of Trihydroxyaporphines (Fig. 5)

The presence of the A-ring of aporphines, and the addition of substituents to it can extend explorations of the dimensions and other characteristics of the dopamine receptor. Accordingly the synthesis of a series of (–)2,10,11-trihydroxyaporphines was undertaken and their in vivo (Neumeyer et al. 1981b) and in vitro characteristics were evaluated. The enantiomers (6aR and 6aS) of 2,10,11-trihydroxyaporphine (16a, THA) were synthesized from thebaine and bulbocapnine and evaluated pharmacologically (Neumeyer et al. 1982). The (–)1,2,11-trihydroxyaporphine (17) was prepared from the naturally occurring aporphine alkaloid (R)-pukateine (15) using boron tribromide (Ram and Neumeyer 1984).

2.4 Synthesis of N-Haloalkylaporphines and ^{3}H-N-Chloroethylnorapomorphine

Substitution of the N-alkyl side chain on APO has had a profound effect on the dopamine receptor binding and on the pharmacological effects of such aporphines. It has been suggested that compounds containing an alkylating moiety, such as the chloroethyl group, might be useful to label membrane DNA receptor components irreversibly. Phenoxybenzamine (PBZ), a haloalkylamine, can bind to DA (D-1) sensitive adenylate cyclase in rat striatal homogenates (Walton et al. 1978). The synthesis of (Neumeyer et al. 1980) and demonstration that (–) N-(2-chloroethyl) norapormorphine, (4c, NCA) causes persistent DA receptor blockade suggested that this inhibition of DA receptor function may involve covalent bonding of a receptor site (Baldessarini et al. 1980, Costall et al. 1980a,b). The chemical and pharmacological studies with NCA support the hypothesis that this compound in aqueous solution is an alkylating analogue of APO, and its prolonged duration of action and biochemical DA antagonism is due to a covalent modification of DA receptors. The solvolysis of NCA was shown to be mediated through a cyclic aziridinium intermediate (Cohen and Neumeyer 1983). The synthesis of highly reactive and chemically pure 8,9-^{3}H-NCA (Guan et al. 1984) made it possible for Seeman and his group to label macromolecular species associated with DA receptors.

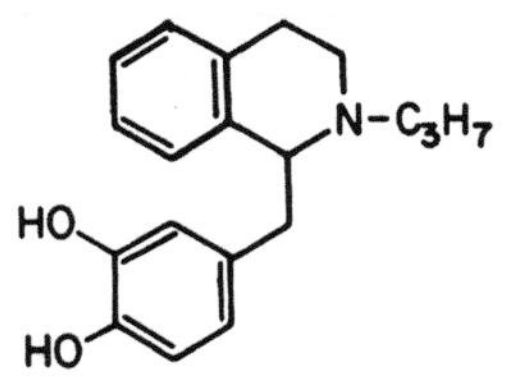

18 *1-(3,4-Dihydroxybenzyl)-2-n-propyl-1,2,3,4-tetrahydroisoquinoline*

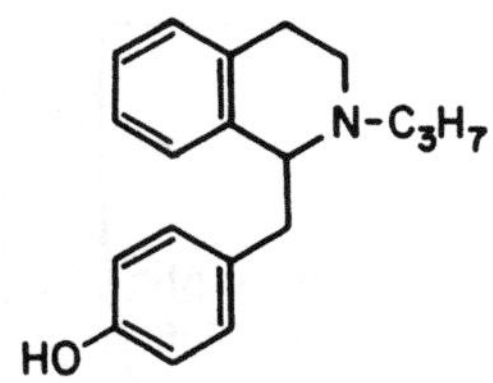

19 *1-(4-Hydroxybenzyl)-2-n-propyl-1,2,3,4-tetrahydroisoquinoline*

Fig. 6. Benzylisoquinolines

2.5 Synthesis of Benzylisoquinolines (Fig. 6)

The syntheses of 1-(3,4-dihydroxybenzyl)-2-propyl-1,2,3,4-tetrahydroisoquinoline (18) and 1-(4-hydroxybenzyl)-2-propyl-1,2,3,4-tetrahydroisoquinoline (19) were carried out in order to examine the structure-activity relationships when the rigid aporphine ring system is replaced by the more flexible benzyltetrahydroisoquinoline moiety which still contained all of the necessary substituents of the potent NPA molecule (4b) (Neumeyer et al. 1977).

3 Characteristics of Putative Dopamine Receptor Types

Based on radioligand binding of DA-agonists and antagonists, as well as effects on adenylate cyclase or endocrine functions in tissue of intact or selectively lesioned animals, several tentative schemes involving at least four DA receptor types have been proposed (Seeman 1980, Kebabian and Calne 1979, Creese et al. 1981, Schwartz et al. 1983, Sokoloff et al. 1980, Goldberg et al. 1978).

DA-agonist "receptors" were first defined by association with low-affinity (μM) stimulation of adenylate cyclase in forebrain, a DA-receptor associated phenomenon now often referred to as DA receptor type "D-1" (Kebabian and Calne 1979). A somewhat more extensively studied type of DA receptor in brain (mainly calf or rat neostriatum), commonly called "D-2" was defined initially by its lack of association with the stimulation of adenylate cyclase (Kebabian and Calne 1979) and later characterized by its high affinity for many DA-antagonists and lower affinity for DA agonists (200–1000 nM) (Seeman 1980). Some D-2 sites (especially in pituitary) may be associated with inhibition of adenylate cyclase by DA agonists (Creese et al. 1982) (Fig. 7).

In a series of biochemical, anatomical, and behavioral experiments reported in 1975, Carlsson (Carlsson 1975) postulated the concept of the DA "autoreceptor" to account for observations that (–)apomorphine (APO) inhibited synthesis of DA in synaptosomal or slice preparations of rat striatum in low concentrations, even after nigrostriatal lesions; this inhibition was reversed by DA antagonists. In

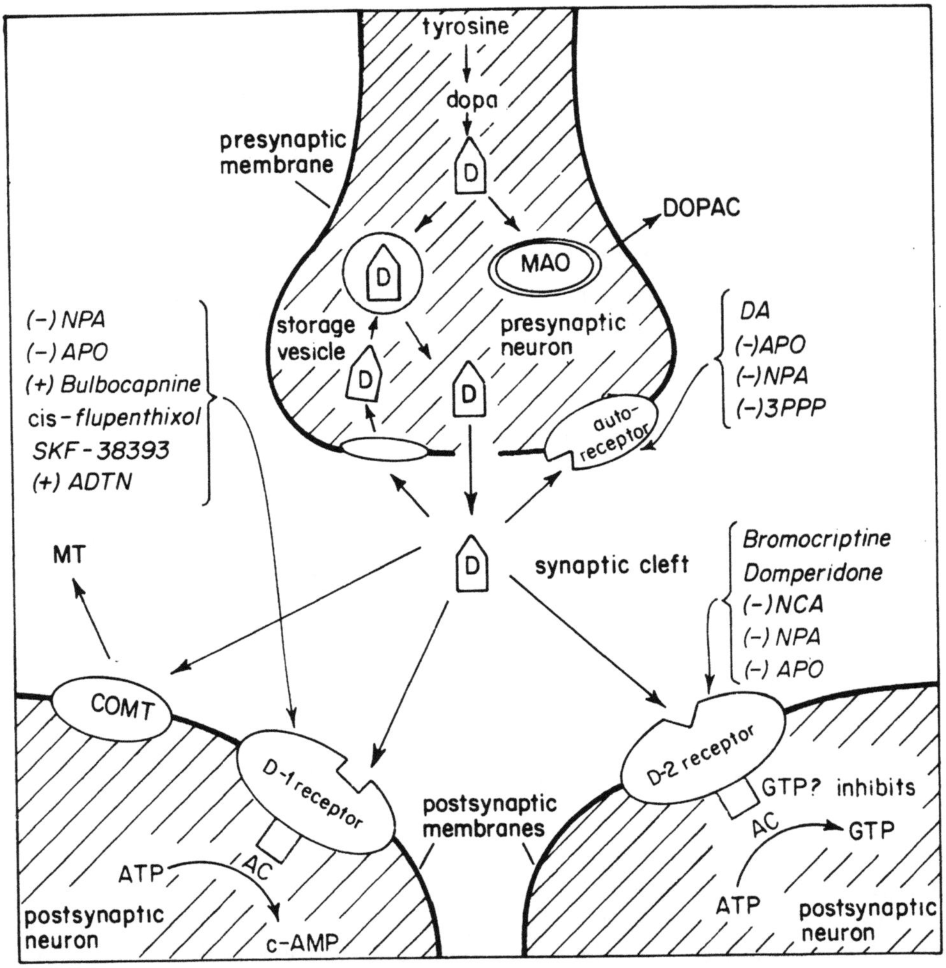

Fig. 7. Dopamine *(D)* is stored in the transmitting neuron terminal and released into the synaptic cleft, where it may bind to an autoreceptor and thus inhibit release of more dopamine. Dopamine may also bind to a D-1 receptor associated with stimulating adenylate cyclase *(AC)*, or to a D-2 receptor that inhibits AC. Dopamine can be metabolized to dihydroxyphenylacetic acid *(DOPAC)* by monoamine oxidase *(MAO)*, or to 3-methoxytyramine *(MT)* by catechol-O-methyltransferase *(COMT)*. Agents acting at D-1, D-2, or autoreceptors (D-3?) are indicated, although it is not implied that these agents act exclusively at these sites

addition, APO inhibited tyrosine hydroxylation in striatal projections by a mechanism that did not depend on the intracellular uptake of APO but as largely neuroleptic-sensitive and presumably mediated by membrane surface receptors. A dose of APO or its N-propyl analog (NPA) necessary to decrease locomotion in rodents in vivo paralleled an in vivo decrease in DA production and was a tenth of that required for behavioral stimulation in rats (Carlsson 1975, Argiolas et al. 1982). Thus, presumably presynaptic "autoreceptors" may be stimulated selectively by low doses of some DA agonists while postsynaptic receptors respond at higher doses. Electrophysiological investigations (Bunney et al. 1973) further indicate that DA "autoreceptors" probably occur on nigral cell bodies or dendrites as well as on striatal nerve terminals and that these may be important in regulating neuronal activity that may parallel DA synthesis and release. Proposed "autoreceptors" on DA cell bodies are ca. 10 times more sensitive to both APO (i.v.) and DA (iontophoretically applied) than striatal postsynaptic receptors are (Bunney et al. 1973, Aghajanian and Bunney 1977), NPA is reportedly even more selective for presynaptic sites as it inhibited DA synthesis in rat striatum in very low ($\mu g kg^{-1}$) doses (Argiolas et al. 1982); some binding of NPA may occur postsynaptically (Leff et al. 1981) and this APO analog as well as an orally, longer acting precursor of it have behaviorally excitatory effects similar to higher doses of APO when given in $mg kg^{-1}$ doses (Baldessarini et al. 1982, Sperk et al. 1982).

3.1 Aporphines as Central Dopamine Agonists and Antagonists

One potential strategy aimed at novel therapeutically useful anti-DA drugs is the development of agents selective for presynaptic autoreceptors. For example, beneficial effects of low doses of (–)APO and (–)NPA have been reported against psychosis or dyskinesias (Tolosa and Sparber 1975, Corsini et al. 1981, Tamminga et al. 1983, Lal 1981). These seemingly paradoxical antipsychotic or antidyskinetic effects are consistent with Carlsson's proposal of DA "autoreceptors" which can decrease dopaminergic transmission by decreasing the synthesis and release of DA. This view is further supported by recent clinical observations that cerebrospinal fluid (CSF) (Levy et al. 1982) and plasma (Cutler et al. 1982) levels of the major metabolite of DA homovanillic acid (HVA), were decreased in schizophrenics given a small dose of APO.

Research efforts are currently underway to develop potentially clinically useful agents which may interact selectively with DA autoreceptors. For example, Carlsson and colleagues have reported evidence of stereoselective presynaptic DA agonist activity of 3-(-hydroxyphenyl)-N-n-propylpiperidine ("3-PPP") (Hjorth et al. 1981). More thorough study of the enantiomers of 3-PPP has led to the characterization of the R(+)enantiomer as a classical dopaminergic agonist stimulating both auto- and postsynaptic receptors. The S(–)enantiomer has been found to stimulate DA autoreceptors and block postsynaptic DA receptors (Hjorth et al. 1983). Thus, in the racemic mixture, postsynaptic stimulation by the (+)enantiomer is presumably negated by blockade due to the (–)enantiomer, thereby resulting in a selective autoreceptor profile of the racemate. A finding somewhat analogous to these results with (+) and (–) 3-PPP was that (+)APO which was found to antagonize the effects of (−)APO (Riffee et al. 1982). Another interesting development is the realization that R(−)NPA is an extremely potent agonist of DA autoreceptors in vivo (Argiolas et al. 1982). Its proposed prodrug methylenedioxy derivative, which we have developed, appears to be the first orally active aporphine with DA-agonistic properties in animals at high doses that is also capable of producing possibly neuroleptic-like motor-inhibitory and anti-arousal effects in rodents at low ($\mu g kg^{-1}$) doses (Baldessarini et al. 1982, Campbell et al. 1982, Sperk et al. 1982).

3.1.1 Structural Requirements of Aporphine Congeners for High Affinity Dopamine Agonist Binding at Central Dopamine Receptors

The structural requirements of a variety of aporphine congeners and structurally related DA agonists prepared in our laboratories led to an hypothesis relating the mode of binding of such agonists to DA receptors (Fig. 8 and Table 1) (Arana et al. 1983b, Neumeyer et al. 1981c, 1983a,b).

The appropriate configuration at the chiral center (carbon 6a) of aporphines, the location of the hydroxyl groups, particularly the hydroxyl groups meta to the aminoethyl fragment, the receptor boundaries, and a proposed interaction of the nitrogen atom to a nucleophilic center on the receptor surface or "acceptor" site, are factors which appear to control binding and pharmacological activity at central DA agonist receptors.

Substitution of the N-alkyl side chain on APO has a profound effect on the DA receptor binding and on the pharmacological effects of such aporphines (Neumeyer et al. 1983a). Increasing the size of the N-alkyl substituent increases the selectivity for ^{3}H-spiperone (^{3}H SPR) binding site, a property which may reflect greater lipophilicity or altered selectivity for D-2 receptors (Neumeyer et al. 1981c). The analysis of ^{3}H-agonist displaceable binding in calf caudate membranes is presented in Table 1. The biochemical analysis of these radioligand binding assay methods shows saturable, displaceable, highly specific characteristics in a subsynaptosomal membrane fraction of calf caudate tissue. Using IC_{50} values calculated from displacement curves for over 100 compounds (not all reported here), the high-affinity (defined by using 0.5nM of each ^{3}H-agonist ligand) binding demonstrates that ^{3}H-APO and ^{3}H-ADTN are useful in labelling a putative DA agonist receptor in calf caudate membranes. The high correlation between the IC_{50}'s of these two ligands, as shown in Table 1, indicates that they may label the same species of putative DA receptors in bovine caudate. The ten most potent compounds in this series also have been shown to be potent agonists of striatal adenylate cyclase, although the potency in the radioreceptor system is in the nanomolar range, whereas it is in the micromolar range in the enzyme assay. The results presented describe a radioligand binding assay which is pharmacologically consistent with a dopamine agonist receptor, as revealed with both ^{3}H-APO and ^{3}H-ADTN, and define the salient structural requirements for high-affinity binding to this putative receptor (Arana et al. 1981, 1983a, b).

Table 1. Inhibition of binding against ^{3}H-apomorphine and ^{3}H-ADTN

		IC_{50} (nM)	
		^{3}H APO	^{3}H-ADTN
Group 1.	**Monohydroxyaporphines**		
(–) 1b	(–)11-Hydroxy-N-n-propyl-noraporphine	5.4	7.0
(–) 1a	(–)11-Hydroxyaporphine	7.8	12.0
(±) 1b	(±)11-Hydroxy-N-n-propyl-noraporphine	36	40
(±) 2a	(±)10-Hydroxyaporphine	112	150
(±) 2b	(±)10-Hydroxy-N-n-propyl-noraporphine	700	516
(±) 3b	(±)8-Hydroxy-N-n-propyl-noraporphine	ca. 10,000	4,000
Group 2a.	**Dihydroxyaporphines: catechols**		
(–) 4b	(–)10,11-Dihydroxy-N-n-propylnoraporphine (–) NPA	2.5	2.0
(±) 4b	(±)10,11-Dihydroxy-N-n-propylnoraporphine (±)NPA	5.0	3.5
(–) 4a	(–)10,11-Dihydroxyaporphine (–)APO	1.0	4.9
(–) 4d	(–)Norapomorphine	14.4	13
(–) 4e	(–)10,11-Dihydroxy-N-hydroxy-ethylnoraporphine	25	60
(+) 4b	(+)10,11-Dihydroxy-N-n-propyl-noraporphine (+)NPA	66	70
(+) 6	(+)1,2-Dihydroxyaporphine	124	145
(–) 6c	(–)N-(Chloroethyl)norapo-morphine (NCA)	394	468
(–) 6	(–)1,2-Dihydroxyaporphine	750	1,140
Group 2b.	**Dihydroxyaporphines: non-catechols**		
(–) 7	(–)2,11-Dihydroxyaporphine	50	114
(–) 8a	(–)2,10-Dihydroxyaporphine	295	184
Group 3.	**Protected Catechols**		
	(+)Bulbocapnine	70	230
(–) 10a	(–)2,11-Dihydroxy-10-methoxy-aporphine (Morphothebaine)	204	
(–) 13b	(–)10,11-Methylenedioxy-NPA (MDO-NPA)	850	342
(–) 15	(–)1,2-Methylenedioxy-11-hydroxyaporphine (pukateine)	70	371
(–) 10b	(–)2,11-Dihydroxy-10-methoxy-N-n-propylnoraporphine	1,030	420
(–) 13c	(–)10,11-Propylidene-2-dioxy-aporphine	2,079	

Table 1 (cont.) Inhibition of binding aginst ^{3}H-apomorphine and ^{3}H-ADTN

		IC_{50} (nM)	
		^{3}H APO	^{3}H-ADTN
(–) 14b	(–)2-Hydroxy, 10,11-methylenedioxy-N-n-propylnoraporphine	236	586
(–) 14a	(–)2-Hydroxy, 10,11-methylenedioxyaporphine	583	627
(–) 13f	(–)10,11-Butylidene-2-dioxy-n-N-propylnoraporphine	1,381	
(–) 13a	(–)10,11-Methylenedioxyaporphine (MDO-APO)	1,310	899
(–) 9a	(–)10-Methoxy-11-hydroxyaporphine (Apocodeine)	2,400	1,351
(–) 13d	(–)10,11-Diphenylmethylenedioxyaporphine	10,000	
(–) 12a	(–)2,10-Dimethoxy-11-hydroxyaporphine	--	ca. 7,200
Group 4.	Trihydroxyaporphines		
(–) 16a	(–)2,10,11-Trihydroxyaporphine (–)THA	10	11
(–) 16b	(–)2,10,11-Trihydroxy-N-propylnoraporphine (TNPA)	25	5.5
(–) 16e	(–)2,10,11-Trihydroxy-N-ethylnoraporphine	25	10
(–) 16f	(–)2,10,11-Trihydroxy-N-allylnoraporphine	24	28
(–) 16c	(–)2,10,11-Trihydroxynoraporphine	30	44
(+) 16a	(+)2,10,11-Trihydroxyaporphine (+) THA	5,000	1,600

3.1.2 Behavioral Activity of Aporphines in Rats with 6-Hydroxydopamine (6 OHDA) Lesions of Caudate or Nucleus Accumbens

The behavioral actions of several aporphines were examined (Jackson et al. 1983) in rats with selective unilateral 6-hydroxydopamine-induced destruction of nigrostriatal dopamine neurons, and in rats with bilateral 6 OHDA-induced destruction of mesolimbic dopamine neurons. Dopaminomimetics such as (R)-APO in these animal models elicit circling behavior and locomotor activity respectively. In animals with unilateral nigrostriatal lesions (–)-2,10,11-trihydroxy-N-n-propylnoraporphine [(–)16b, TNPA)] and (–)-10,11-methylenedioxy-N-n-propylnoraporphine [(13b), MDO-NPA] elicited weak, but prolonged, contraversive circling, whereas (–)2,10,11-trihydroxyaporphine (16a, THA) was inactive. In animals with bilateral destruction of mesolimbic dopamine neurons TNPA and MDO-NPA elicited a strong stimulation of locomotor activity, while THA was inactive. The results suggest that TNPA and MDO-NPA, but not THA, exert central dopaminomimetic effects in vivo.

The results are also consistent with previous data indicating that N-propyl substitution of aporphines causes a relative enhancement of activity in animal models which emphasize effects at mesolimbic rather than striatal dopamine receptors. Such compounds could provide a more rational basis for drug therapy.

3.1.3 Motor Inhibition Induced by Aporphine Derivatives in the Mouse

The ability of several aporphine and benzylisoquinoline derivatives to inhibit mouse spontaneous locomotor activity at low doses, and at higher doses to have a reduced motor inhibitory effect, was used to determine whether motor inhibitory and facilitatory potentials could be dissociated, and the optimal structure required to cause these behavioral changes (Bradbury et al. 1983). Ability to displace ^{3}H-ADTN from its binding sites in rat striatal tissue was used as a broad measure of the abilities of these compounds to bind to the "dopamine receptor". The order of potency for "low dose" inhibition of mouse spontaneous locomotion was (–)N-n-propylnorapomorphine [(–)NPA, 4b] > (–)apormorphine [(–)APO, 4a] > (–)2,10,11-trihydroxy-N-n-propylnoraporphine [(–)TNPA, 16b] > (–)2,10,11-trihydroxyaporphine [(–)THA 16a] > (–)10,11-dihydroxy-N- (hydroxyethyl)noraporphine [(–)4e] > norapomorphine [(–)4d] > (±)10-hydroxy-N-n-propylnoraporphine (2b) > 1-(3,4-dihydroxybenzyl)- 2-n-propyl-1,2,3,4-tetrahydroisoquinoline (18). The latter two compounds were only weakly active and (±)8-hydroxy-N-n-propylnoraporphine (19b) and 1-(4-hydroxybenzyl)-2-n-propyl-1,2,3,4-tetrahydroisoquinoline (19) were inactive. The reduction in motor inhibitory effect as dose of agonist was increased (indicative of facilitation of locomotion) was only observed with (–)NPA and (–)APO, whereas (–)TNPA and (–)THA caused non-specific changes which interfered with motor performance or the doses required were so large as to render testing impractical. The potent inhibitory actions of (–)NPA, (–)APO, (–)TNPA, and (–)THA were antagonized by pretreatment with spiperone but not with prazosin or yohimbine. Specific binding of 2.0 nM ^{3}H ADTN was displaced by nanomolar concentrations of (±)-ADTN and aporphine derivatives. Isoapomorphine (5) and the two benzylisoquinolines were ineffective at 10^{-6} M. The best structure for those derivatives examined was shown in all tests to be (–)NPA (Bradbury et al. 1983).

3.1.4 Alkylating Aporphines as Irreversible DA Receptor Antagonists

The synthesis and demonstration that R(–)N-(2-cholorethyl)-10,11-dihydroxynoraporphine (N-chloroethylnorapomorphine, NCA, 4c) may cause persistent DA receptor blockade, suggested that this inhibition of DA receptor function may involve covalent bonding of a receptor binding site (Neumeyer et al. 1981d). NCA blocked DA-sensitive adenylate cyclase activity (D-2) in a non-competitive and apparently irreversible manner. This effect was prevented by coincubation with DA or APO but not with norepinephrine. Analogs of NCA with low affinity for DA receptor sites defined by binding of 1 nM ^{3}H APO (Arana et al. 1981) also had much weaker effects against the DA-sensitive cyclase activity. These included the 10-0-methylated derivative (analog of apocodeine), and a 10,11-diester derivative of NCA (Baldessarini et al. 1980).

Additional pharmacological actions of NCA both at central and peripheral DA receptor sites have been described recently (Mustafa et al. 1982, Lehmann and Langer 1982, Lehmann et al. 1983, Goosey and Daggett 1982, Helmeste 1983, Schmidt et al. 1984, Szabo and Neumeyer 1983c). NCA and recently available ^{3}H-NCA (Guan et al. 1984) should aid in the further characterization and isolation of DA receptors and in evaluation of their disposition and turnover in vivo. Moreover, NCA may also represent a novel class of long-acting DA antagonist.

3.2 Stereoselectivity of DA Receptor Agonists and Antagonists

Most DA antagonists in current clinical use block DA receptors non-selectively and are associated with a wide range of antipsychotic, neurological, and endocrine actions. The recent realization that R(−)N-n-propylnorapormorphine [(R)NPA] is an extremely potent agonist of DA autoreceptors in vivo (Argiolas et al. 1982) suggested its comparison with the (+) enantiomer (S)NPA for DA-like activity. Pharmacological activities of (S) and (R)NPA were evaluated (Neumeyer et al. 1983b) with nM concentrations of ^{3}H-labelled APO, ADTN, and spiperone (SPR) for competition for binding to a membrane-rich subsynaptosomal fraction of calf caudate nucleus. Similar experiments have been carried out with the enantiomers of APO and other aporphines (Arana et al. 1983a, b, 1984; Neumeyer et al. 1981a). The efficacy of (R) and (S)NPA in stimulating DA-sensitive adenylate cyclase from homogenates of rat corpus striatum and pieces of carp retina was also evaluated. Three behavioral effects in the rat (stereotyped behavior, sedation, and catalepsy) were also examined. Only (R)-NPA induced catalepsy at low doses; (S)-NPA failed to antagonize this action of the R isomer. The effects of (R) and (S)-NPA on adenylate cyclase paralleled the behavioral effects and radioreceptor binding assays in that the R(−)isomer was strongly preferred at such D-1 receptors. The (S)-enantiomer of NPA was, however, the weakly preferred configuration for rat liver catechol-O-methyltransferase. A DA-receptor model that accomodates the configuration of NPA and related aporphines is shown in Fig. 8.

Fig. 8. Interáction of dopamine (**A**); (R)-APO (**B**); (R)-NPA (**C**); (R)2,10,11-trihydroxyaporphine (**D**); (R)11-hydroxy-N-n-propylnoraporphine (**E**); (S)-NPA (**F**) at the DA receptor. The model accounts for the receptor-binding characteristics of aporphines and related rigid analogs of DA. The features include M_1 and M_2 putative binding sites on the receptor surface complementary to OH-binding sites on the agonist molecule; an M_3-binding site that represents an interaction by OH substituents in position 2 of the aporphine A ring which diminishes affinity; and an electronegative site B complementary to a positively charged nitrogen atom on the agonist molecule. The optimal distance of the OH binding site complementary to M_1 is 6.5−7.3Å from the nitrogen atom (transα-rotameric conformation of DA). Substituents on the nitrogen should be accommodated by the lipophilic cavity on the receptor surface. A Hydrogen atom shown only at carbon-6a (chiral center) of aporphines accounts for enantioselectivity. An obstruction on the receptor precludes the appropriate interaction in the 6aS configuration (Fig. 8F). IC_{50} (nM) values were determined for calf caudate nuclei

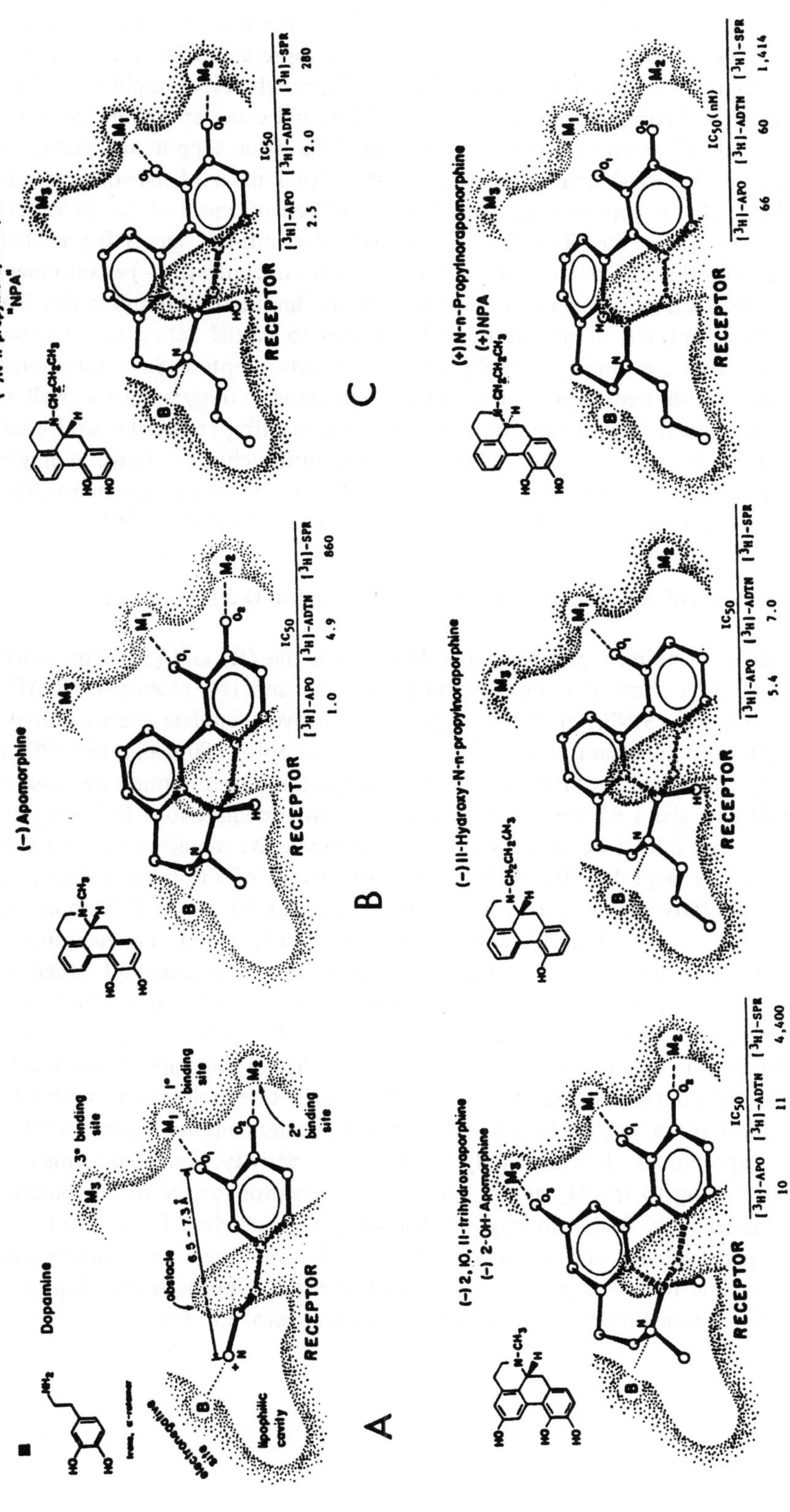

Fig. 8. (Legend see p. 160)

We have recently found evidence of a potentially important neuropharmacologic difference between the enantiomers of aporphines represented by the S(+) and R(−) enantiomers of apormorphine (APO) and N-propylnorapomorphine (NPA). The results of evaluating enantiomers of APO, NPA, and other aporphines as models of D-1, and D-2 receptor types in mammalian CNS tissue support the preferred interactions of the R(−) enantiomers. In addition, preliminary behavioral experiments indicate that it appears very likely that the (+) enantiomers of the two aporphines, and especially the S(+)NPA, may be selective anti-DA agents for arousal mechanisms, possibly associated with the limbic system. In contrast, (−) enantiomers seem to be less selective and can evidently stimulate limbic or extrapyramidal DA receptors; their anti-DA effects seem to be limited to partial action at a narrow range of low doses presumably partially selective for autoreceptors. Since traditional antipsychotic agents have nonselective anti-DA effects in extrapyramidal as well as limbic DA receptors, they are almost all associated with extrapyramidal neurological side effects as the price for their clinically useful antipsychotic actions. It might thus be possible to develop clinically useful and more selective, less neurotoxic limbic DA-antagonists based on the actions we have observed with S(+)NPA.

3.2.1 Interactions of Aporphine Enantiomers with D-1 and D-2 DA Receptors

Recent studies (Goldman and Kebabian 1984) with the (R) and (S) enantiomers of APO and NPA show that both interact with D-1 and D-2 receptors. (R)APO as well as (R) and (S)NPA stimulate dopamine sensitive adenylate cyclase activity of carp retina, a biochemical model of the D-1 receptor. Similarly (R)APO as well as (R) and (S)NPA stimulate the D-2 receptor in the intermediate lobe of the rat pituitary gland whereas (S)APO blocks the intermediate lobe D-2 receptor. The dopaminergic inhibition of the release of immunoreactive α-melanocyte-stimulating hormone-like peptides (IR-αMSH) from melantroph of the intermediate lobe of the rat pituitary gland is used as a physiological model of the D-2 dopamine receptor (Cote et al. 1982, Goldman et al. 1983) (Table 2). Further studies with (S)-bulbocapnine O-methyl ether and (S)-apomorphine dimethyl ether indicated that these aporphines are able to antagonize either the D-1 or the D-2 dopamine receptor (J. Kebabian, personal communication). What is of particular interest in these findings is that the intrinsic antagonistic activity of these aporphines resides in the (S) enantiomer and that the requirement for a catechol system in the 10- or 11-position (or a single 11-hydroxy group), a requirement for agonist activity in such (R)aporphines, does not apply for antagonist activity. One could thus conclude that binding to the M_1 and M_2 sites on the receptor surface of our putative receptor model (Fig. 8) is not required for antagonist activity. These differences in the intrinsic activity and apparent affinity of the aporphine enantiomers with and without free hydroxyl groups is in accord with the hypothesis that dopamine agonist and antagonist receptors are distinct pharmacological entities.

Table 2. Efficacy of aporphine enantiomers upon the D-1 and the D-2 dopamine receptors. (Goldman and Kebabian 1984)

	D-1 Receptor		D-2 Receptor	
Compound	Action	Affinity (nM)	Action	Affinity (nM)
R(−)APO	Partial agonist (55%)	2.3±0.6	Agonist (100%)	14.0± 6.0
S(+)APO	Antagonist	12.8±2.5	Antagonist	495 ±99
R(−)NPA	Agonist	5.0±1.7	Agonist (100%)	0.3± 0.1
S(+)NPA	Partial agonist (34%)	6.7±2.0	Agonist (100%)	16.7± 6.7

Data for the D-1 receptor were obtained in experiments determining adenylate cyclase activity in cell-free homogenates of carp retina. For D-1 agonists, the value in parenthesis is the maximal response to the indicated agent expressed as a percentage of the maximal response to dopamine (100 μM); affinity is the concentration of the indicated agent causing a half-maximal enhancement of enzyme activity. For S(+)APO, affinity is the inhibitory constant (K_i).
Data for the D-2 receptor were obtained in experiments determining the release of IR-αMSH from enzymatically dispersed intermediate lobe cells. For the D-2 agonists, the value in parenthesis is the maximal agonist-induced inhibition of L-isoproterenol-stimulated release of IR-αMSH expressed as a percentage of the inhibition due to 1 μM LY 171555: affinity is the concentration of the indicated agent causing half-maximal inhibition of L-isoproterenol-stimulated release of IR-αMSH. For S(+)APO, affinity is the inhibition constant (K_i).

3.3 Vascular Effects of Aporphines in the Perfused Rat Kidney

The dopaminomimetic activity of apomorphine on the renal vascular bed has been well documented (Goldberg et al. 1978, Imbs et al. 1974, Crumley et al. 1976). The renal vascular effects of several aporphines were studied on the isolated perfused rat kidney in the presence of 10^{-5} phenoxybenzamine and 10^{-5} sotalol and after contraction of the vascular bed with prostaglandin $F_{2\alpha}$ (10^{-7} -3 x 10^{-6}M). Under these conditions (R)APO showed renal dopaminomimetic activity, i.e. renal vasodilation competitively inhibited by (+) butaclamol (10^{-8}M) but not by (−) butaclamol (3 x 10^{-8}M). It had an apparent affinity 25 times higher but a markedly lower intrinsic activity than DA. (R)-NPA and (R)TNPA (16b) were less potent and the mono-10-hydroxylated aporphine (2b) was completely inactive. (S)-bulbocapnine also showed weak dopaminomimetic activity in this test system (Schmidt et al. 1984). In other studies, Goldberg reported (S)-bulbocapnine to be an antagonist of DA_1 receptors (Goldberg and Kohli 1983). (R)-NCA (4c) was also found to have renal vascular dopaminolytic activity. It was found to irreversibly block the dopamine receptor of the vascular bed in the isolated kidney. However, the inhibition was incomplete, even when 10^{-5} M (R)NCA was used, indicating the relatively low affinity of (R)-NCA for the renal vascular dopamine receptor (Schmidt et al. 1984).

3.4 Dopamine Agonist Aporphines as Anticonvulsants in Two Models of Reflex Epilepsy

The anticonvulsant action of various aporphine derivatives acting at dopamine receptors has been investigated in two genetically determined animal models – DBA/2 mice with sound-induced seizures and baboons *Papio papio* with photically induced seizures (Anlezark et al. 1983). Protection against the clonic and tonic phases of the seizure response in DBA/2 mice was seen for 15–60 min after (R) 2,10,11-trihydroxy-N-n-propylnoraporphine (TNPA, 16b) (1.25 $mgkg^{-1}$) and (R) MDO-NPA (13b) (0.625–1.25 $mgkg^{-1}$). Short-lasting protection (up to 30 min) was seen following (R)2,10-11-trihydroxy-N-ethylnoraporphine (TNEA, 16b) (1.25–6.25 $mgkg^{-1}$). Changes in audiogenic seizure susceptibility were accompanied by piloerection, ptosis and loss of spontaneous locomotor and exploratory behavior. No protection was seen after (R)-norapomorphine (4d) (0.05–18.75 $mgkg^{-1}$). In baboons, marked reductions in photomyoclonic responses were seen following (R) MDO-NPA (13b) (0.25 $mgkg^{-1}$, lasting up to 2 h); TNPA (0.5–2.5 $mgkg^{-1}$, lasting up to 7 h); (R)-THA (16a) (5 $mgkg^{-1}$, lasting 2 h). Little change in responsiveness followed administration of (R)norapomorphine (1.25 or 6.25 $mgkg^{-1}$). Changes in photosensitivity were accompanied by yawning and pupil dilation. (R)- MDO-NPA (0.5–6.25 $mgkg^{-1}$) was also administered orally in baboons. Moderate protection against photically induced myoclonus was seen following 6.25 $mgkg^{-1}$. The order of potency of the compounds in DBA/2 mice and photosensitive baboons was the same. These results demonstrate that aporphine derivatives with longer duration of action than apomorphine and a low incidence of side effects can reduce seizure susceptibility in two models of reflex epilepsy. The potency of (R)APO against syndromes of photically induced epilepsy in man has already been demonstrated (Lal 1981, Anlezark et al. 1981). Orally active, long-acting aporphine derivatives may prove to have further use in the clinic.

3.5 Aporphines as Duodenal Anti-ulcerogens in the Rat

Data are accumulating that indicate the possible involvement of central and peripheral dopamine binding sites in the pathogenesis of duodenal ulceration. Structure activity studies with duodenal ulcerogens implicated dopamine as a putative mediator and/or modulator in duodenal ulceration (Szabo and Neumeyer 1983). A marked change in the incidence and intensity of cysteamine-induced duodenal ulcer was demonstrated by the administration of dopamine agonists or antagonists (Szabo 1979). Dopamine agonists (e.g. bromocriptine or lergotrile) administered either as a pre- or post-treatment, decreased the intensity of the acute und chronic duodenal ulcers and diminished the output of gastric acid in rats given cysteamine or propionitrile (Szabo 1979, Szabo and Neumeyer 1983).

On the other hand, dopamine antagonists (e.g. haloperidol or pimozide) aggravated the duodenal ulcers and mortality after cysteamine administration. The chemically induced duodenal ulceration was associated with changes in the sensitivity and number of dopamine receptors in the gastric and duodenal mucosa and muscularis propria (Szabo and Neumeyer 1983).

In rats given cysteamine, (R)-MDO-NPA at either 50 or 100 μg/100g caused significant prevention of experimental duodenal ulcers and abolished mortality. The ulcer index (which is the composite of ulcer incidence and intensity) was significantly decreased only by 100 μg of (R)-NPA. A single high dose of either (R)MDO-NPA or (R)-NPA was not active, while a daily treatment with small quantitites virtually abolished the cysteamine-induced duodenal ulcers. The dopamine antagonist (S)-butaclamol at 100 μg or (R)-NCA at 50 or 100 μg, aggravated the experimental duodenal ulcers and mortality caused by cysteamine (Neumeyer and Szabo 1983). These findings support previous clinical observations (Szabo 1979), which indicate that untreated dopamine deficiency disorders, such as Parkinson's disease, have an excess of peptic ulcer, while states with dopamine excess or receptor hyperactivity (e.g. schizophrenia) are accompanied by decreased prevalence of peptic ulcer. In preliminary secretory studies, both MDO-NPA and NPA significantly suppressed the enhanced gastric acid output induced by cysteamine in the chronic gastric fistula rat. Gastric acid secretions were collected and titrated at hourly intervals for 7 h after cysteamine (15 mg/100 g p.o.). Gastric acid output of control group was 1043 ± 102 μEq/7 h, versus pretreatment with 0.1 mg/100 g p.o. MDO-NPA in which gastric acid output was reduced to 288 ± 67 μEq/7 h ($P < 0.001$).

The dopamine agonist MDO-NPA seems to exert prominent duodenal antiulcerogenic effect. Its action is about 200 times more potent than the histamine H_2-receptor antagonist cimetidine, tenfold more active than other dopamine agonists (e.g., bromocriptine, lergotrile), and its potency is identical to naturally occurring prostaglandins which also inhibit this experimental duodenal ulcer (Robert et al. 1974). Thus, MDO-NPA is a prodrug with orally effective and prolonged activity at dopamine receptors (e.g. in duodenum and/or brain) and these properties may serve as tools to elucidate new factors in the pathogenesis of duodenal ulceration. The drug or one of its analogs may also have clinical utility for the prevention and/or treatment of duodenal ulcer disease.

4 Conclusions

In summary, a series of mono-, di- and trihydroxylated aporphines and two benzylisoquinoline derivatives have been synthesized and characterized. Our studies have demonstrated that aporphines continue to be useful ligands for the elucidation of dopaminergic mechanisms in such in vitro paradigms as receptor binding. A receptor model which accounts for the structure-activity relationships of high-affinity DA agonist binding in calf caudate tissue as reflected by ^{3}H-APO and ^{3}H-ADTN has been refined. Such studies, when combined with other in vitro paradigms, with other radioligands and enzyme stimulation assays, as well as in vivo systems such as behavioral effects in rodents, and vascular effects in the perfused rat kidney should aid in the further characterization of both central and peripheral receptors and in the design of clinically useful DA agonists and antagonists.

Acknowledgements. This research was supported in part by NIH Grants NS-15439 and NS-18178 at Northeastern University. JLN received support on a Distinguished Professor Award from Northeastern University. The assistance of Ms. Noralie Barnett in the preparation of the figures and Ms. Patricia Bohannon for preparation of the manuscript is greatly appreciated.

References

Aghajanian GK, Bunney BS (1977) Pharmacological characterization of dopamine "autoreceptors" by microiontophoretic single-cell recording studies. Adv Biochem Psychopharmacol 16:433–438

Anlezark GM, Marrosu F, Meldrum BS (1981) Dopamine agonist in reflex epilepsy. In: Morselli PL, Lloyd KG, Meldrum BS (eds) Neurotransmitters, seizures and epilepsy. Raven Press, New York, pp 251–262

Anlezark GM, Blackwood DHR, Meldrum BS, Ram VJ, Neumeyer JL (1983) Comparative assessment of dopamine agonist aporphines as anticonvulsants in two models of epilepsy. Psychopharmacology 81: 135–139

Arana GW, Baldessarini RJ, Harding M (1981) Binding of ^{3}H-apormorphine: Pharmacology. Biochem Pharmacol 30:3171–3179

Arana GW, Baldessarini RJ, Lamont JS, Amlicke D, Neumeyer JL (1983a) Pharmacology of high-affinity binding of ^{3}H (±)2-amino-6,7-dihydroxy-1,2,3,4-tetrahydronaphthalene (ADTN) to bovine caudate nucleus fractions. Biochem Pharmacol 32:2873–2880

Arana GW, Baldessarini RJ, Neumeyer JL (1983b) Structure-activity relationships of aporphines at central dopamine receptors. Acta Pharm Suec Suppl 2:23–36

Arana GW, Lamont JS, Baldessarini RJ, Teicher M, Neumeyer JL, Cohen BM (1984) Binding of ^{3}H-apomorphine to an aporphine binding site as well as to dopamine sites in bovine caudate nucleus tissue. Neuropharmacology 23:885–892

Argiolas A, Merea G, Serra G, Merlis MR, Fadda F, Gessa GL (1982) N-n-propylnorapomorphine: An extremely potent stimulant of dopamine autoreceptors. Brain Res 231:109–116

Baldessarini RJ, Kula NS, Arana GW, Neumeyer JL, Law S-J (1980) Chloroethylnorapomorphine, a proposed long-acting dopamine antagonist: Interactions with dopamine receptors of mammalian forebrain in vitro. Eur J Pharmacol 67:105–110

Baldessarini RJ, Neumeyer JL, Campbell A, Sperk G, Ram VJ, Arana GW, Kula NS (1982) An orally effective, long-acting dopaminergic prodrug: (–)10,11-methylenedioxy-N-propylnoraporphine. Eur J Pharmacol 77:87–88

Bradbury AJ, Costall B, Naylor RJ, Neumeyer JL (1983) Motor inhibition induced by aporphine derivatives in the mouse. J Pharm Pharmacol 35:494–499

Bunney BS, Aghajanian GK, Roth RH (1973) Comparison of effects of L-dopa, amphetamine and apomorphine on firing rate of rat dopaminergic neurons. Nature (London) 245: 123–125

Campbell A, Baldessarini RJ, Ram VJ, Neumeyer JL (1982) Behavioral effects of (–)10,11-methylenedioxy-N-n-propylnoraporphine, an orally effective, long-acting agent active at central dopamine receptors, and analogous aporphines. Neuropharmacology 21: 953–961

Carlsson A (1975) Receptor-mediated control of dopamine metabolism in pre- and post-synaptic receptors. In: Usdin E, Bunney WE Jr (eds). Marcel Dekker, New York, pp 49–65

Cohen S, Neumeyer JL (1983) Aporphines 50. Kinetics and solvolysis of N-(2-chloroethyl) norapomorphine, an irreversible dopamine receptor antagonist. J Med Chem 26:1348–1353

Corsini GU, Pitzalis GF, Bernardi F, Bocchetta A, DelZompo M (1981) The use of dopamine agonists in the treatment of schizophrenia. Neuropharmacology 20:1309–1313

Costall B, Fortune DH, Granchelli FE, Law S-J, Naylor RJ, Nohria V (1980a) On the ability of N-chloroethylaporphine derivatives to cause irreversible inhibition of dopamine receptor mechanisms. J Pharm Pharmacol 32:571–576

Costall B, Fortune DH, Law S-J, Naylor RJ, Neumeyer JL, Nohria V (1980b) (–)N-chloroethylnorapomorphine inhibits striatal dopamine function via irreversible receptor binding. Nature (London) 285:571–573

Cote TE, Eskay RL, Frey EA, Grewe CW, Munemura M, Stoof JC, Tsuruta K, Kebabian JW (1982) Biochemical and physiological studies of the beta-adrenoceptor and the D-2 dopamine receptor in the intermediate lobe of the rat pituitary gland. Neuroendocrinology 35:217–224

Creese I, Sibley DR, Hamblin M, Leff SR (1981) Dopamine receptors: Subtypes, regulation and localization. Fed Proc 40:147–152

Creese I, Morrow AL, Leff SE, Sibley DR, Hamblin MW (1982) Dopamine receptors in the central nervous system. Int Rev Neurobiol 23:255–301

Crumley H, Hinshaw W, Pinder R, Goldberg LI (1976) Dopamine-like renal and mesenteric vasodilation by apomorphine, 6-propyl-norapomorphine and 2-amino-6,7-dihydroxy-1,2,3,4-tetrahydronaphthalene. Nature (London) 259–584

Cutler NR, Jeste DV, Karoum F, Wyett RF (1982) Low-dose apomorphine reduces serum homovanillic acid concentration in schizophrenic patients. Life Sci 30:735–756

Davis PS, Maksoud HA, Trainor TM, Vouros P, Neumeyer JL (1985) Stereospecific microbiological 10-0-Demethylation of R(–)-10,11-Dimethoxyaporphines. Biochem Biophysical Research Comm (in Press)

Ernst AM (1966) Mode of action of apomorphine and dextroamphetamine on gnawing compulsion in rats. Psychopharmacologia 10:316

Ernst AM, Smelik PG (1967) Site of action of dopamine and apomorphine on compulsive gnawing behaviour in rats. Experientia 22:837

Goldberg LI, Kohli JD (1983) Differentiation of dopamine receptors in the periphery. In: Kaiser C, Kebabian JW (eds) Dopamine receptors. Am Chem Soc, Washington, p 101

Goldberg LI, Volkman PH, Kohli JD (1978) Vascular dopamine receptors. Ann Rev Pharmacol Toxicol 18:57–79

Goldman ME, Kebabian JW (1984) Apomorphine enantiomers – Interactions with D-1 and D-2 dopamine receptors. Mol Pharmacol 25:18–23

Goldman ME, Beaulieu M, Kebabian JW, Eskay RL (1983) α-Melanocyte-stimulating hormone-like peptides in the intermediate lobe of the rat pituitary gland: characterization of content and release in vitro. Endocrinology 112:435–441

Goosey MW, Doggett NS (1982) The effects of (–)N-chloroethylnorapomorphine on ^{3}H apomorphine binding to striatal dopamine receptors. Biochem Pharmacol 31:3693–3694

Gottlieb R, Neumeyer JL (1976) Aporphines 22. Electrochemical synthesis of aporphines via cathodic cyclization of iodobenzylisoquinolinium salts. J Am Chem Soc 98:7108–7109

Granchelli FE, Filer CN, Soloway AH, Neumeyer JL (1980) Aporphines 27. Mechanistic aspects of the rearrangement of thebaine and codeine analogues in methanesulfonic acid. Improved method for synthesis of N-alkylated aporphines. J Org Chem 45:2275–2278

Guan J-H, Neumeyer JL, Filer CN, Ahern DG, Lilly L, Watanabe M, Seeman P (1984) Aporphines 58. N-(2-Chloroethyl [8,9^3 H]-norapomorphine, an Irreversible Ligand for Dopamine Receptors: Synthesis and Application.J Med Chem 27:806–810

Helmeste DM (1983) Dissociation between behavioral effects of (–)-N-(2-chloroethyl)-norapomorphine and D_2 dopamine-like activity. Eur J Pharmacol 95:277–278

Hjorth S, Carlsson A, Wikstrom H, Lindberg P, Sánchez D, Hacksell U, Arvidsson L-E, Svensson U, Nilsson JLG (1981) 3-PPP, a new centrally acting DA-receptor agonist with selectivity for autoreceptors. Life Sci 28:125–138

Hjorth S, Carlsson A, Clark D, Svenson K, Wikstrom H, Sánchez D, Lindberg P, Hacksell V, Arvidsson L-E, Johansson A, Nilsson JL (1983) Pharmacological manipulation of central dopamine (DA) autoreceptors – biochemical and behavioural consequences. Acta Pharm Suec Suppl 1:130–138

Imbs JL, Schmidt M, Schwarz J (1974) Effets de l'apormorphine sur l'action vasodilatatrice rénale de la dopamine. J Pharmacol 5:579–584

Jackson EA, Neumeyer JL, Kelly PH (1983) Behavioural activity of some novel aporphines in rats with 6-hydroxydopamine lesions of caudate or nucleus accumbens. Eur J Pharmacol 87: 15–23

Kebabian JW, Calne DB (1979) Multiple receptors for dopamine. Nature (London) 277:93–96

Koch MV, Cannon JG, Burkman A (1968) Centrally acting emetics II. Norapomorphine derivatives. J Med Chem 11:977–981

Lal S (1981) Clinical studies with apomorphine. In: Corsini GU, Gessa GL (eds) Apomorphines and other dopaminomimetics, vol II. Clinical pharmacology. Raven Press, New York, pp 1–11

Leff S, Adams L, Hyttel J, Creese I (1981) Kainate lesion dissociates striatal dopamine receptor radioligand binding sites. Eur J Pharmacol 70:71–75

Lehmann J, Langer S (1982) Dopamine autoreceptors differ pharmacologically from post-synaptic dopamine receptors: Effect of (–)-N-(2-chloroethyl)-norapomorphine. Eur J Pharmacol 77:85–86

Lehmann J, Lee CR, Langer SZ (1983) Dopamine receptors modulating ^{3}H acetylcholine release in slices of the cat caudate: Effects of (–)N-(2-chloroethyl)-norapomorphine. Eur J Pharmacol 90:393–400

Levy MI, Davis BM, Mons RC, Mathe AA, Kindler KS, Horváth TB, Hsieh JYK, Davis KL (1982) Apomorphine and schizophrenia: Treatment CSF and neuroendocrine responses. Proc Ann Meet Soc Biol Psychiatry, New Orleans, p 128

Maksoud HM, Kuttab SH, Neumeyer JL, Vouros P (1983) Analysis of N-n-propylnorapomorphine in plasma and tissue by capillary gas chromatography-electron-capture detection. J Chromatogr 274:149–159

Maksoud HM, Neumeyer JL, Vouros P, Davis P (1984) (in preparation)

Manske RHF (1954) In: Manske RHF (ed) The aporphine alkaloids, vol IV. Academic Press, London New York, p 119

Matthiessen A, Wright CRA (1869) Researches into the chemical constitution of narcotine and of its products of decomposition. Proc R Soc 17:337–340

Mustafa AA, Sharma RP, Woodruff GN (1982) Dopamine receptor blocking properties of N-chloroethylaporphine derivatives. Br J Pharmacol 75:39

Neumeyer JL, Neustadt BR, Weintraub JW (1967) Synthesis of aporphines. Tetrahedron Lett 32:3107–3109

Neumeyer JL, Oh KH, Weinhardt KK, Neustadt BR (1969) The chemistry of aporphines. IV. Synthesis of aporphines via Reissert alkylation, photochemical cyclization OR the Pschorr cyclization route. J Org Chem 34:3786–3788

Neumeyer JL, Granchelli FE (1970) Aporphines VI. A novel synthesis of 7-hydroxyaporphine, 7-hydroxynoraporphine and 6a,7-dehydronoraporphine. Tetrahedron Lett 60:5261–5264

Neumeyer JL, Neustadt BR, Weinhardt KK (1970) Aporphines V. Total synthesis of (±)apomorphine. J Pharm Sci 59:1850–1852

Neumeyer JL, Neustadt BR, Oh KH, Weinhardt KK, Boyce CB, Rosenberg FJ, Teiger DJ (1973a) Aporphines 8. Total synthesis and pharmacological evaluation of (±)apomorphine, (±)apocodeine, (±)N-n-propylnorapomorphine and (±)N-n-propylnorapocodeine. J Med Chem 16: 1223–1228

Neumeyer JL, McCarthy M, Battista SP, Rosenberg FJ, Teiger DG (1973b) Aporphines 9. Synthesis and pharmacological evalution of (±)-9,10-dihydroxyaporphine, (±)-isoapomorphine, (+)-, (–)-, and (±)-1,2-dihydroxyaporphine and (+)-1,2,9,10-tetrahydroxyaporphine. J Med Chem 16:1228–1233

Neumeyer JL Granchelli FE, Fuxe K, Ungerstedt U, Corrodi H (1974) Aporphines 11. Synthesis and dopaminergic activity of monohydroxyaporphines. Total synthesis of (±)-11-hydroxyaporphine, (±)-11-hydroxy-N-n-propylnoraporphine. J Med Chem 17:1090–1095

Neumeyer JL, Dafeldecker WP, Costall B, Naylor RJ (1977) Aporphines 21. Dopaminergic activity of aporphines and benzylisoquinoline derivatives. Synthesis of 8-hydroxyaporphines and 1-(hydroxybenzyl)-2-n-propyl-1,2,3,4-tetrahydroisoquinolines. J Med Chem 20:190–196

Neumeyer JL, Law SJ, Baldessarini RJ, Kula NS (1980) Aporphines 30. (–)-N-(2-chloroethyl-norapomorphine) a novel irreversible dopamine receptor antagonist. J Med Chem 23: 594–595

Neumeyer JL, Lal S, Baldessarini RJ (1981a) Historical highlights of the chemistry, pharmacology and early clinical uses of apomorphine. In: Gessa GL, Corsini GU (eds) Apomorphine and other dopaminomimetics, vol I. Basic pharmacology. Raven Press, New York, p 1

Neumeyer JL, Law SJ, Meldrum B, Anlezark G, Watling KJ (1981b) Aporphines 34. (–)-2,10,11-trihydroxy-N-n-propylnoraporphine, a novel dopaminergic aporphine alkaloid with anti-convulsant activity. J Med Chem 24:898–899

Neumeyer JL, Arana GW, Law SJ, Lamont JS, Kula NS, Baldessarini RJ (1981c) Aporphines 36. Dopamine receptor interaction of trihydroxyaporphines. Synthesis, radioreceptor binding and striatal adenylate cyclase stimulation of 2,10,11-trihydroxyaporphines in comparison with other hydroxylated aporphines. J Med Chem 24:1440–1445

Neumeyer JL, Law SJ, Lamont JS (1981d) Apomorphine and related aporphines as probes of the dopamine receptor. In: Gessa GL, Corsini GU (eds) Apomorphine and other dopaminomimetics, vol I. Basic pharmacology. Raven Press, New York, pp 209–218

Neumeyer JL, Arana GW, Ram VJ, Kula NS, Baldessarini RJ (1982) Aporphines 39. Synthesis and dopamine receptor binding and pharmacological activity of (R)-(–)- and S(+)-2-hydroxyapomorphine. J Med Chem 25:990–992

Neumeyer JL, Arana GW, Ram VJ, Baldessarini RJ (1983a) Synthesis and structure activity relationships of aporphines at central dopamine receptors. Acta Pharm Suec Suppl 2:11–24

Neumeyer JL, Reischig D, Arana GW, Campbell A, Baldessarini RJ, Kula NS, Watling KJ. (1983b) Aporphines 48. Enantioselectivity of (R)(–)- and (S)(+)-N-n-propylnorapomorphine on dopamine receptors. J Med Chem 26:516–521

Neumeyer JL, Szabo S (1983) (–)10,11-Methylenedioxy-N-n-propylnoraporphine as orally effective dopamine agonist and duodenal antiulcerogen in the rat. Eur J Pharmacol 88: 273–274

Pschorr R, Jaecke B, Fecht H (1902) Über die Konstitution des Apomorphins. Chem Ber 35: 4377–4392

Ram VJ, Neumeyer JL (1981) Aporphines 35. Synthesis of (R)-(–)- and (S)-(+)-apomorphine from thebaine and (+)-bulbocapnine. J Org Chem 13:2830–2831

Ram VJ, Neumeyer JL, (1982) Aporphines 42. Synthesis of (R)-(–)-11-hydroxyaporphines from morphine. J Org Chem 47:4372–4374

Ram VJ, Neumeyer JL (1984) (in preparation)

Riffee WH, Wilcox RE, Smith RV, Davis PJ, Brubaker A (1982) Inhibition of R(–)-apomorphine-induced stereotypic cage-climbing behavior in mice by S(+)-apomorphine. Adv Biosci 37: 357–362

Robert A, Nezamis JE, Lancaster C, Baldalmenti JN (1974) Cysteamine-induced duodenal ulcers: a new model to test antiulcer agents. Digestion 11:199–214

Saari WS, King SW, Lotti VJ (1973) Synthesis and biological activity of (6aS)-10,11-dihydroxyaporphine, the optical antipode of apomorphine. J Med Chem 16:171–172

Schmidt M, Imbs JL, Neumeyer JL, Giesen EM, Schwartz J (1984) Vascular effects of apomorphine and related compounds in the perfused rat kidney. Eur J Pharmacol 97:75–85

Schwartz JC, Sokoloff P, Matres MP, Protais P, Costentin J, Bouthenet N, Sales N (1983) Characterization of multiple classes of dopamine receptors by biochemical, autoradiographic and behavioral studies. Acta Pharm Suec Suppl 1 1:47–59

Seeman P (1980) Brain dopamine receptors. Pharmacol Rev 32:229–313

Shamma M (1972) The isoquinoline alkaloids – Chemistry and pharmacology. Academic Press, London New York, p 194

Shamma M, Guinaudeau H (1984) Biogenetic pathways for the aporphinoids. Tetrahedron (in press)

Smith RV, Davis PJ (1978) Regiospecific synthesis of isoapocodeine from 10,11-dimethoxyaporphine by using *Cunninghamella elegans*. Appl Environ Microbiol 35:738–742

Sokoloff P, Martres MP, Schwartz JC (1980) Three classes of dopamine receptor (D-2, D-3, D-4) identified by binding studies with ^{3}H-apomorphine and ^{3}H-domperidone. Naunyn-Schmiedebergs Arch Pharmacol 315:89–102

Sperk G, Campbell A, Baldessarini RJ, Stoll A, Neumeyer JL (1982) Tissue levels of N-n-propylnorapomorphine after treatment with (–)10,11-methylenedioxy-N-n-propylnorapomorphine, an orally long-acting prodrug active at central dopamine receptors. Neuropharmacology 21:1311–1316

Szabo S (1979) Dopamine disorder in duodenal ulceration. Lancet 2:880

Szabo S, Neumeyer JL (1983) Dopamine agonists and antagonists in duodenal ulcer disease. In: Kaiser C, Kebabian JW (eds) Dopamine receptors. Am Chem Soc Washington, pp 175–196

Tamminga C, Gotts MD, Miller MR (1983) Dopamine agonist treatment of schizophrenia and neuroleptic-associated dyskinesias. Acta Pharm Suec Suppl 2: 153–158

Tolosa ES, Sparber (1975) Apomorphine in Huntington's chorea: Clinical observations and theoretical considerations. Life Sci 15:1371–1380
Walton KG, Liepmann P, Baldessarini RJ (1978) Inhibition of dopamine-sensitive adenylate cyclase activity by phenoxybenzamine. Eur J Pharmacol 52:231–234

The Chemistry and Pharmacology of Morphinan Alkaloids

A. BROSSI[1]

1 History

The chemistry of morphinan drugs is closely related to that of morphine through recognition of its structure by Robinson and Sugasawa (1933) and Schöpf and Thierfelder (1939). Real progress, however, was made when Grewe and Mondon (1948) reported a novel detailed synthesis of morphinan, the basic unit of the opium alkaloids. This accomplishment, together with the first total synthesis of morphine by Gates and Tschudi (1952), its biosynthesis from reticuline by Barton et al. (1965), the final proof of the stereochemistry of natural morphine proposed by Stork (1952) and Rapoport and Levine (1953) and confirmed by Kalvoda et al. (1955) and Goto's (1964) work on sinomenine, set the stage for an unprecedented effort by Universities, Pharmaceutical Industries and Government Research Laboratories (NIH), to synthesize structures simpler than that of morphine. It was hoped that total synthesis would afford compounds with similar biological features but devoid of unfavourable side-effects and without the addictive properties of morphine. It was expected that the accumulated knowledge would ultimately lead to a practical synthesis of morphine accomplished 30 years later by Rice at the NIH.

2 Correlation of Opium Alkaloids with Morphinan Drugs (Fig. 1)

Thebaine is chemically the most interesting opium alkaloid, readily reduced to dihydrothebainone (Schöpf and Winterhalder 1927), an important relay substance for converting opium alkaloids into morphinan drugs such as levorphanol (Sawa et al. 1961), or for resynthesizing opium alkaloids (Weller and Rapoport 1976). Thebaine also connects with morphinandienones, such as salutaridine (Heinisch and Vieböck 1971). The reduction of thebaine to dihydrothebaine-ϕ (Bentley et al. 1952) and to β-dihydrothebaine (Razdan et al. 1978) made both dihydro analogs available for further transformations (Sawa and Maeda 1964, Razdan and SISA 1981). Oxidation of thebaine with peracids afforded 14-β-hydroxy analogs (Iijima et al. 1977),

[1] Section of Medicinal Chemistry, Laboratory of Chemistry, National Institute of Arthritis, Diabetes, Digestive and Kidney Diseases, National Institute of Health, Bldg. 4, Room 135, Bethesda, Maryland 20205, USA

The Chemistry and Biology of Isoquinoline Alkaloids, ed. by Phillipson et al.

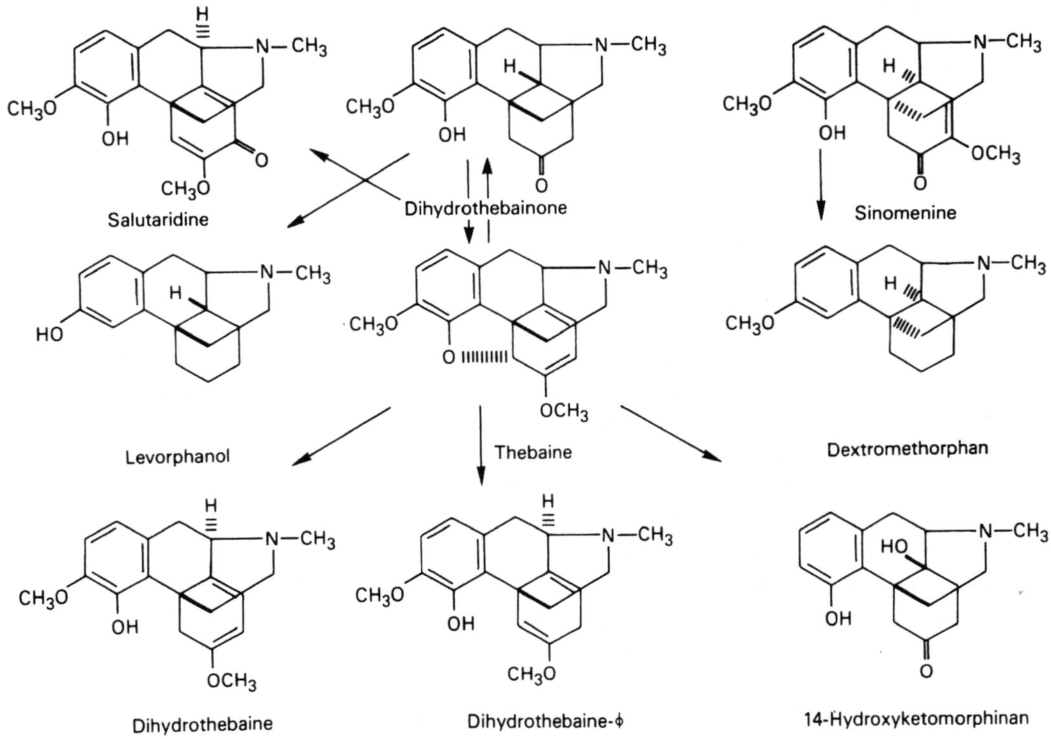

Fig. 1. Correlation of opium alkaloids with morphinan drugs

representing today the most important group of morphinan and opioid drugs and illustrated here by a 14-β-hydroxy substituted ketomorphinan structure (Schmidhammer et al. 1981). The efforts by Goto and Sudzuki (1929) showing that sinomenine and related structures possess a configuration opposite to that of opium alkaloids was confirmed by conversion of sinomenine into dextromethorphan (Sawa et al. 1961) and by the total synthesis of this drug (Corrodi et al. 1959).

3 The Grewe Synthesis of Morphinans

The break-through achieved by Grewe and Mondon by their simple synthesis of (±)-morphinan is shown in Fig. 2. The potential of this synthesis was later explored by cyclization of dihydrobases, prepared from tetrahydroisoquinolines by Birch reduction and treatment with mineral acids (Grewe et al. 1967), to afford 2-hydroxy-substituted ketomorphinans (Grewe and Friedrichsen 1967) and small amounts of the expected (±)-dihydrothebainone. It was shown that the cyclization was greatly dependent on the nature of the N-substituent. The Grewe route to ketomorphinans was further explored (Maeda and Ohsugi 1968, Olieman et al. 1980) and extended to the synthesis of (±)-3-hydroxy-N-methylmorphinan.

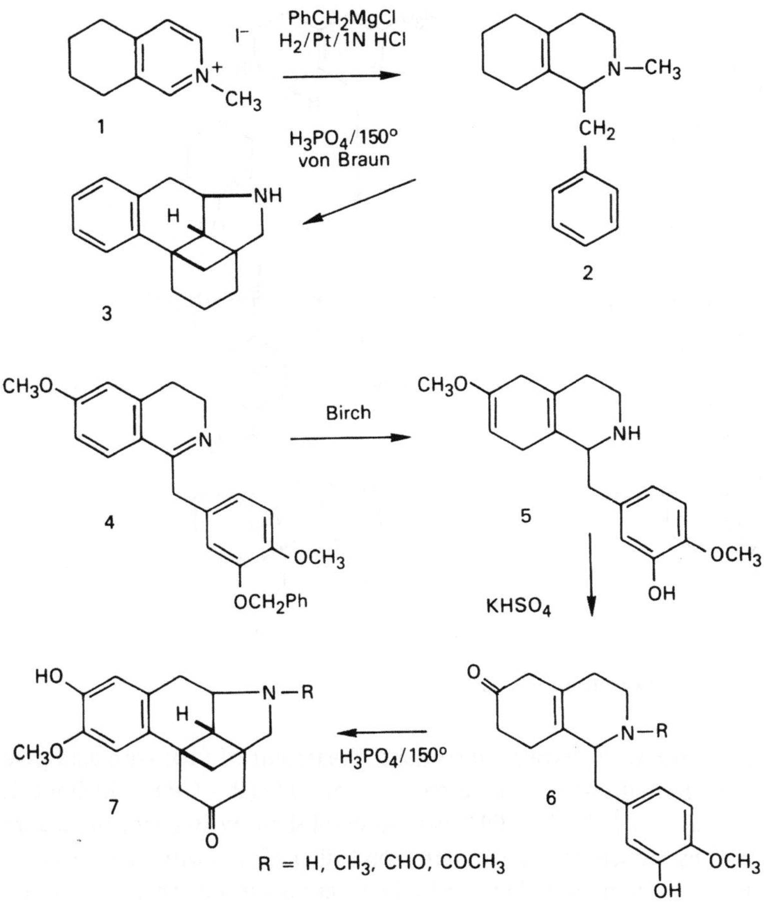

Fig. 2. The Grewe synthesis of morphinans

4 Chemistry Applied to Morphinan Synthesis

Chemical procedures most frequently applied to the synthesis of morphinans are illustrated on Figs. 3 and 4, showing the total synthesis of (-)-N-methylmorphinan (12) (Henecka 1953a, Schmidhammer et al. 1982), the ketomorphinan (13) (Henecka 1953a) and a partial synthesis of "reversed" (-)-dihydrothebainone (Brossi et al. 1982).

Regioselective ether cleavage of the aromatic unconjugated ether group (Brossi and Teitel 1970, Schmidhammer and Brossi 1982) in the isoquinoline (8) afforded after reduction and chemical resolution the tetrahydroisoquinoline (9), (TIQ). Birch reduction of (9) with Li in liquid ammonia and tert.-butanol as a hydrogen donor (Beyerman et al. 1976a), afforded after N-formylation the dihydro-derivative (10). Preferential use of (10) for the Grewe cyclization was suggested by Leimgruber's rule (Leimgruber 1972), demonstrating that this cyclization is facilitated when the amino

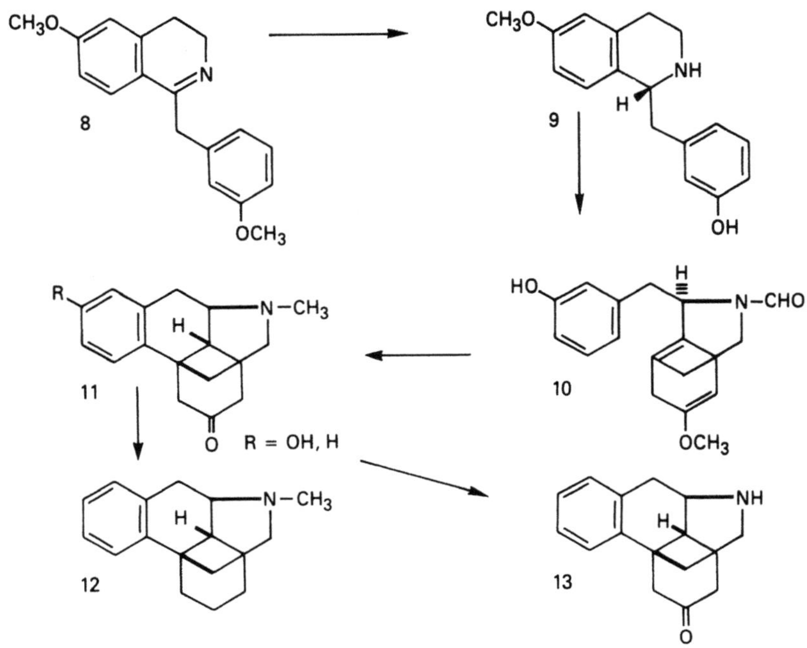

Fig. 3. Chemistry useful in morphinan synthesis

group carries an electron-withdrawing substituent. Treatment of (10) with acid gave after hydrolysis and N-methylation the ketomorphinan (11) (R = OH) in high yield. Phenolic deoxygenation of (11) (R = OH) was accomplished by Bognar's procedure (Bognar et al. 1974) by reducing the phenyltetrazolyl ether of (11) with Pd/C in acetic acid. Other methods to eliminate phenolic hydroxy groups in morphinans were reported by reduction of diphosphate esters (Manmade et al. 1982) or phenyl ethers (Sawa and Tada 1968) with Li in liquid ammonia, both requiring protection of the carbonyl group. The carbonyl group in the ketomorphinan (11) (R = H) was removed by a Wolff-Kishner reduction affording (12). N-Demethylation of (11) (R = H) was accomplished with trichloroethoxycarbonyl chloride (Rozwadowska et al. 1980), a variant of the von Braun demethylation affording, after treatment of the carbamate with Zn/NH_4Cl in refluxing methanol, the ketomorphinan (13). Opening of the epoxide ring in an opioid, prepared from dihydromorphinone (Reden et al. 1979) with Zn/NH_4Cl in methanol, afforded after standard N,O-functionalization the "reversed" dihydrothebainone, as shown in Fig. 4.

5 The Roche Contribution

The report by Schnider and Grüssner at Roche, Basel, that (±)-3-hydroxy-N-methylmorphinan, prepared with Grewe chemistry, afforded after optical resolution antipodes with different pharmacological profiles (Schnider and Grüssner 1951) was fol-

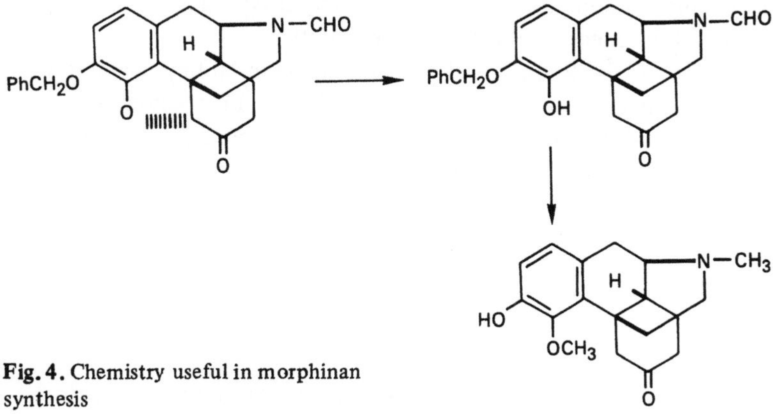

Fig. 4. Chemistry useful in morphinan synthesis

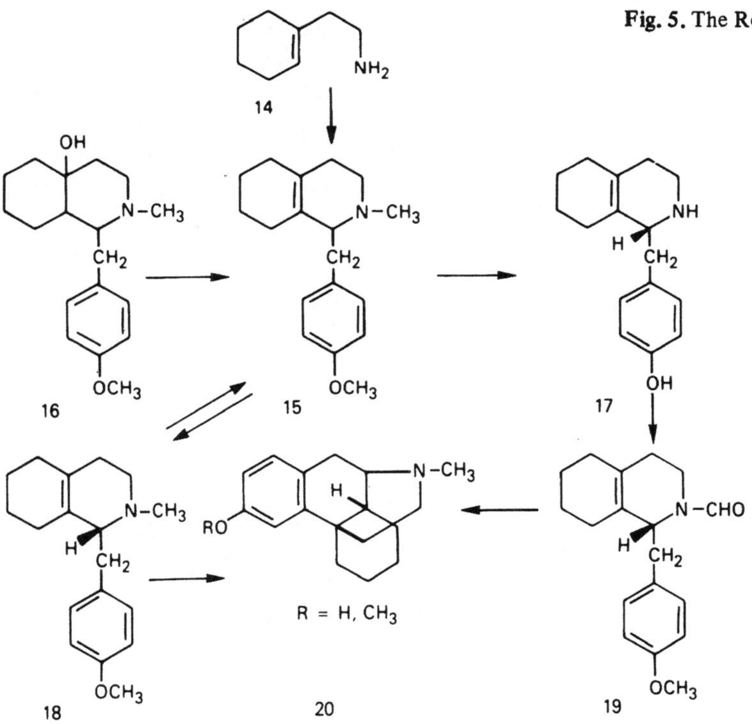

Fig. 5. The Roche contribution

lowed with a concerted effort to make them available by easier methods. Figure 5 summarizes the results obtained, which are discussed in detail in Hellerbach's excellent review on morphinan analgesics (Hellerbach et al. 1966).

Cyclohexenylethylamine (14) (Schnider and Hellerbach 1950) provided a facile route to octahydroisoquinolines, such as (15), and its optically active analogs (17), (18) and (19) with morphine configuration. Racemization of (18) was achieved by refluxing the unwanted optical isomer in presence of a metal catalyst (Hindely 1961).

Another route to octahydroisoquinolines from the 10-hydroxy substituted isoquinoline (16), as reported by Henecka (1953b) lacks versatility.

Further improvements of the Roche synthesis were made in Nutley where it was found that the Grewe cyclization of the optical active N-formyl protected isoquinoline (19) took place below 100°, and afforded the methyl ether (20) (R = CH_3) in very high yield after hydrolysis of the N-formyl group and N-methylation (Leimgruber 1972).

6 Isomorphinans

Morphinans with trans-B/C rings are called isomorphinans. They are accessible by syntheses developed by Gates (Gates and Webb 1958) and by Belleau at Bristol in Canada (Monkovic et al. 1975a). Figure 6 shows that the Gates ketoamide (21) was first converted into the dehydromorphinan (22) by LAH reduction and N-methylation. Catalytic reduction, optical resolution and ether cleavage afforded (23), shown here as the (-)-isomer. 14-Hydroxy-substituted trans-morphinans as represented by (26), were obtained by an elegant synthesis starting with 7-methoxy-2-tetralone. The carbamate (24) was converted with peracid into the expoxide (25) which upon treatment with NaH in benzene in the presence of amyl alcohol afforded (26) ($R_1 = CH_3$,

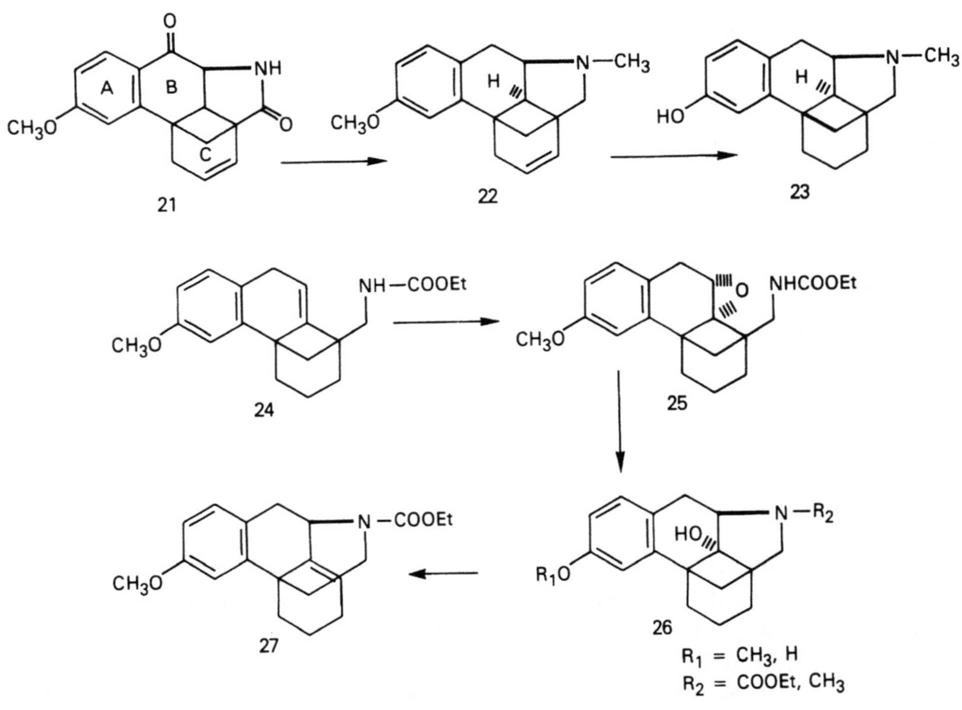

Fig. 6. Isomorphinans

R_2 = COOEt). Reduction of (26) with LAH and ether cleavage with boron tribromide yielded (±)-(26) (R_1 = H, R_2 = CH_3). Dehydration of the carbamate (26) with phosphorous oxychloride in pyridine afforded the interesting 8,14-dehydromorphinan (27), connecting the trans series with the 14-hydroxy-substituted cis-series. Only few isomorphinans were resolved into optical isomers and adequately tested.

7 The Bristol Synthesis of 14-β-Hydroxymorphinans

The amine (28), prepared from 7-methoxytetralone-2, provided an entry into the important class of 14-hydroxymorphinans of the B/C cis-series (Monkovic et al. 1973, 1975b). This is shown in Fig. 7 by a synthesis of the drug oxilorphan. Bromination of (28) in chloroform afforded the bromohasubanan derivative (29) which underwent ring enlargement when heated with sodium bicarbonate in DMF at 130° to afford the dehydromorphinan (30) converted into the cyclopropylcarbonyl derivative (31). Epoxidation of (31) with peracid gave (32) and reduction with LAH afforded the methyl ether of (±)-(33). Ether cleavage with boron tribromide and optical resolution with d-tartaric acid completed the synthesis of (-)-(33), introduced as a narcotic antagonist under the name oxilorphan. A serious drawback of the Bristol synthesis is the optical resolution step at a very late stage yielding unwanted optical isomers of no use.

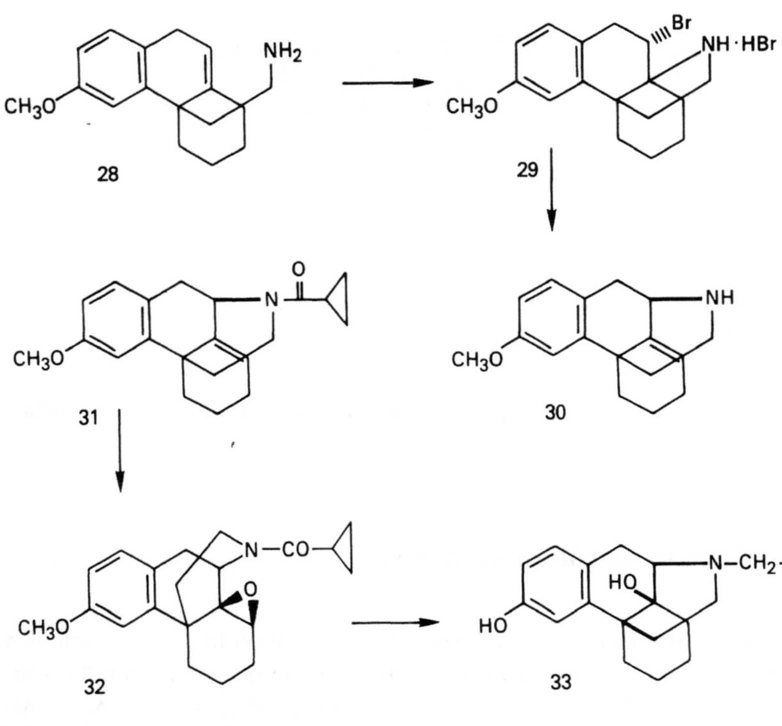

Fig. 7. The Bristol synthesis of oxilorphan

8 The Beyerman Total Synthesis of (-)-Dihydrothebainone

The Beyerman synthesis was first completed in the N-formyl series (Beyerman et al. 1976a), including studies of an efficient conversion of such compounds into dihydrothebainone (Beyerman et al. 1976b). Since the Grewe synthesis afforded mainly 2,3-dioxygenated ketomorphinans (Maeda and Ohsugi 1968), Beyerman used appropriately substituted TIQ with three oxygen functions in the benzylic part. It was hoped that Birch reduction, Grewe cyclization and removal of the 2-hydroxy group would ultimately yield the desired ketone. Such a synthesis starting from optically active TIQ with the required configuration is shown in Fig. 8 (Lie et al. 1978a). This elegant approach was successfully completed and afforded, in addition to (-)-dihydrothebainone, several novel trioxy-substituted ketomorphinans required for structure-activity studies.

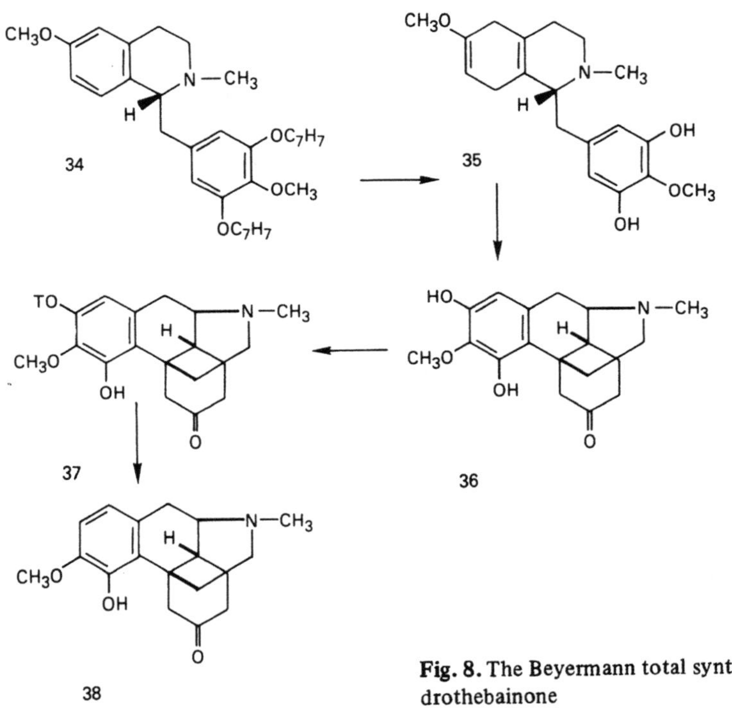

Fig. 8. The Beyermann total synthesis of natural dihydrothebainone

9 Morphinandienones and Morphinandiones

Morphinandienones, recognized as intermediates in the biosynthesis of opium alkaloids (Barton et al. 1965), were obtained in vitro by processes mimicking the plant synthesis (Schwartz and Mami 1975, Szántay et al. 1982, White et al. 1983). When N-formylnorreticuline (39) (R_1 = CHO, R_2 = H) was brominated at C-6′ to (39)

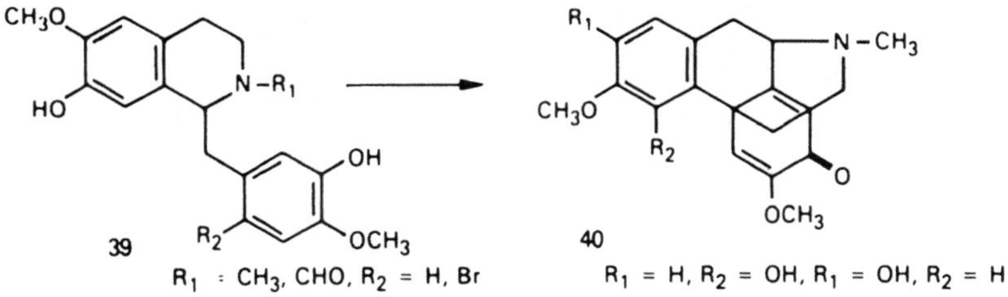

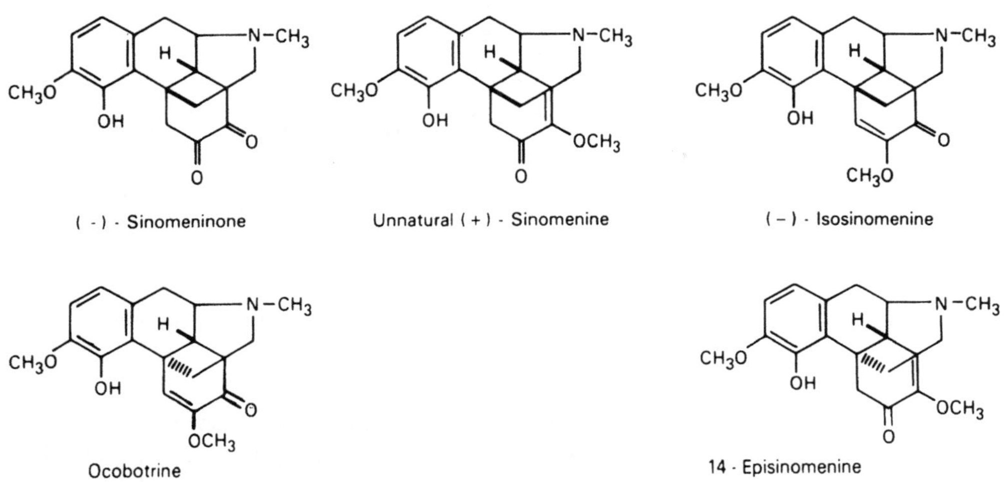

Fig. 9. Morphinandienones and morphinandiones

(R_1 = CHO, R_2 = Br), the oxidative coupling reaction afforded after deformylation and N-methylation (±)-palladine (40) (R_1 = OH, R_2 = H) instead of (±)-salutaridine (40), (R_1 = H, R_2 = OH) (Blasko et al. in press).

The variety of morphinan alkaloids of this class is illustrated in Fig. 9, by the unnatural (-)-sinomeninone, its two enol ethers (Minamikawa and Brossi 1978) and ocobotrine and 14-eposinomenine, alkaloids of the B/C transseries (Vecchietti et al. 1976). This group of alkaloids will undoubtedly further increase and sooner or later selected representatives will become available for pharmacological screening.

10 Sinomenine

The alkaloid from *Sinomenium acutum* has a (-)-rotation but belongs to a group of alkaloids with the antipodal configuration of opium alkaloids. Considerable work with sinomenine was carried out in Japan, particularly by Goto and his colleagues (Goto 1964) and its conversion into (+)-dihydrothebainone, sinomeninone and isosinomenine was achieved (Goto and Nambo 1930) (Fig. 10). Although no clinically useful drug has emerged from this effort, the development of dextromethorphan as a non-narcotic antitussive agent by Roche was clearly stimulated by sinomenine's

Fig. 10. Sinomenine

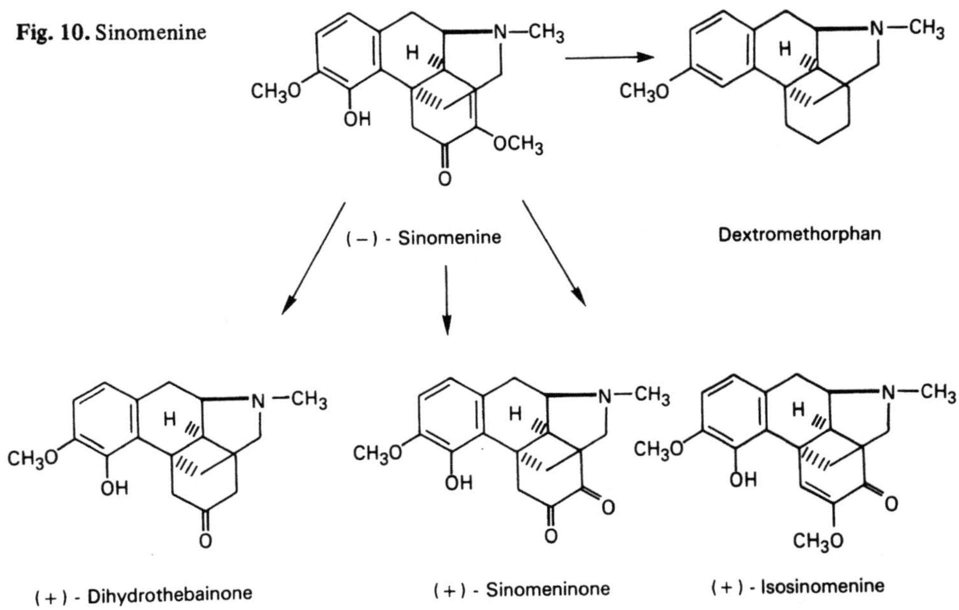

structure. The availability of (+)-dihydrothebainone, later accomplished with better procedures (Iijima et al. 1978a), made it possible to investigate several unnatural (+)-opioids as neurochemical tools (Jacquet et al. 1977, Iijima et al. 1978b). It is interesting to note that sinomenine, only recently tested in an antitussive screen (Nakamura personal communication) proved superior to dextromethorphan and codeine, suggesting that antitussive agents may still be discovered in the (+)-series of morphinan alkaloids (Kerekes et al. in press).

11 The Rice Synthesis of (-)-Dihydrothebainone

Rice reports in Chapter 12 a practical synthesis of optically active ketomorphinans with natural or unnatural configuration (Rice 1980, 1983a). The crucial step was the successful cyclization of the optically active β,γ-unsaturated ketone (41), substituted

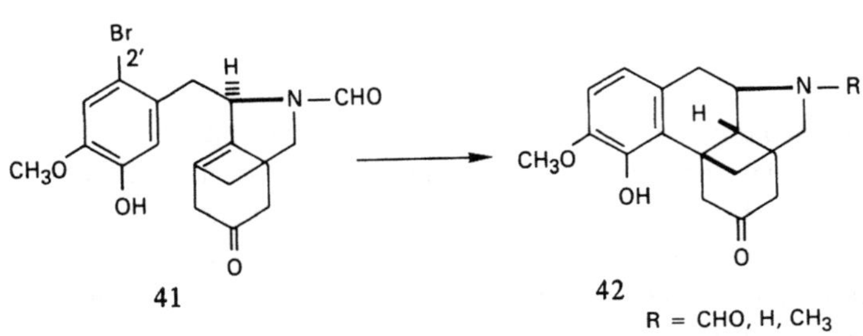

Fig. 11. The Rice synthesis of (-)-dihydrothebainone

at C-2′ with bromine, with superacid to the optically active N-formyl-bromonordihydrothebainone (42) (R = CHO), as shown in Fig. 11. Catalytic debromination and acid hydrolysis afforded nordihydrothebainone (42) (R = H) and after N-methylation dihydrothebainone (42) (R = CH_3). The Rice synthesis has been scaled-up and is workable, providing for the first time an entry into the largely unexplored class of unnatural morphinan alkaloids.

12 Aromatic Differently Substituted Morphinans

The synthesis of 2-hydroxy- and aromatic unsubstituted ketomorphinans was discussed previously (Fig. 4). The 4-hydroxymorphinan (44) (R = CH_3), shown in Fig. 12 was first obtained from morphine via 3-deoxydihydromorphinone (43) (Manmade et al. 1982, Hsu et al. 1979) and later in racemic form by total synthesis from (45) (Hsu et al. 1982, Brossi et al. 1982) or equivalents of (45) (Hsu et al. 1982). The ketomorphinan (44) (R = H, CH_3) has been obtained recently by an improved total synthesis in the form of both optical isomers (Kerekes et al. in press). The N-formyl morphinan (46) prepared from (43) (Rozwadowska et al. 1980) allowed an entry into 1-hydroxymorphinans via the quinone (47), followed by elimination of the 4-oxygen function (Schmidhammer and Brossi 1983). Most ketomorphinans were decarbonylated by the Wolff-Kishner reaction affording morphinans prepared differently (Simon et al. 1981). The so-called 1-hydroxymorphinan prepared by a Grewe synthesis was in fact a hydroxy-substituted apomorphinan (Schmidhammer et al. 1983a).

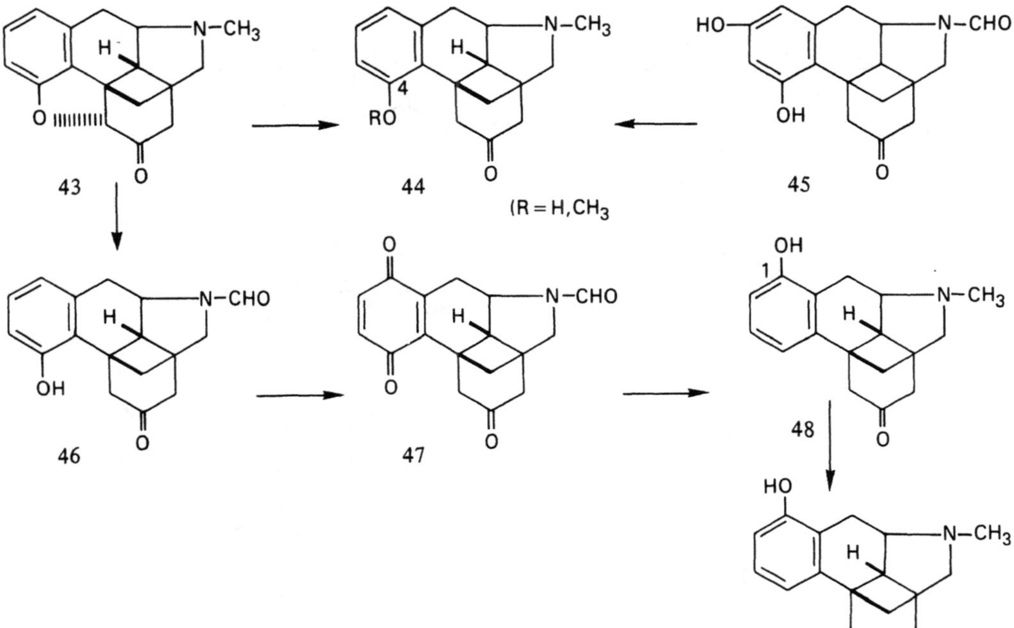

Fig. 12. Aromatic differently substituted 6-ketomorphinans

Several aromatic di-oxygenated morphinans and ketomorphinans were prepared and biologically evaluated (Schmidhammer et al. 1983b). Only the 3,4-disubstituted morphinans (Rahman and Brossi 1977) and their 6-keto-analogs (Brossi et al. 1982) showed potent antinociceptive properties.

13 Metabolic Conversions of Morphinan Drugs

Detailed information on the metabolism of clinically tested morphinan analgesics and antitussives was obtained by chromatographic analysis (Hellerbach et al. 1966, Brossi et al. 1955) and by mass-spectral, NMR, GC and X-ray analysis (Mannering and Schanker 1958, Blount et al. 1973). The data revealed that the major pathway of degradation of N-alkylated morphinan drugs, independent of configuration and aromatic substitution, involved N-dealkylation leading to nor-compounds. With dextromethorphan O-demethylation also occurred, (51), being the major metabolite (Brossi et al. 1955). Phenolic metabolites and phenolic drugs are often excreted as conjugates, particularly in dogs (Blount et al. 1973). Interestingly, the 6-β-alcohol (54), shown in Fig. 13, was found to be a minor metabolite of levallorphan (Mannering and Schanker 1958, Blount et al. 1973), whereas the 7-β-alcohol (55) was a minor metabolite of (-)-2-hydroxy-N-cyclopropylmethylmorphinan (Vane et al. 1978). Metabolites of (+)-2-hydroxy-substituted morphinans, such as (52), were found to be the catechol (53) (R = H) and its O-methyl ether (53) (R = CH_3) (Vane et al. 1978). It should be mentioned that tert. butyl ether of levorphanol did not metabolize to levorphanol and was metabolically stable (Mohacsi et al. 1982).

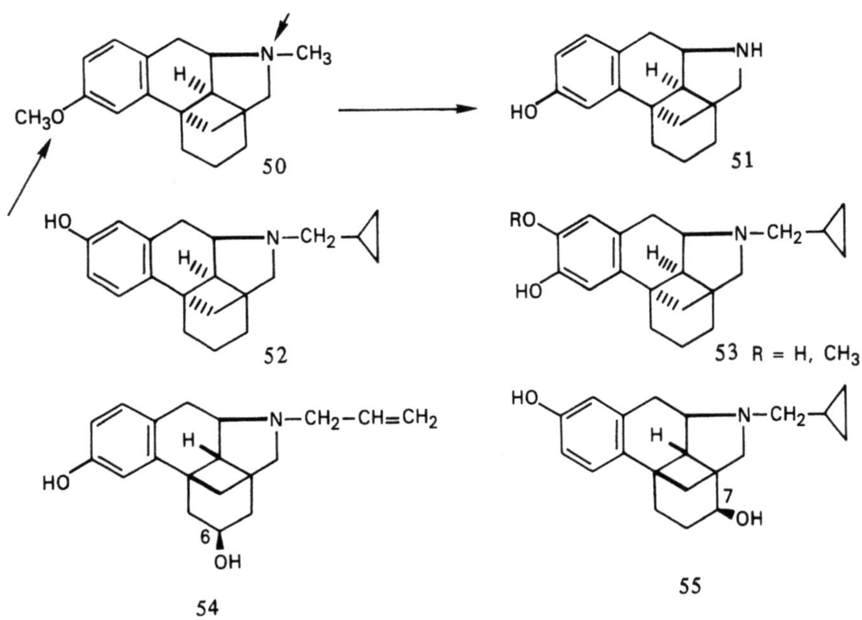

Fig. 13. Metabolic conversions of morphinans drugs

14 Pharmacology of Morphinan Drugs

Affinity-binding to opiate receptors from vertebrate brain in the presence and absence of NaCl (Pert et al. 1973, Pert and Snyder 1974), with simultaneous evaluation of the drugs in the classical screening in vivo (hot-plate, tail-flick, writhing tests (Jacobson 1978, Hite 1981), afforded data which allowed grouping of these compounds into agonists, mixed agonist-antagonists, and antagonists. Based on comparative studies, often incomplete, and frequently obtained with racemic mixtures, the following conclusions can be drawn:

Opioids with agonist activity have N-methyl or N-phenethyl-substitution. Opioids with antagonist activity are obtained by replacing the N-methyl group with a propyl, allyl, propargyl, cyclopropylmethyl, cyclobutylmethyl, dimethylallyl or furylmethyl group (Merz et al. 1974). The presently marketed morphinan drugs are shown in Fig. 14. The newest drug, butorphanol, has been tested clinically and found to be a highly potent mixed agonist-antagonist, showing in animal screening low physical dependence liability and fewer hallucinogenic side effects in comparison to levallorphan or cyclorphan (Gordon 1974).

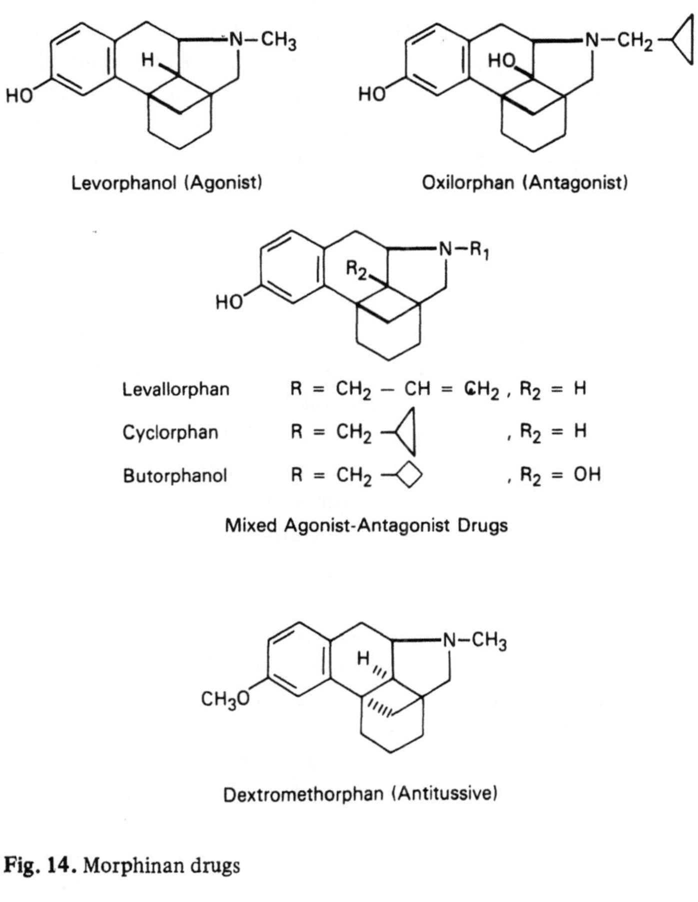

Fig. 14. Morphinan drugs

14.1 Stereochemistry

Narcotic agonists and antagonists bind to the μ-opiate receptor enantioselectively and only the optical isomers with the configuration of morphine are recognized. Dextromethorphan is at present the only active (+)-morphinan drug with stereochemistry antipodal to that of opioids. Dextromethorphan is a non-narcotic antitussive agent and the best of many analogs prepared at Roche (Hellerbach et al. 1966). Recent developments in our laboratory suggest that other (+)-morphinans may have similar biological properties (Kerekes et al. in press).

Isolevorphanol with a trans B/C ring was reported to be several times more potent than morphine (Gates and Webb 1958), whereas isocyclorphan was analgesically much less potent than cyclorphan, but retained the antagonistic effects (Belleau 1982).

14.2 14-OH-Opioids

The introduction of a hydroxy group at C-14 afforded cis and trans isomers (β and α in optically active compounds). Studies carried out mostly with racemic mixtures are incomplete (Belleau 1982). A great number of 14-β-OH substituted ketomorphinans were evaluated (Rahman and Brossi 1977, Seki et al. 1964, Schmidhammer et al. 1981) and it was found that β-hydroxylation at C-14 did not drastically alter the opioid agonist activity. Conversion of a 14-β-OH into a 14-β-OMe group, on the other hand, afforded compounds with very high antinociceptive potency accompanied by considerable opioidtype side-effects (Schmidhammer et al. in press).

14.3 Aromatic Substitution

Modification of biologically active molecules by chemists was often guided by convenience. The development of (-)-3-phenoxy-N-methylmorphinan as a codeine substitute at Roche (Mohacsi et al. 1979), or (-)-3-acetyl-N-methylmorphinan as another narcotic agonist (Mohacsi et al. 1982), are typical examples.

The results summarized in Fig. 15 show that aromatic substitution in N-methylmorphinans is important and optimal in 3-hydroxy-N-methylmorphinan (levorphanol). Aromatic substitution is equally important in the ketomorphinan series where high agonist potency was observed with the aromatic unsubstituted compound reaching the potency of levorphanol with the 4-methoxy-substituted analog. The 3,4-dimethyl ether derivative was also found to be highly potent, but the potency decreased considerably when one of the two methoxy groups was cleaved, or the dimethoxy arrangement replaced by a methylenedioxy group (Raman and Brossi 1977). Reduction of the carbonyl group in agonist ketomorphinans to alcohols abolished the activity almost completely (Jacobson et al. 1981). From the data available on aromatic differently substituted morphinans (Simon et al. 1981) and ketomorphinans (Schmidhammer et al. 1983a), it can be concluded that the area marked by carbon atoms C-3 to C-6 in the morphinan structure plays an important role in the process leading to antinociception. Work with antagonist structures will show whether this molecular hole is of similar importance in this series of analgesics.

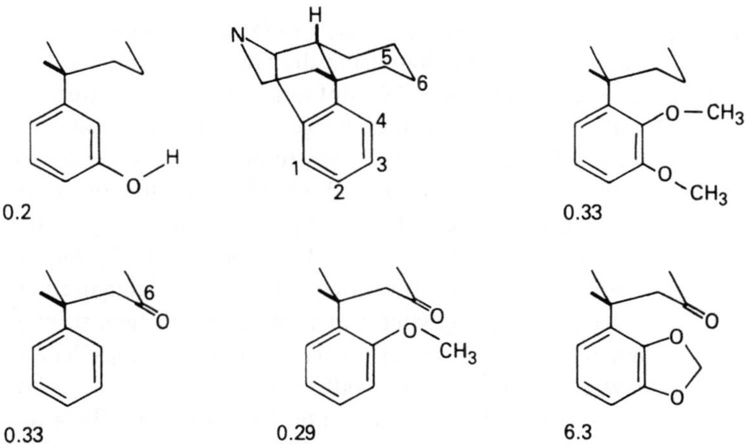

Fig. 15. Analgesic potencies (ED_{50} by hot plate assay, sc mg/kg, mice, morphine = 1.0)

14.4 Other Substitutions in the Morphinan Skeleton

Substitution at C-10 of N-methylmorphinans with hydroxy or carbonyl groups, resulted in considerable loss of antinociception (Häfliger et al. 1956, Rapoport and Masamune 1955). Similar structural modifications have, however, to be extended to antagonist compounds before definitive conclusions can be made. Several 14-OMe-N-methylketomorphinans, substituted at C-8 (Polazzi et al. 1980) or C-7 (SISA 1983) with alkyl groups were prepared by lengthy syntheses, but these compounds do not offer any clinical advantage over already existing drugs. It remains to be seen whether the recently synthesized 14-exo-ethenomorphinans (Ghosh et al. 1983), focussing attention to the area around the C-6/C-7 carbons of morphinans, will provide useful information.

N-Substitution of morphinans with chemically unique side chains was obtained by the addition of acrylonitrile to norlevorphanol (Jacobson et al. 1979) affording a N-cyanoethylated morphinan which did bind well to opiate receptors, had little acute toxicity, and did not substitute for morphine in morphine-dependent monkeys. This result suggests that further modification of the N-side chain in morphinans and opioids may lead to interesting compounds.

15 Mode of Action of Opioids

We still do not know how opioid agonists and antagonists function on a molecular level and what molecular features, if any, are responsible for side effects and addiction. Ongoing work with highly selective irreversible ligands pledging the binding to different types of receptors (Rice et al. 1983) may reveal useful information about these fundamental questions. It is hoped that precise studies will address the metabolic degradation of opioids in a more quantitative manner.

The interesting suggestion made by Belleau et al. (1974) and Feinburg et al. (1976) that opioids may adopt a conformation at the drug-receptor complex with the nitrogen substituent in an equatorial or axial conformation, with an apparent requirement for an equatorial configuration of the nitrogen substituent in opioids is still not answered satisfactorily (Hori et al. 1983). It has to be questioned seriously whether the data recently provided by Kobylecki et al. (1982) with quaternary derivatives, chemically and biologically quite different species, usefully support the aforementioned conclusion. It is difficult to understand why an N-propyl group in antagonist opioids should have the N-propyl substituent in an axial configuration. The old theory by Beckett et al. (1956) suggesting that agonists and antagonists may N-dealkylate differently, implemented in Belleau's theory on clastic binding (Belleau and Morgan 1974), is still a valuable concept. Steric rather than electronic effects imposed by the substitution on nitrogen may be more important. The N-dealkylation process, involving highly specific enzymes and occurring on a drug-protein complex, may afford nor-compounds with opioid but not with pure antagonist compounds, and offer a possible rationale for the biological differences of these two classes of drugs.

Studies on receptor recognition involving morphine derivatives were based, in part, on the assumption that these structures would not change when the crystal was put into solution, even though there is only limited experimental support for this case (Okuda et al. 1964, Glasel 1982). It has now been shown that the structure of morphine differs between the crystalline state and in aqueous solution (Brown et al. 1983), suggesting that the receptor in the central nervous system probably recognizes morphine and its analogues with conformations different to that present in the crystal.

Acknowledgement. I would like to thank Mrs. Kathy Carter for the technical assistance in the preparation of this manuscript.

References

Barton DHR, Kirby GW, Steglich W, Thomas GM, Battersby AR (1965) Investigations on the biosynthesis of morphine alkaloids. J Chem Soc:2423–2438

Beckett AH, Casy AF, Harper NJ (1956) Analgesics and their antagonists: some steric and chemical considerations. J Pharm Pharmacol 6:874–884

Belleau B (1982) The chemical regulation of biological mechanisms. In: Creighton AM, Turner S (eds) R Soc Chem, Spec Publ no 42. Burlington House, London, pp 200–221

Belleau B, Morgan P (1974) Clastic binding on the opiate receptor. J Med Chem 17:908–909

Belleau B, Conway T, Ahmed FR, Hardy FR (1974) Importance of the nitrogen lone electron pair orientation in stereospecific opiates. J Med Chem 17:907–908

Bentley KW, Robinson R, Wain AE (1952) The reduction of thebaine and dihydrothebainone by sodium and ammonia. J Chem Soc:958–966

Beyerman HC, Lie TS, Maat L, Buurman E, Bijsterveld EJM, Sianige HJM (1976a) A convenient synthesis of codeine and morphine. Recl Trav Chim Pays-Bas Belg 95(1):24–25

Beyerman HC, Bommel L van, Maat L, Olieman C (1976b) Recl Trav Chim Pays-Bas Belg 95:312

Beyerman HC, Leeuwen FF van, Lie TS, Maat L, Olieman C (1976c) Chemistry of opium alkaloids, part VI, A practical technique for laboratory Birch reductions. Recl Trav Chim Pays-Bas Belg 95(10):238–240

Blasko G, Börnyei G, Barczai-Beke M, Pechy P, Szantay C (1984) J Org Chem (in press)
Blount JF, Mohacsi E, Vane FM, Mannering GJ (1973) Isolation, X-ray analysis and synthesis of a metabolite of (-)-3-hydroxy-N-allylmorphinan. J Med Chem 16:352–355
Bogńar R, Gaál Gy, Kerekes P, Horváth G, Kovacs MT (1974) Hydroxyl group elimination in the morphine series. Org Prep Proc Int 6(6):305–311
Brossi A, Teitel S (1970) Partial O-demethylation of aromatic substituted 3,4-dihydroisoquinolines. Helv Chim Acta 53:1779–1787
Brossi A, Häfliger O, Schnider O (1955) Oxy-morphinane 6. Mitt. Die papierchromatographische Bestimmung von Morphinanderivaten und die Verfolgung ihrer Ausscheidung beim Hund. Arzneim. Forsch 5:62–66
Brossi A, Atwell L, Jacobson AE, Rozwadowska MD, Schmidhammer H, Flippen-Anderson JL, Gilardi R (1982a) Structure-activity relationship of oxygenated morphinans VII: 5-methylated and 14-hydroxysubstituted agonists and antagonists of the 4-hydroxy- and 3,4-dioxygenated 6-morphinanone series. Helv Chim Acta 65:2394–2404
Brossi A, Hsu FL, Rice KC (1982b) A simplified synthesis of (±)-4-hydroxy-N-methylmorphinan-6-one. J Org Chem 47:5214–5216
Brown CE, Roerig SC, Fujimoto JM, Burger VT (1983) The structure of morphine differs between the crystalline state and aqueous solution. J Chem Soc Chem Commun:1506–1508
Corrodi H, Hellerbach J, Züst A, Hardegger E, Schnider O (1959) Hydroxy-morphinane. Die Konfiguration der Morphinane. Helv Chim Acta 42:212–217
Feinburg AP, Creese I, Snyder SH (1976) The opiate receptor: a model explaining structure-activity relationships of opiate agonists and antagonists. Proc Natl Acad Sci USA 73:4213–4219
Gates M, Tschudi G (1952) The synthesis of morphine. J Am Chem Soc 74:1109–1110
Gates M, Webb WG (1958) The synthesis and resolution of 3-hydroxy-N-methylisomorphinan. J Am Chem Soc 80:1186–1194
Ghosh AC, Portlock DE, Dalzell HC, Malberg C, Herligy P, Razdan RK (1983) Diels-Alder reaction of ß-dihydrothebaine and its 4-phenyl ether with methyl vinyl ketone: synthesis of 6,14-exoethenomorphinans. J Org Chem 48:4137–4139
Glasel JA (1981) A comparison of solution solid state and theoretical conformations of morphine. Biochem Biophys Res Commun 102:703–709
Gordon M (1974) Annu Rep Med Chem 9:38–50
Goto K (1964) Sinomenine – an optical antipode of morphine alkaloids. Kitasato Inst, Tokyo
Goto K, Nambo T (1930) Sinomenine and disinomenine. Part XVI. On isobromo-sinomenine (or bromo-sinomeneine). Bull Chem Soc Jpn 5:165–169
Goto K, Sudzuki H (1929) Sinomenine and disnomenine. Part XI. On the position of the double linking in sinomenine. Bull Chem Soc Jpn 4:244–254
Grewe R, Friedrichsen W (1967) Cyclization of octahydroisoquinoline derivatives by morphinan ring closure: synthesis of dihydrothebainone. Chem Ber 100:1550–1558
Grewe R, Mondon A (1948) Synthesen in der Phenanthren-Reihe, vi.M.H: Synthese des Morphinans. Chem Ber 81:279–286
Grewe R, Fischer H, Friederichsen W (1967) Partial Birch reduction of 1-benzylisoquinoline derivatives. Chem Ber 100:1–8
Häfliger O, Brossi A, Chopard-dit-Jean LH, Walter M, Schnider O (1956) Hydroxy-Morphinane. Über ein Photooxydationsprodukt von (+)-3-Methoxy-N-methyl-morphinan. Helv Chim Acta 39:2053–2062
Heinisch G, Viebóck F (1971) Preparation and reactions of 4-O-methylsalutaridine methoperchlorate. Monatsschr Chem 102:775–780
Hellerbach J, Schnider O, Besendorf H, Pellmont B (1966) In: Synthetic analgesics, part II(A). Morphinans, organic chemistry, vol VIII. Pergamon Press, New York, pp 1–112
Henecka H (1953a) Neue Synthesen in der Morphinanreihe. Ann Chem 583:110–129
Henecka H (1953b) Neue Synthesen in der Morphinanreihe. Liebigs Ann 583:110–128
Hindley NC (1961) Hoffmann-La Roche, Br Pat 832025
Hite GJ (1981) In: Foye WO (ed) Principles in medicinal chemistry. Lea and Febiger, Philadelphia, pp 261–302

Hori M, Katoaoka T, Shimizu H, Imaf E, Suzuki Y, Kawamura N, Fujimura H, No Zak M, Niwa M (1983) Agonist-antagonist properties of 5,7-ethano-4,5,5a,6,7,11b-hexahydro-2,6,7-trimethyl-1H-benzo(g)homoquinolin-9-ol and 4,6-ethano-3,4,4a,5,6,10b-hexahydro-2,5,6-trimethylbenzo(f)quinolin-8-ol. Chem Pharm Bull 31:2520-2522

Hsu FL, Jacobson AE, Rice KC, Brossi A (1979) Partial synthesis of 3-deoxydihydromorphine from (-)-4-hydroxy-6-keto-N-methylmophinan. Heterocycles 13:259–261

Hsu FL, Rice KC, Brossi A (1982) Total synthesis of (±)-3-deoxy-7,8-dihydromorphine, (±)-4-methoxy-N-methylmorphinan-6-one and 2,4-dioxygenated (±)-congeners. Helv Chim Acta 65: 1576–1589

Iijima I, Rice KC, Brossi A (1977) The oxidation of thebaine with m-chloroperbenzoic acid. Studies in the (+)-morphinan Ser 3. Helv Chim Acta 60:2135–2137

Iijima I, Minamikawa J, Jacobson AE, Brossi A, Rice KC (1978a) Studies in the (+)-morphinan series. 5, Synthesis and biological properties of (+)-naloxone. J Med Chem 21:398–400

Iijima I, Minamikawa J, Rice KC, Jacobson AE, Brossi A (1978b) Studies in the (+)-morphinan, Ser 4: A markedly improved synthesis of (+)-morphine. J Org Chem 43:1462–1463

Jacobson AE (1978) In: Barnett G, Trsic M, Willette R (eds) Quasar, Res Monogr 22. Natl Inst Drug Abuse, pp 129–145

Jacobson AE, Rice KC, Reden J, Lupinacci L, Brossi A, Streaty RA, Klee W (1979) Paradoxical effects of N-cyanoalkyl substituents upon the activities of several classes of opioids. J Med Chem 22:328–331

Jacobson AE, Schmidhammer H, Hsu FL, Rozwadowska MD, Atwell L, Aceto MD, Harris LS, Katz JL, Woods JH, Medzirhadsky F (1981) Structure-activity relationships of oxygenated morphinans III: an exploration of the effect of the aromatic oxygen and 6-keto group on antinociceptive activity; receptor affinity; and narcotic antagonism. NIDA Res Monogr 41: 86–92

Jacquet YF, Klee WA, Rice KC, Iijima I, Minimikawa J (1977) Stereospecific and nonstereospecific effects of (+)- and (-)-morphine: evidence for a new class of receptors? Science 198: 842–845

Kalvoda J, Buchschacher P, Jeger O (1955) Über die absolute Konfiguration des Morphins und verwandter Alkaloide. Helv Chim Acta 38:1847–1856

Kerekes P, Chang C-S, Brossi A (1984) in preparation

Kobylecki RJ, Lane AC, Smith CFC, Wakelin LPG, Cruse WBT, Egert E, Kennard O (1982) N-methylnalorphine: Definition of N-allyl conformation for antagonist at the opiate receptor. J Mec Chem 25:1278–1280

Leimgruber W (1972) Hoffmann-La Roche, Nutley, USA Pat 3624429

Lie TS et al. (1978a) Chemistry of opium alkaloids, part XI. Synthesis of racemic and chiral codeine and morphine via the dihydrothebainones. Recl Trav Chim Pays-Bas Belg 98:419–420

Lie TS et al. (1978b) Chemistry of opium alkaloids, part X, Synthesis of racemic and optically active codeine and morphine via the N-formylnordihydrothebainones. Recl Trav Chim Pays-Bas Belg 97:127–130

Maeda R, Ohsugi E (1968) Synthesis of racemic 3-methoxy-6-oxo-N-methylmorphinan. Chem Pharm Bull (Tokyo) 16:897–908

Manmade A, Marshall JL, Minns RA, Dalzell H, Razdan RK (1982) Total synthesis of (±)-3-deoxy-7,8-dihydromorphine. J Org Chem 47:1717–1721

Mannering GJ, Schanker LS (1958) Metabolic fate of levo-3-hydroxy-N-allylmorphinan (Levallorphan). J Pharmacol Exp Ther 124:296–304

Merz H, Langbain A, Stockhaus K, Walther G, Wick H (1974) Narcotic antagonism. In: Braude MC, Harris LS, May EL, Smith JP, Villareal JE (eds) Advances in psychopharmacology, vol VIII. Raven Press, New York, pp 91–107

Minamikawa J, Brossi A (1978) Studies in the (+)-morphinan, Ser IV. Dimers of natural (-) and unnatural (+)-sinomenine. Heterocycles 10:79–84

Mohacsi E (1983) Hoffmann-La Roche. Nutley, USA Pat 4374139

Mohacsi E, Leimgruber W, Baruth H (1979) Problems of drug dependence. Proc 41st Ann Sci Meet 77

Mohacsi E, Leimgruber W, Baruth H (1982) Synthesis and pharmacology of metabolically stable tert-butyl ethers of morphine and levorphanol. J Med Chem 25:1264–1266

Monkovic I, Conway TT, Wong H, Perron YG, Pachter IJ, Belleau B (1973) Total synthesis and pharmacological activities of N-substituted 3,14-dihydroxymorphinans 1. J Am Chem Soc 95:7910–7912

Monkovic I, Wong H, Belleau B, Pachter IJ, Perron YG (1975a) Synthetic morphinans and hasubanans IV; Total synthesis of 3,14-dihydroxyisomorphinans; 3-methoxy-,$\Delta^{8,14}$-morphinans; and 9,α-hydroxy-3-methoxyhasubanan. Can J Chem 53:2515–2523

Monkovic I, Wong H, Pircio AW, Perron YG, Pachter IJ, Belleau B (1975b) Oxilorphan and butorphanol (L-17-cyclobutylmethyl-3,14-dihydroxymorphinan); Potent narcotic antagonists and nonaddicting analgesics in the 3,14-dihydroxymorphinan, Ser 5. Can J Chem 53:3094–3102

Okuda S, Yamaguchi S, Kawazoe T, Tsuda K (1964) Studies on morphine alkaloids. I. Nuclear magnetic resonance spectral studies on morphine alkaloids (I). Chem Pharm Bull 12(1): 104–112

Olieman C, Nagelhout Ph, Groot AD de, Maat L, Beyerman HC (1980) Chemistry of opium alkaloids, part XIV. Synthesis of racemic 3-hydroxy-N-methyl-6-oxomorphinan. Recl Trav Chim Pays-Bas Belg 99:353–355

Pert CB, Snyder SH (1974) Opiate receptor of agonists and antagonists affected differentially by sodium. Mol Pharmacol 10:868–879

Pert CB, Pasternak G, Snyder SH (1973) Opiate agonists and antagonists discriminated by receptor binding in brain. Science 182:1359–1361

Polazzi JO, Schut RN, Kotick MP, Howes JH, Osgood PF, Razdan RK, Villareal JE (1980) Analgesic narcotic antagonist 2. 8-alkymorphinan-6-ones. J Med Chem 23:174–179

Rahman MF, Brossi A (1977) Preparation and analgesic activity of some 3,4-disubstituted N-methylmorphinans of the (-) series. Heterocycles 6:881–885

Rapoport H, Levine JB (1953) Stereochemical studies in the morphine series. The relative configuration at carbon thirteen and fourteen. J Am Chem Soc 75:5329–5334

Rapoport H, Masamune S (1955) The stereochemistry of 10-hydroxycodeine derivatives. J Am Chem Soc 77:4330–4335

Razdan RK, SISA (1981) USA Pat 4272540

Razdan RK, Portlock DE, Dalzell HC, Malmberg C (1978) Synthesis of ß-dihydrothebainone. J Org Chem 43:3604–3606

Reden J, Reich MF, Rice KC, Jacobson AE, Brossi A (1979) Deoxymorphines: Role of the phenolic hydroxyl in antinociception and opiate receptor interactions. J Med Chem 22: 256–259

Rice KC (1980) Synthetic opium alkaloids and derivatives. A short total synthesis of (±)-dihydrothebainone, (±)-dihydrocodeinone and (±)-nordihydrocodeinone as an approach to a practical synthesis of morphine, codeine and congeners. J Org Chem 45:3135–3137

Rice KC (1983a) A practical total synthesis of natural and unnatural codeine, morphine, thebaine, their opium-derived congeners and antagonists. Proc Jpn Pharm Soc XX:64–66

Rice KC, Jacobson AE, Burke TR Jr, Bajwa BS (1983) Irreversible ligands with high selectivity toward δ or μ opiate receptors. Science 220:314–316

Robinson R, Sugasawa S (1933) Preliminary synthetic experiments in the morphine group, Part V. Completion of the synthesis of a laudanosoline dimethyl ether related to sinomenine. J Chem Soc 280–281 and references therein

Rozwadowska MD, Hsu FL, Jacobson AE, Rice KC, Brossi A (1980) (-)-4-Hydroxy-N-formylmorphinan-6-one; a versatile intermediate for the synthesis of 3-deoxyopioids. Can J Chem 58:1855–1859

Sawa YK, Maeda S (1964) Elimination of the 4-hydroxyl group of the alkaloids related to morphine I. Tetrahedron 20:2247–2253

Sawa YK, Tada H (1968) Elimination of the 4-hydroxyl group of the alkaloids related to morphine-XI. Synthesis of (-)-14-hydroxy-3-methoxy-N-methylmorphinan derivatives. Tetrahedron 24:6185

Sawa YK, Tsuji N, Maeda S (1961) Elimination of the 4-hydroxyl group of the alkaloids related to morphine. I. Tetrahedron 15:144–153

Schmidhammer H, Brossi A (1982) Synthesis of (-)- and (+)-2-hydroxy-6-keto-N-methylmorphinans, their O-methyl ethers and 2-deoxy congeners. Can J Chem 60:3055–3060

Schmidhammer H, Brossi A (1983) Synthesis of (-)-1-hydroxy-N-methylmorphinan-6-one and its O-methyl ether from (-)-4-hydroxy-N-formylmorphinan-6-one. J Org Chem 48:1469–1471

Schmidhammer H, Jacobson AE, Atwell L, Brossi A (1981) Structure-activity relationship of oxygenated morphinans V. Narcotic agonist and antagonist activity in the 14-hydroxymorphinan series, preliminary communication. Helv Chim Acta 64:2540–2543

Schmidhammer H, Jacobson AE, Brossi A (1982) Morphinans and 6-ketomorphinans unsubstituted in the aromatic ring; high analgesic activity of (-)-6-keto-N-methylmorphinan, IV. Heterocycles 17:391–394

Schmidhammer H, Brossi A, Flippen-Anderson JL, Gilardi R (1983a) Apomorphinans from isoquinolines: Grewe cyclization of 1-(2-hydroxybenzyl)-N-methyloctahydroisoquinoline and its O-methyl ether. Helv Chim Acta 66:2437–2442

Schmidhammer H, Jacobson AE, Brossi A (1983b) Chemical and biological study of aromatic oxygenated 6-ketomorphinans. Med Res Rev 3:1–19

Schmidhammer H, Aeppli L, Atwell L, Fritsch F, Holck M, Jacobson AE, Nebuchla M, Sperk G (1984) J Med Chem (in press)

Schnider O, Grüssner A (1951) Oxy-morphinane (3. Mitteilung). Optisch aktive 3-Oxymorphinane. Helv Chim Acta 34:2211–2217

Schnider O, Hellerbach J (1950) Synthese von Morphinanen (2. Mitteilung). Helv Chim Acta 33:1437–1448

Schöpf C, Thierfelder K (1939) Synthetische Versuche in der Benzylisochinolin-Reihe III Laudanosolin-3,7-dimethyläther ans Laudanosin. Liebigs Ann 537:143–156 and references therein

Schöpf C, Winterhalder L (1927) Die Konstitution der Morphiumalkaloide. Liebigs Ann 452: 232–267

Schwartz MA, Mami IS (1975) A biogenetically patterned synthesis of the morphine alkaloids. J Am Chem Soc 97:1239–1240

Seki I, Takagi H, Kobayashi S (1964) Pharmacological studies on morphine derivatives III. On the chemical structure-activity relationships of 14-hydroxymorphine derivatives. J Pharm Soc Jpn 84:280–289

Simon LD, Simon FR, Mohacsi E, Berger L, Simon EJ (1981) Effect of the position of the phenolic group in morphinans on their affinity for opiate receptor binding. Life Sci 28:2769–2772

SISA (1983) Cambridge, US Pat 4370333

Stork G (1952) The reduction of thebaine. J Am Chem Soc 74:767–773

Szántay C, Bárczai-Beke M, Péchy P, Blaskó G, Dörnyei G (1982) Studies aimed at the synthesis of morphine. 3. Synthesis of (±)-salutaridine via phenolic oxidative coupling of (±)-reticuline. J Org Chem 47:594–596

Vane FM, Ellis DH, Rao J, Blount JF, Mohacsi E, Szuna A, Kamm JJ (1978) Metabolism of the enantiomers of 2-hydroxy-N-cyclopropylmethylmorphinan in dog and man. Biomed Mass Spectrom 5:498–507

Vecchietti V,, Casagrande C, Gerrari G (1976) Ocobotrine and 14-episinomenine, new trans-morphinane alkaloids of *Ocotea brachybotra*. Tetrahedron Lett:1631–1633

Weller DD, Rapoport H (1976) A practical synthesis of codeine from dihydrothebainone. J Med Chem 19:1171–1175

White JD, Caravatti G, Kline TB, Edstrom E, Rice KC, Brossi A (1983) Biomimetic total synthesis of (-)-codeine. Tetrahedron 39:2393–2397

The Development of a Practical Total Synthesis of Natural and Unnatural Codeine, Morphine and Thebaine

K. C. RICE[1]

1 Introduction

In this report, a brief overview of the history of the chemistry and pharmacology of the opium alkaloids will be presented in order to place recent progress in opiate total synthesis in proper perspective. After a brief description of how past, present and possible future developments influenced the development of the NIH opiate total synthesis, some recent progress in this area will be described.

2 Historical

(-)-Morphine (1) is the major pharmacologically active component of opium, which is the air-dried sap of the unripe seed capsule of the poppy *Papaver somniferum.* The use of opium as a remedy for various ills may be documented in the Smith and Ebers Papyri about 1500 B.C. (Terry and Pellens 1928). The traditional production of opium still employed today consists of incising the poppy capsule, and collecting and drying the extruded sap; this procedure affords an approximately 20-fold increase in alkaloid content over that of the dried poppy capsule. The opium produced in this manner contains 10–16% morphine (1), 0.8–2.5% codeine (2), 0.5–2.0% thebaine (3) (Windholz 1984), and traces of neopine (4) (Bentley 1954). Interestingly, this purification technique has changed little since early descriptions of opium harvest were given (Kritikos and Papadaki 1967). When it is recalled that the major effort of the physician in the centuries preceding the present was the relief of pain and treatment of symptoms rather than cause, the value accorded to opium as a medicinal producing pain relief, euphoria and sleep is easily understood. The effective practice of medicine, even in its earliest and most rudimentary form, has thus required, as it now does, a continuous source of opium since the discovery of this remarkable substance in the Bronze Age. Indeed, Sydenham wrote in about 1680 that "among the remedies which it has pleased Almighty God to give man to relieve his sufferings, none is so universal and so efficacious as opium." (Terry and Pellens 1928). This is still the case today.

1 Section on Medicinal Chemistry, Laboratory of Chemistry, National Institute of Arthritis, Diabetes, Digestive and Kidney Diseases, National Institute of Health, Bldg 4, Rm 135, Bethesda, Maryland 20205, USA

The Chemistry and Biology of Isoquinoline Alkaloids, ed. by Phillipson et al.

Fig. 1

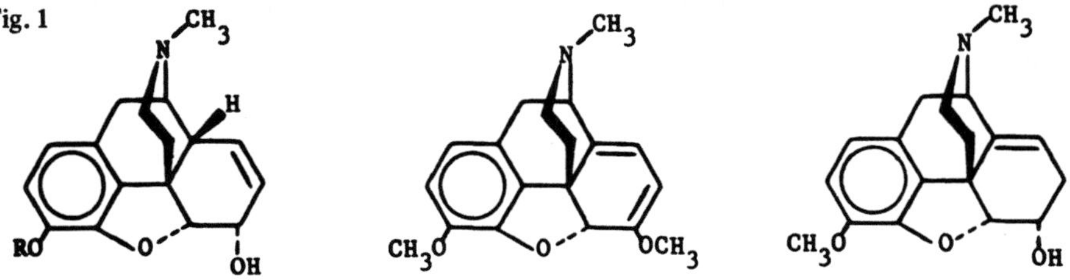

1: R = H, (-)-Morphine

2: R = CH_3, (-)-Codeine

3: (-)-Thebaine

4: (-)-Neopine

3 Isolation and Structure of Morphine, Codeine and Thebaine

Compounding and formulation of opium has produced innumerable preparations, of which laudanum, first introduced by Paracelsus in the 16th century, was used in the U.S. as late as the last decade (Jaffe and Martin 1980). Investigations into the composition of opium by the Parisian Derosne in 1803 (Derosne 1803) afforded "salt of opium" as the first crystalline component to be isolated, however the identity of this material is uncertain, although it is believed to have been morphine meconate or narcotine. Two years later, the German pharmacist Sertürner described isolation of pure crystalline morphine (Sertürner 1806) which, as the first well-characterized organic base to be isolated from a plant, marked the beginning of alkaloid chemistry. Sertürner recognized that this substance, which he named after Morpheus, the Greek God of Dreams, behaved as a base and showed similar effects in dogs and humane to those of the opium from which it was derived (Sertürner 1806, 1817). The isolation of the only other morphinan constituents of medicinal or commercial value today followed with the discovery of codeine by Robiquet (1832) and thebaine by Pelletier (1835). The large-scale separation of morphine and codeine for many years employed the Gregory Process, first described in 1833 (Gregory 1833). Neopine, the 8,14-double bond isomer of codeine and a very minor constituent of opium with a pharmacological profile similar to that of codeine, was described in 1911 (Dobie and Lauder 1911).

Structural elucidation of the morphine alkaloids proceeded slowly in the hands of the infant science of chemistry with correct elemental analysis of morphine (Laurent 1847) and simple chemical transformations such as acetylation of morphine to heroin (Wright 1874) and methylation to codeine (Grimaux 1881). These studies marked the beginning of one of the most massive degradative assaults ever mounted for structural elucidation of a natural product, an effort which may never again be equalled. This effort, brilliantly carried forward by Pschorr, Knorr and others (Holmes 1952), led to the correct structures for thebaine, morphine and thus codeine, since the relation of the latter to morphine was already known (Gulland and Robinson 1925). Formulation of this structure for morphine led to efforts at total synthesis in order to confirm it, particularly the attachment of the ethylamine bridge at C_9–C_{13}, which was the weakest point of the Gulland-Robinson structure of 1925.

4 First Synthesis and Stereochemistry

Numerous innovative approaches were utilized toward this goal, with a biomimetic attempt appearing as early as 1931 (Robinson and Sugasawa 1931). By 1950, and predominantly as a result of the brilliant and independent research of Gates, Grewe, and Ginsburg, it was clear that this goal was not far away (Holmes 1952). At about the same time a total synthesis of racemic Δ^6-β-dihydrodesoxycodeine methyl ether (5) was described (Gates and Tschudi 1950) which conclusively established the attachment of the ethylamine bridge at C_{13} in codeine and morphine. This material was subsequently transformed to (-)-dihydrothebainone [(-)-6], a key intermediate in the first total synthesis of natural codeine and morphine which was reported in 1952 (Gates and Tschudi 1952, 1956). This unequivocal and masterful synthesis completely confirmed the Gulland-Robinson structure and was followed by a second synthesis of (-)-(6), and thus natural codeine and morphine, (Ginsburg and Elad 1954). Semisynthetic unnatural (+)-codeine and (+)-morphine were also prepared from natural (-)-sinomenine (7) (Goto and Yamamoto 1954) utilizing the methodology of Gates in the late steps. The relative stereochemistry of morphine, previously determined chemically (Holmes 1952), was confirmed by X-ray analysis in 1955 (MacKay and Hodgkin 1955). The absolute stereochemistry was elegantly established by Jeger in the same year through chemical degradation of dihydrocodeinone (Kalvoda et al. 1955), thus revealing the complete structure of morphine as (1) after nearly 150 years of investigation.

Fig. 2

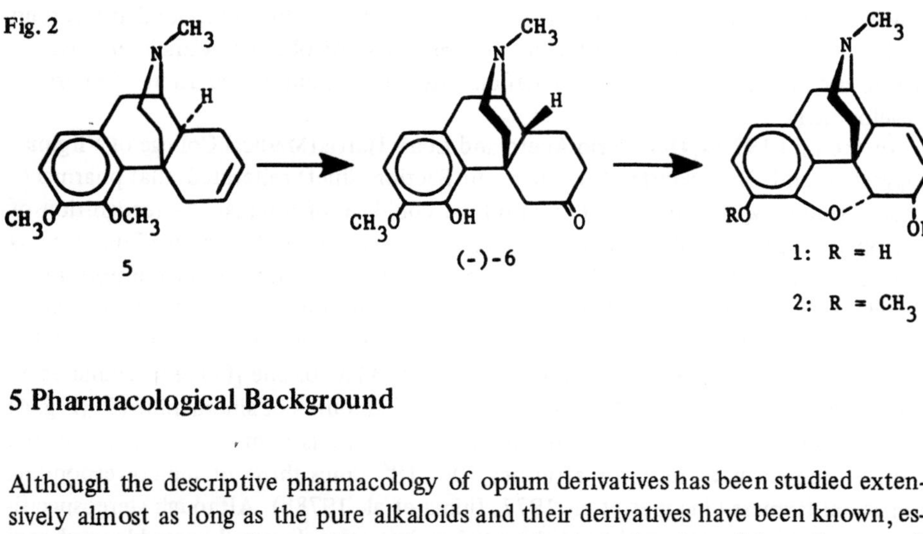

5 Pharmacological Background

Although the descriptive pharmacology of opium derivatives has been studied extensively almost as long as the pure alkaloids and their derivatives have been known, essentially nothing was known about the true mechanism of action of these drugs until the discovery of opiate receptors in the mammalian central nervous system (CNS) in 1973. We know now that these receptors, discovered almost simultaneously by Snyder, Simon and Terenius (Pert and Snyder 1973, Simon et al. 1973, Terenius 1973), are saturable, high affinity, stereospecific binding sites through which the effects of opiates are expressed. The endorphins, some of which resemble morphine biochemically, are endogenous peptides that were discovered in the CNS in 1975 by Hughes et al. (1975). It is now clear that these peptides are the natural ligands for the receptors which exist in at least three subtypes, (mu, delta and kappa) that were proposed as early as 1967 (Martin 1967) on the basis of studies in the chronic spinal dog. The function and role of the multiple types of opiate receptors in the overall modulation of the CNS is far from clear, but it is highly likely that interaction of these receptors with their ligands controls many aspects of the human perception of pain, pleasure and mood (Snyder 1983, 1984).

Fig. 3

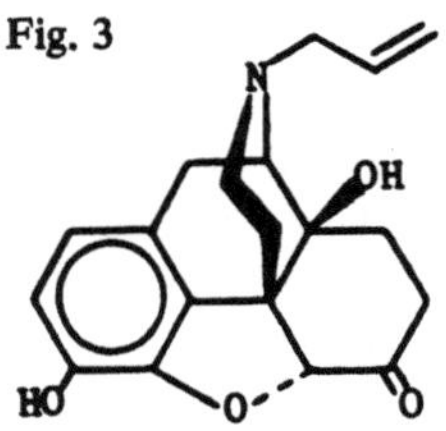

(-)-8: (-)-Naloxone

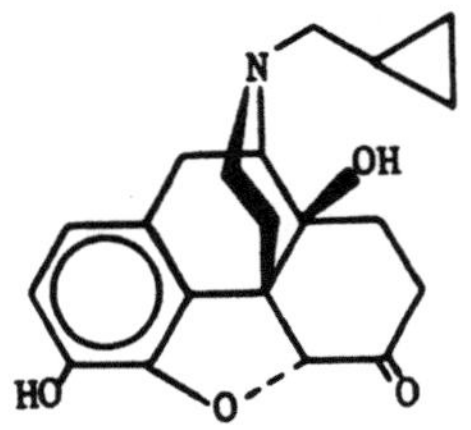

(-)-9: (-)-Naltrexone

The potent narcotic antagonists (-)-naloxone [(-)-(8)] (Narcan) and (-)-naltrexone [(-)-(9)] act at the receptor level and are extensively utilized in opiate research for detection of opiate receptor mediated effects and the former is also used clinically as a life-saving antidote in cases of narcotic overdose. Formerly, naloxone was thought to have little or no pharmacological effect other than narcotic antagonism. Recently, however, highly significant new data have indicated that this compound and certain of its surrogates may be of value in the treatment of obesity, Alzheimers disease, cardiovascular shock, neurological deficit after stroke, and modulation of neuronal development.

In the mid 1970's, Drs. Mario Aceto and Louis Harris (Medical College of Virginia) suggested, and Dr. Everette May (then our Section Chief) reiterated, that pharmacological studies with unnatural (+)-morphine could be of interest. As recognition of the opiate receptor-endorphin system began to clarify in 1976, it seemed likely to us that unnatural (+)-codeine and (+)-morphine (and congeners) could be important research tools for study of the receptor-mediated and non-specific effects of opiates. These compounds (Iijima et al. 1977, 1978a, Jacquet et al. 1977, Iijima et al. 1978a) as well as unnatural (+)-oxymorphone (10) and (+)-naloxone [(+)-(8)] (Iijima et al. 1978b) were synthesized at NIH from the limited amounts of sinomenine (7) (Fig. 4) available and have proved to be invaluable research tools in many systems, since the natural isomers have receptor affinities 10^3-10^4 times those of the corresponding unnatural isomers (Jacquet et al. 1977, Iijima et al. 1978b). Although stereospecificity of opiate receptors had been detected earlier using dextrophan and levorphanol (chiral 3-hydroxy-N-methylmorphinans), significant differences have been found between the morphinans and the opiates (Jacquet et al. 1977). Recently, a high-affinity site has been postulated which specifically binds the 4,5-epoxymorphinan structures of the opiates, but not the 3-hydroxymorphinans (Grevel and Sadee 1983).

The elucidation of the structure of opiate receptor subpopulations, and of their role in the CNS, is a complex problem under investigation in many laboratories worldwide. Insight into some aspects of the problem is being gained using a number

Fig. 4

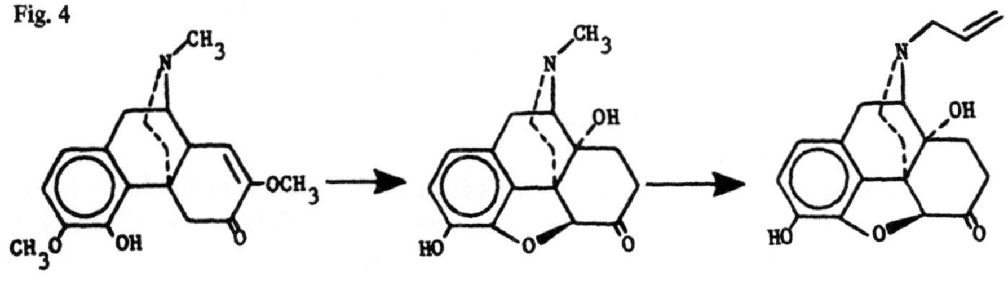

7: (-)-Sinomenine 10: (+)-Oxymorphone (+)-8: (+)-Naloxone

of biochemical, pharmacological and immunological approaches. For example, active opiate receptors have been solubilized (Simonds et al. 1980), alkylating opiates and opiate-like ligands specific for the mu (Portoghese et al. 1980, Pasternak and Hahn 1980, Rice et al. 1983) and delta receptor (Rice et al. 1983) subpopulations are available, and identification of a M_r 58,000 subunit of the delta receptor has been made (Klee et al. 1982). Recent developments in this area (Snyder 1983, 1984), as well as investigations into the possible involvement of opiate binding sites in regulation of the immune system, have recently been reviewed (Weber and Pert 1984).

6 Recent Synthetic Efforts

Synthetic studies in opiate total synthesis have continued unabated since the Gates synthesis. The first biomimetic total synthesis was reported by Barton in 1963 (Barton et al. 1963, 1967), albeit in very low yield. Since this time, numerous reports utilizing Grewe (Grewe and Friedrichsen 1967, Lie et al. 1979, Rice 1980, 1982), biomimetic (Szántay et al. 1983, Schwartz and Zoda 1981, White et al. 1983) and other approaches (Moos et al. 1983, Evans and Mitsch 1982) have been described. The impetus for development of a practical total synthesis of opiates at NIH was provided by two considerations, which were the unavailability of sinomenine (7) for synthesis of new, and previously known unnatural opiates in larger quantity, and the shortage of opium for production of medical opiates which occurred during 1973–1975 (Schwartz 1975, 1980). The latter event necessitated release of about half of

Fig. 5

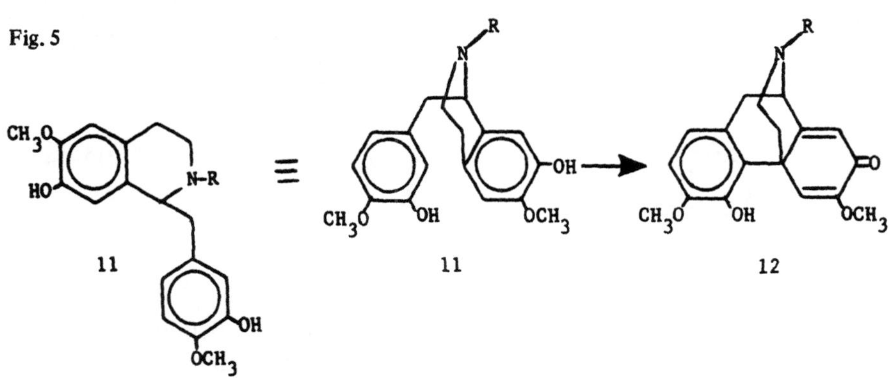

11 11 12

the U.S. strategic material supply of opium for domestic production and strongly emphasized the general desirability of obtaining a synthetic route which could provide a source of medical opiates independent of natural sources. As with complete dependence on foreign sources of any natural product, a variety of factors could cause recurrence of such a problem in the future.

Initial investigations in opiate total synthesis at NIH focused on the biomimetic route, shown above and recently studied extensively by Szántay et al. (1983) and Schwartz and Zoda (1981) in which a reticuline derivative (11) is oxidatively cyclized to a salutaridine derivative (12) which would be convertible to thebaine and hence to codeine and morphine. Traditionally, this approach has been complicated by control of the product distribution ratio and the tendency of the desired morphinandienones to undergo secondary transformations. An effective synthesis of reticuline derivatives (Rice and Brossi 1980) was developed as a prerequisite to this approach at NIH, and a simple synthesis of racemic and chiral pavinans and isopavinans (Rice et al. 1980) resulted from these studies, but attempts at development of an oxidative cyclization which would be useful for large scale synthesis were unsuccessful.

7 The NIH Total Synthesis (Rice 1980, 1981)

A second approach attempted by the writer based on Grewe cyclization proved successful for large-scale opiate synthesis and a number of factors considered in its development are briefly discussed below. It has been shown independently by Grewe and Friedrichsen (1967) and Morrison et al. (1967) that Grewe cyclization of [(±)-13] could be utilized to provide racemic dihydrothebainone [(±)-6] with the required oxygenation pattern for conversion to the natural opium alkaloids; however, because of nonregioselective cyclization the yield was only 3%, with the isomeric morphinan [(±)-14] being the major product. The strategy adopted in the present work was to

Fig. 6

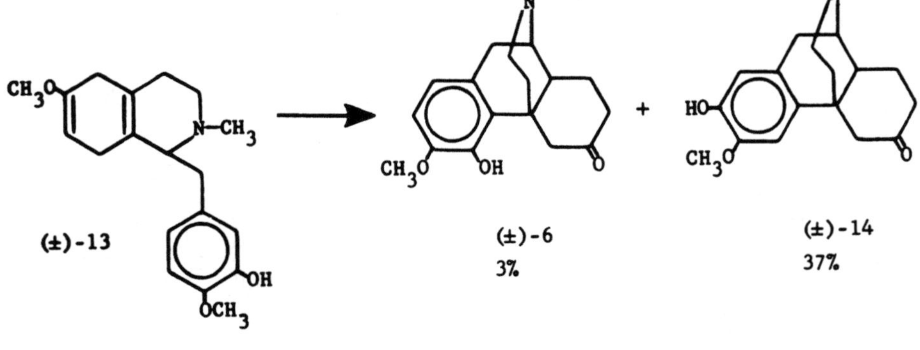

direct the regioselectivity of cyclization with a readily introduced and removed blocking group. This approach had been unsuccessfully attempted (Beyerman et al. 1976, DeGraw et al. 1974) earlier with [(±)-15], prepared in low yield from [(±)-24],

Fig. 7

(±)-15

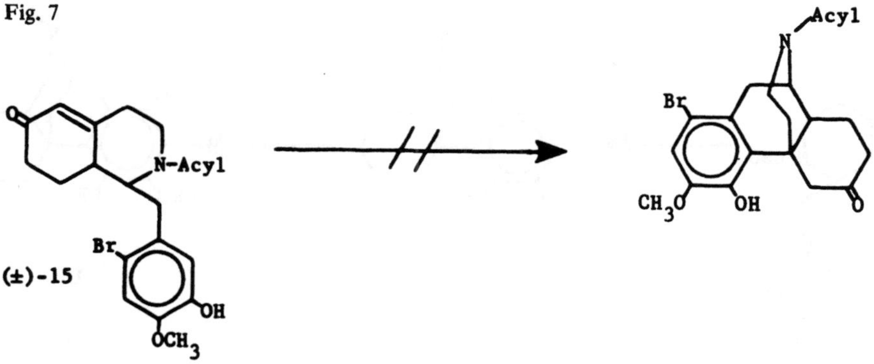

in order to obtain useful yields of a dihydrothebainone derivative with the proper oxygenation pattern for conversion to the opium alkaloids. Since 1-bromo derivatives are obligatory intermediates in the oxide bridge closure of 4-hydroxymorphinans and this function is easily removed by catalytic hydrogenation, a bromine substituent appeared ideal. The work of Leimgruber (1972) and later our own (Rice et al. 1980), had shown that N-acyl amines undergo Grewe or Grewe-like cyclization in a much more facile manner than the corresponding amines. Thus, it seemed prudent to utilize the readily removable N-formyl substituent in the morphinan-forming reaction which would afford, after hydrolysis, the corresponding N-normorphinan. A possible morphinan precursor appeared to be (16), that might cyclize to 1-bromo-N-formylnor-dihydrothebainone (17) from which the projected key intermediate 1-bromo-N-nor-dihydrothebainone (18) would be obtained after subsequent hydrolysis. If oxide bridge closure in the N-nor series could be executed with (18) to give (19), and could be followed by debromination, then both nordihydrocodeinone (20) and dihydrocodeinone (21) would be rendered equally available, provided debromination and concomitant reductive alkylation could be accomplished. Because substitution of N-alkyl functions such as cyclopropyl and cyclobutylmethyl for N-methyl produces dramatic alterations in the pharmacological profiles of morphinans (Braude 1973) ranging from essentially pure narcotic antagonists to clinically valuable agonist-antagonists with little or no addiction potential in man, the optional availability of either the N-nor or the N-methyl intermediates was a very important consideration in designing a versatile opiate synthesis and was expected to greatly enhance the utility of the projected route. Since conversion of dihydrothebainone to dihydrocodeinone, first described by Schöpf and Pfeifer (1930), was recently utilized in the development of a highly efficient method for conversion of dihydrothebainone to codeine and thebaine by Weller and Rapoport (1976), and high yielding transformation of codeine to morphine had been described by the author (Rice 1977), an efficient synthesis of dihydrothebainone or dihydrocodeinone would afford a practical total synthesis of codeine, morphine and thebaine. Such a synthesis would also render the complete spectrum of medically valuable opium derivatives available since this entire group consists of codeine, morphine, their transformation products and the drugs based on thebaine. For the NIH requirements, such a route should be capable of providing optional access to both the natural and unnatural series. In addition, prochiral or chiral

Fig. 8

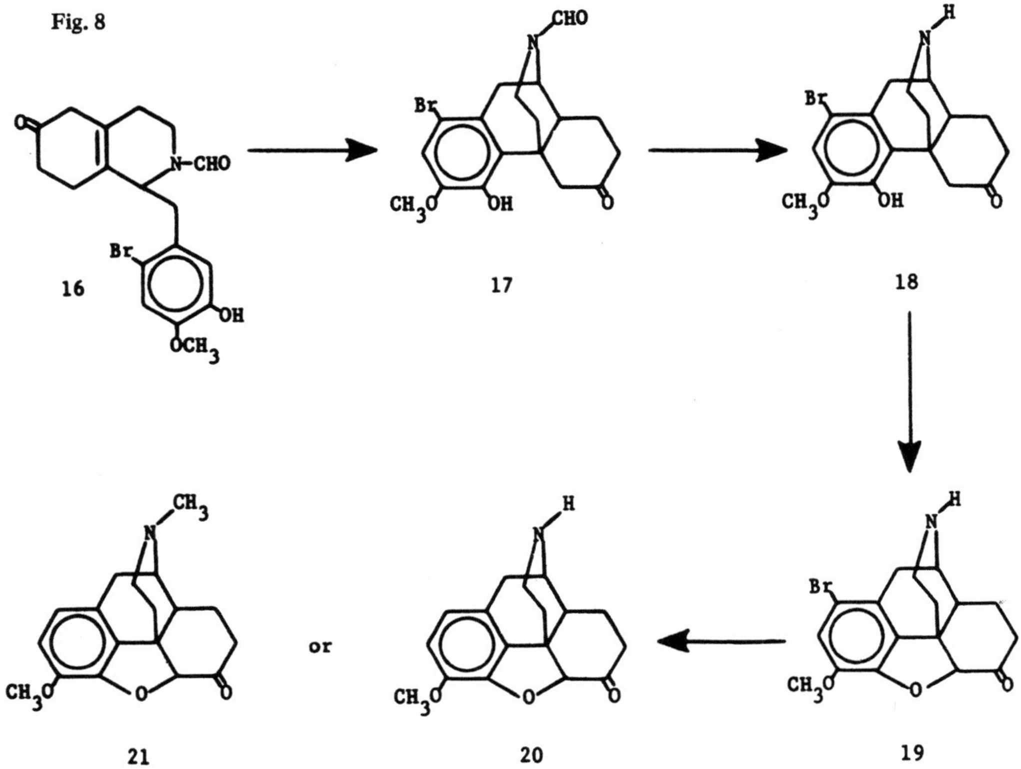

intermediates with undesired stereochemistry should be directable to the desired stereochemistry. It was also hoped that unprotected phenolic intermediates could be utilized where applicable; this would obviate the need for the additional steps required for protection and deprotection. These requirements have now been met and were accomplished as briefly described below. In the following discussion, the term "isolated intermediate" refers to any intermediate which was filtered and washed; crude intermediates obtained by solvent extraction and evaporation are not included.

As reported in preliminary form (Rice 1980), the route was initially developed in the racemic series with unprotected phenolic intermediates and utilized commercially available m-methoxyphenethylamine (22) as starting material. Thermal condensation of this amine with acid (23) gave a 95% isolated yield of the corresponding amide, which was converted to racemic tetrahydroisoquinoline [(±)-24] employing the method utilized earlier in our norreticuline synthesis (Rice and Brossi 1980). Birch reduction of [(±)-24], previously described by Grewe and Friedrichsen (1967), then afforded the hexahydro derivative (25) in 74% yield from m-methoxyphenethylamine (22). Selective N-formylation of (25) afforded formamide (26), which gave ketal (27) when treated with ethylene glycol-methanesulfonic acid in THF. Bromination with N-bromoacetamide of ketal (27) generated in situ gave bromoketal (28) which, without isolation, was deketalized to β,γ-unsaturated ketone (16) in 90% yield from formamide (26). Bromine-directed Grewe cyclization of (16) to (±)-1-bromo-N-formylnordihydrothebainone (17) was accomplished in 60% isolated yield by treatment with triflic acid. Attempted cyclization of the corresponding α,β-unsaturated ketone (15) failed to provide more than traces of (17) under any conditions examined and confirmed earlier failures (Beyerman et al. 1976, DeGraw et al. 1974) to cyclize (15). Acid hydrolysis of (17) afforded 1-bromonordihydrothebainone (18) as a centrally important intermediate in essentially quantitative yield. Hydrogenation of unisolated (18) over palladium in the presence of formaldehyde then provided (±)-dihydrothebainone [(±)-6] in quantitative yield based on (17). Closure of the oxide bridge of

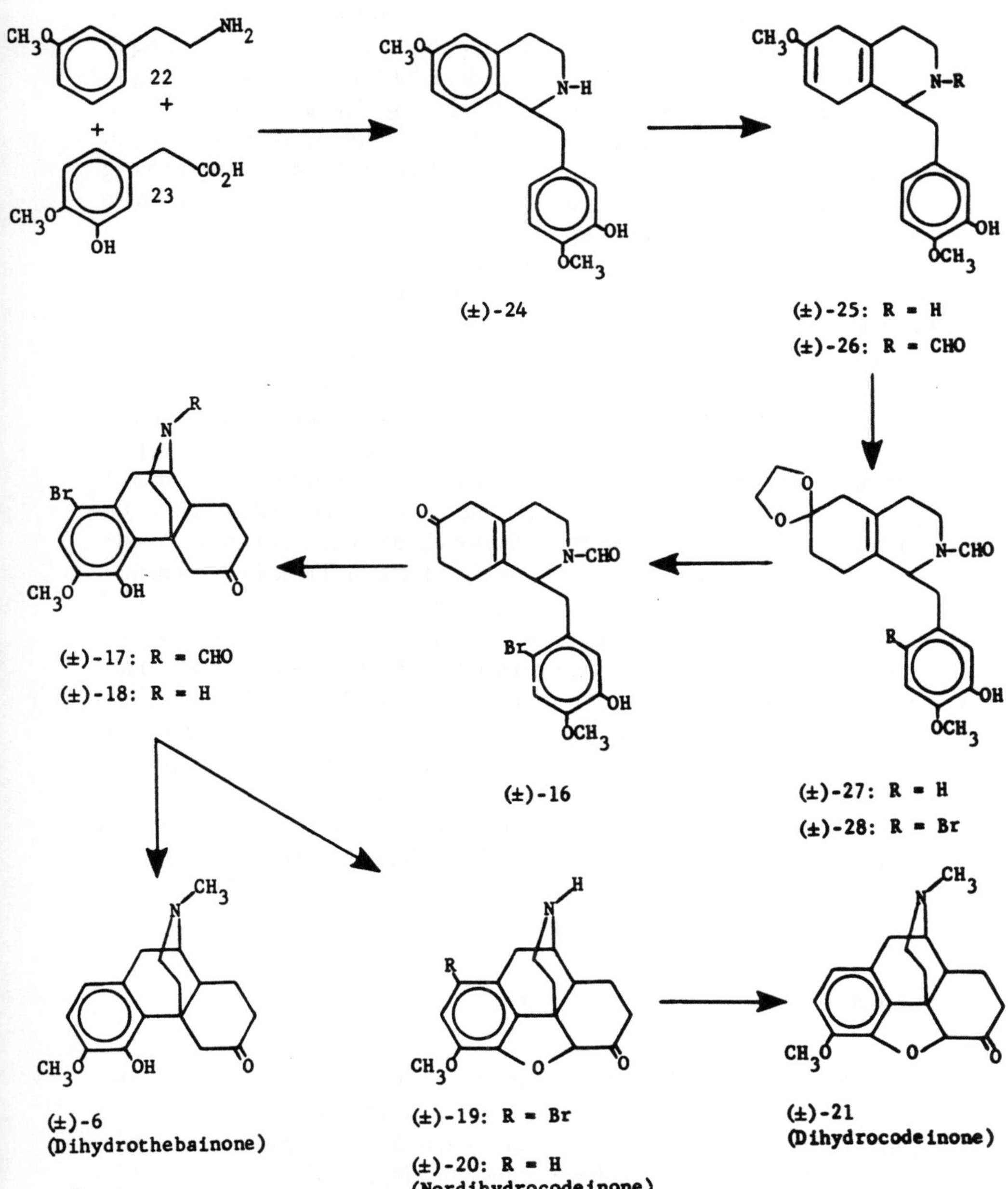

Fig. 9

(18), unprecedented in the nor series, was easily accomplished by treatment with a slight excess of bromine followed by aqueous base as in the N-methyl series. Hydrogenation of the resulting (±)-1-bromonordihydrocodeinone [(±)-19] without isolation gave (±)-nordihydrocodeinone [(±)-20] in 80% yield from (17). When formaldehyde was included, concomitant reductive alkylation and debromination occurred to give (±)-dihydrocodeinone [(±)-21] in similar yield. The overall yields of racemic dihydrothebainone, nordihydrocodeinone and dihydrocodeinone based on m-methoxyphenethylamine were 37, 30 and 29%, respectively, with only six isolated intermediates which were directly obtained sufficiently pure for further transformation. Since completion of this work, a facile optical resolution of the racemic tetrahydroisoquinoline [(±)-24] using the tartaric acids has been developed (Rice 1982) and the optical purity of the resulting enantiomers demonstrated by derivatization with chiral α-methylbenzylisocyanate and HPLC analysis of the resulting diastereoisomers. In addition, an efficient racemization (Rice et al. unpublished) has permitted recycling of the undesired optical isomer of (24) and other modifications at various stages have been introduced. The Grewe conditions for Birch reduction of (24) have now been modified to provide a facile and highly reproducible reduction of chiral (24) to the corresponding chiral (25) in 94% yield. The ready availability of both enantiomers of tetrahydroisoquinoline (24) permitted extension of this route to both the natural and unnatural opiate series which has now been well worked out and scaled up to prepare nearly 100 g quantities of chiral 1-bromonordihydrocodeinone hydrobromide in one batch from m-methoxyphenethylamine. The synthesis has been repeated with essentially the same results by Carroll and his coworkers (Rice et al. unpublished) at the Research Triangle Institute. Assuming two cycles of racemization-recycling of the undesired enantiomer of (24) and a theoretical yield of 50% in the optical resolution, it is now possible to synthesize either natural or unnatural codeine, thebaine and morphine in 20–25% overall yield from m-methoxyphenethylamine with only 6–8 isolated intermediates, including the tartrate salt of the chiral tetrahydroisoquinoline (24). In addition to unnatural codeine, morphine and thebaine, the route has been used to prepare multigram amounts of unnatural (+)-naltrexone [(+)-9] (Rice unpublished) (+)-oxymorphone (10) (Rice et al. unpublished) and (+)-naloxone [(+)-8] (Rice et al. unpublished) of which the latter two compounds have been shown to be biochemically indistinguishable (Klee and Streaty unpublished) at the receptor level from the samples prepared earlier from natural sinomenine (7). Synthesis of a substantial number of other unnatural opium derivatives for use as neuropharmacological research tools and for pharmacological examination in other than opiate systems is in progress and will be reported in due course.

8 Summary

The development of opiate chemistry and pharmacology since isolation of morphine by Sertürner has passed the important milestones of structural elucidation, the first total synthesis, discovery of the opiate receptors and their endogenous ligands, solubilization of active opiate receptors, identification of a M_r 58,000 subunit of the delta receptor, and synthesis of alkylating opioid ligands specific for opiate receptor subpopulations. Furthermore, the recently detected pharmacological properties of previously known narcotic antagonists may well lead to new treatments for a number of conditions. In total synthesis, great progress has been made in both biomimetic and Grewe approaches since the masterful first synthesis by Gates and Tschudi. As described here, a relative simple methodology has now been developed for large scale total synthesis of both the natural and unnatural isomers of all medically valuable opium-derived morphinan derivatives. The route potentially provides independence from natural sources of these drugs, may be of value in their commercial production, and certainly renders the unnatural enantiomers available in any required quantity for detailed pharmacological studies.

As in other areas of research, advances in opiate chemistry are made possible by building on results of earlier workers and the present work is no exception. Certainly the morphinan forming cyclization of Grewe, the Schöpf conversion of dihydrothebainone to dihydrocodeinone, and transformation of the latter to codeine and thebaine by Weller and Rapoport are stepping stones which were essential for development of the results described here in their present form. Progress in opiate pharmacology clearly follows a similar course to that of opiate chemistry, and it seems quite likely that major advances in our understanding of the opiate receptor-endorphin system will continue to be forthcoming and will eventually lead to manipulation of this system in humans with dramatic results.

Acknowledgements. The author appreciates helpful discussions with Drs. Arnold Brossi, Arthur Jacobson and Ulrich Weiss of NIH, Drs. F.I. Carroll, George Brine and Anita Lewin of Research Triangle Institute, and Dr. Mike Wilson of McNeil Laboratories during various phases of this work. In addition, special thanks are extended to Dr. Everette May for his advice and encouragement during the entire development of the NIH total synthesis.

References

Barton DHR, Kirby GW, Steglich W, Thomas GM (1963) The biosynthesis and synthesis of morphine alkaloids. Proc Chem Soc 203–204

Barton DHR, Bhakuni DS, James R, Kirby GW (1967) Phenol oxidation and biosynthesis, part XII. Stereochemical studies related to the biosynthesis of the morphine alkaloids. J Chem Soc (C):128–132

Bentley KW (1954) The chemistry of the opium alkaloids. Clarendon Press, Oxford, p 124

Beyerman HC, Buurman E, Maat L, Olieman C (1976) Synthesis of racemic and of (+) and (-)-1-methyldihydrothebainone. Recl Trav Chim Pays-Bas Belg 95:184–188

Braude MC (1973) Narcotic antagonists. Raven Press, New York

DeGraw JI, Christensen JC, Brown VH, Cory MJ (1974) Investigation of the Grewe codeine method. Attempts to achieve a practical synthesis. Heterocycl Chem 11:363–367

Derosne JF (1803) Memoire sur l'opium. Ann Chim 45:257–285

Dobie JJ, Lauder A (1911) Hydroxycodeine: A new alkaloid from opium. J Chem Soc 99:34–35

Evans DA, Mitsch CH (1982) Studies directed towards the total synthesis of morphine alkaloids. Tetrahedron Lett 23:285–288

Gates M, Tschudi G (1950) The synthesis of racemic β-Δ^6-dihydrodesoxycodeine methyl ether. J Am Chem Soc 72:4839–4840

Gates M, Tschudi G (1952) The synthesis of morphine. J Am Chem Soc 74:1109–1110

Gates M, Tschudi G (1956) The synthesis of morphine. J Am Chem Soc 78:1380–1393

Ginsburg D, Elad D (1954) The synthesis of morphine. J Am Chem Soc 76:312–313

Goto K, Yamamoto I (1954) (+)-Codeine and (+)-morphine from sinomenine. Proc Jpn Acad 30: 769–773

Gregory W (1833) Über eine neue Methode zur Abscheidung des Morphiums aus dem Opium. Ann 7:261–265

Grevel J, Sadee W (1983) An opiate binding site in the rat brain is highly selective for 4,5-epoxymorphinans. Science 221:1198–1201

Grewe R, Friedrichsen W (1967) Die Cyclisierung von Octahydro-isochinolinderivaten durch Morphinan-Ringschluß. Synthese des Dihydrothebainones. Chem Ber 100(5):1150–1558

Grimaux E (1881) Sur une nouvelle série de bases derivées de la morphine. CR Acad Sci 93:591

Gulland JM, Robinson R (1925) Constitution of codeine and thebaine. Mem Proc Manchester Liter Philos Soc 69:79 [Chem Abstr 20:765 (1926)]

Holmes HL (1952) The morphine alkaloids. In: Manske RHF, Holmes HL (eds) The alkaloids, vol II, chap 8, parts 1,2. Academic, London New York

Hughes J, Smith TW, Kosterlitz HW, Fothergill LA, Morgan BA, Morris HK (1975) Identification of two related pentapeptides from the brain with potent opiate agonist activity. Nature (London) 258:577–579

Iijima I, Rice KC, Silverton JV (1977) Studies in the (+)-morphinan Ser I. An alternate conversion of (+)-dihydrocodeinone into (+)-codeine. Heterocycles 6(8):1157–1165

Iijima I, Minamikawa J-I, Jacobson AE, Brossi A, Rice KC, Klee W (1978a) Studies in the (+)-morphinan Ser 5. Synthesis and biological properties of (+)-naloxone. J Med Chem 21:398–400

Iijima I, Minamikawa J-I, Rice KC, Jacobson AE, Brossi A (1978b) Studies in the (+)-morphinan Ser 4. A markedly improved synthesis of (+)-morphine. J Org Chem 43:1462–1463

Jacquet YF, Klee W, Rice KC, Iijima I, Minamikawa J-I (1977) Stereospecific and nonstereospecific effects of (+)- and (-)-morphine: Evidence for a new class of receptors? Science 198: 842–845

Jaffe JH, Martin WR (1980) Opioid analgesics and antagonists. In: Gilman AG, Goodman LS, Gilman A (eds) The pharmacological basis of therapeutics, 6th edn. Macmillan, New York, p 509

Kalvoda J, Buchschacher P, Jeger O (1955) Über die absolute Konfiguration des Morphins und verwandter Alkaloide. Helv Chim Acta 38:1847–1856

Klee WA, Simonds WF, Sweat FW, Burke TR, Jacobson AE, Rice KC (1982) Identification of a M_r 58000 glycoprotein subunit of the opiate receptor. FEBS Lett 150:125–128

Kritikos PG, Papadaki SP (1967) The history of the poppy and of opium and their expansion in antiquity in the eastern Mediterranean area. Bull Narcot 19(3):17–38

Laurent A (1847) Sur la composition des alcalis organiques et de quelques combinaisons azotées. Ann Chim Phys 19:359–377

Leimgruber W, Mohasci E (1972) Derivatives of Morphinan. US Pat 3634429

Lie TS, Maat L, Beyerman HC (1979) Chemistry of opium alkaloids, part XI. Synthesis of racemic and chiral codeine and morphine via the dihydrothebainones. Recl Trav Chim Pays-Bas Belg 98(6):419–420 and references cited therein

MacKay M, Hodgkin DC (1955) A crystallographic examination of the structure of morphine. J Chem Soc 3261–3267

Martin WR (1967) Opioid antagonists. Pharmacol Rev 19:463–521

Moos WH, Gless RD, Rapoport H (1983) Codeine analogues. Synthesis of 4a-aryldecahydroisoquinolines containing nitrogen ring functionality and of octahydro-1H-indeno [1,2,3-ef] isoquinolines. A total synthesis of codeine. J Org Chem 48:227–238

Morrison GC, Waite RO, Shavel J (1967) An alternate route in the synthesis of morphine. Tetrahedron Lett:4055–4060

Pasternak GW, Hahn E (1980) Long-acting opiate agonists and antagonists: 14-Hydroxydihydromorphinone hydrazones. J Med Chem 23:674–676

Pelletier J (1835) Nouvelles recherches pour servier á l'histoire de l'opium et de ses principes. J Pharmazie (s. 2) 21:555–582

Pert CB, Snyder SH (1973) Opiate receptor: Demonstration in nervous tissue. Science 179: 1011–1014

Portoghese PS, Larson DL, Sayre LM, Fries DS, Takemori AE (1980) A novel opioid receptor site directed alkylating agent with irreversible narcotic antagonistic and reversible agonistic activities. J Med Chem 23:233–234

Rice KC (1977) High-yield conversion of codeine to morphine. J Med Chem 20:164–165

Rice KC (1980) Synthetic opium alkaloids and derivatives. A short total synthesis of (±)-dihydrothebainone, (±)-dihydrocodeinone, and (±)-nordihydrocodeinone as an approach to a practical synthesis of morphine, codeine and congeners. J Org Chem 45:3135–3137 and references cited therein

Rice KC (1982) Synthetic opium alkaloids and derivatives 2. Efficient total synthesis of (-)-dihydrocodeinone and derivatives.. In: Harris LS (ed) Problems of drug dependence 1981. NIDA Res Monogr 41, Washington DC, p 99 and references cited therein

Rice KC, Brossi A (1980) Expedient synthesis of racemic and optically active N-norreticuline and N-substituted and 6'-bromo-N-norreticulines. J Org Chem 45:592–601

Rice KC, Ripka WC, Reden J, Brossi A (1980) Pavinan and isopavinan alkaloids. Synthesis of racemic and natural thalidine, bisnorargemonine and congeners from N-norreticuline. J Org Chem 45:601–607

Rice KC, Jacobson AE, Burke TR Jr, Bajwa BS, Streatly RA, Klee W (1983) Irreversible ligands with high selectivity toward δ or μ opiate receptors. Science 220:314–316

Robiquet PJ (1832) Nouvelles observations sur les principaux products de l'opium. Ann Chim Phys 51:225–267

Robinson R, Sugasawa S (1931) Preliminary synthetical experiments in the morphine group. Part 1. J Chem Soc 3163–3172

Schöpf C, Pfeifer T (1930) Teilsynthesen in der Morphinreihe: I. Der Übergang aus der Dihydrothebainon- in die Dihydrokodeinon-Reihe. Ann 483:157–169

Schwartz MA (1980) Prescription drugs in short supply, chap 2. Marcel Dekker, New York. See also: Proceedings of the AMA symposium on supplies of opium for medical use, March 3–4, 1975

Schwartz MA, Zoda MF (1981) Biomimetic approaches to morphine alkaloids. Total synthesis of (±)-2-hydroxycodeine and (±)-noroxycodone. J Org Chem 46:4623–4625 and references cited therein

Sertürner FWA (1806) Darstellung der reinen Mohnsäure (Opiumsäure) nebst einer chemischen Untersuchung des Opiums mit vorzüglicher Hinsicht auf einen darin neu entdeckten Stoff und die dahin gehörigen Bemerkungen. J Pharm Ärzte Apotheker Chem 14/1:47–93

Sertürner FWA (1817) Über das Morphium, eine neue salzfähige Grundlage, und die Mekonsäure, als Hauptbestandtheile des Opiums. Ann Phys 55:56–90

Simon EJ, Hiller JM, Edelman I (1973) Stereospecific binding of the potent narcotic analgesic [^{3}H]etorphine to rat-brain homogenate. Proc Natl Acad Sci USA 70:1947–1949

Simonds WF, Koski G, Streaty RA, Hjelmeland LM, Klee WA (1980) Solubilization of active opiate receptors. Proc Natl Acad Sci USA 77:4623–4627

Snyder SH (1983) Neurotransmitter receptor binding and drug discovery. J Med Chem 26:1667–1672

Snyder SH (1984) Drug and neurotransmitter receptors in the brain. Science 244:22–31

Szántay C, Blaskó G, Bárczai-Beke M, Dörnyei G, Péchy P (1983) Studies on the synthesis of morphine VI. Recent progress on the biomimetic synthesis of morphine. Planta Med 48:207–211 and references cited therein

Terenius L (1973) Characteristics of the "receptor" for narcotic analgesics in synoptic plasma membrane fraction from rat brain. Acta Pharmacol Toxicol 33:377–384

Terry CE, Pellens M (1928) The opium problem, chap 2. Bur Soc Hyg, New York

Weber RJ, Pert C (1984) Opiatergic modulation of the immune system. In: Muller EE, Genazzani AR (eds) Central and peripheral endorphins, basic and clinical aspects. Raven Press, New York, p 35–42

Weller DD, Rapoport H (1976) A practical synthesis of codeine from dihydrothebainone. J Med Chem 19:1171–1175

White JD, Caravatti G, Kline TB, Edstrom E, Rice KC, Brossi A (1983) Biomimetic total synthesis of (-)-codeine. Tetrahedron 39:2393–2397

Windholz M (1984) The Merck Index, 10th edn. Merck, Rahway NJ, no 6728

Wright CRA (1874) On the action of organic acids and their anhydrides on the natural alkaloids Part I. J Chem Soc 27:1031–1043

Biomimetic and Total Synthesis of Monoterpenoid Isoquinoline Alkaloids

R.T. BROWN[1]

1 Introduction

Among the large number of alkaloids containing an isoquinoline nucleus is a small but important group derived from dopamine and a C-9 unit of general type (1). A typical member of the ipecac alkaloids is emetine (3) considered to be formed by condensation of a second dopamine molecule with protoemetine (2). Some years ago this C-9 unit was shown to be monoterpenoid in origin as in the analogous indole alkaloids, the biosynthetic sequence essentially corresponding to that outlined in Fig. 1 (Battersby and Parry 1971, Nagakura et al. 1978). The ultimate non-nitrogenous precursor is the iridoid glucoside secologanin (4). When we discovered a major source of this material in *Lonicera* spp., we began a programme of partial syntheses of alkaloids using secologanin as starting material, ipecac and related bases being early targets. In general our approach has been "biomimetic", following wherever possible a likely biosynthetic pathway in an attempt to control the plethora of functional groups and chiral centres. Nevertheless, we have also been able to explore alternative routes and use various chemical strategies to achieve syntheses of both known and potential natural products which are discussed in the first part of this chapter. In the second part we will report on the application of chemical lessons learnt from the above transformations to total syntheses of isoquinoline alkaloids via synthetic analogues of secologanin.

2 Ipecac Alkaloids from Secologanin

Because of the clinical importance of emetine in particular, the ipecac alkaloids have been targets for several syntheses which have involved problems in stereochemical control and in resolution of racemic intermediates (Brossi et al. 1971). Provided chirality could be conserved at C-2 and C-7, many difficulties could be avoided by using the natural precursor secologanin as starting material, and this aim has been realised (Brown et al. 1979).

By a sequence of acetylation, oxidation, methylation, and catalytic hydrogenation, secologanin was converted into methyl 3,4-dihydrosecoxyloganin tetra-acetate

1 Department of Chemistry, University of Manchester, Manchester M13 9PL, United Kingdom

The Chemistry and Biology of Isoquinoline Alkaloids, ed. by Phillipson et al.

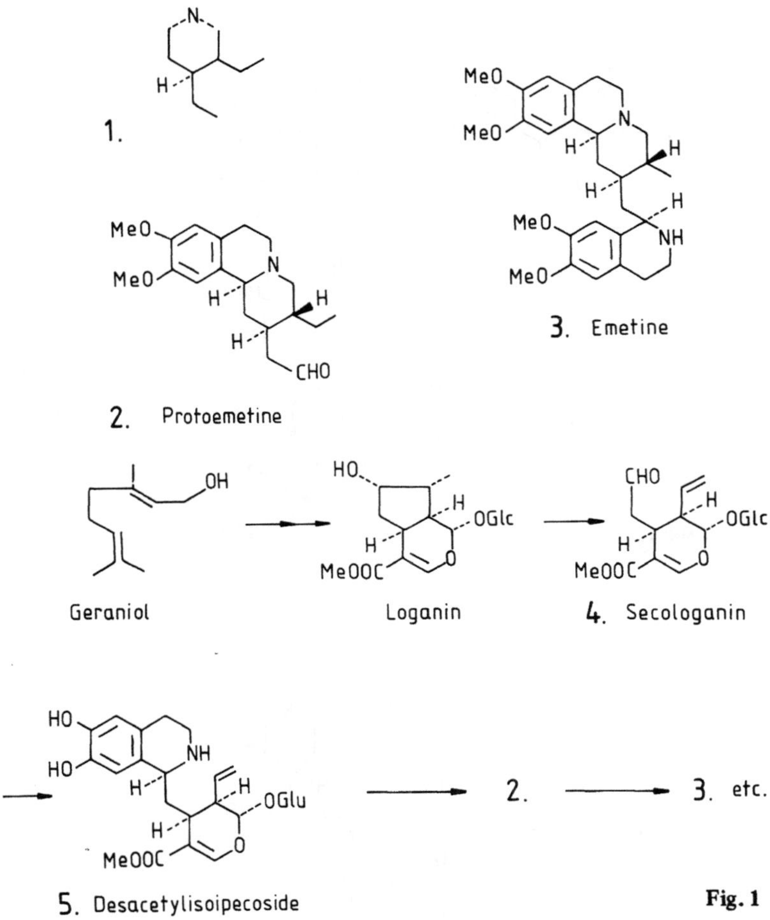

Fig. 1

(6). (Fig. 2) Zemplen deacetylation and removal of the sugar with β-glucosidase in pH5 buffer then afforded the aglucone (7). Following a procedure developed in our laboratory, this was reductively aminated by sequential treatment with 2-3′,4′-dimethoxyphenethylamine and sodium cyanoborohydride to give the tetrahydronicotinate (8) as the major product in 42% overall isolated yield. That this procedure did not result in significant inversion at either chiral centre was established by equilibrating the aglucone C-2 isomers via the ring-opened form with sodium methoxide in methanol for several hours. Repetition of the reductive amination now gave as the major product a less polar epimer which was not observed in the original reaction. After much experimentation it was eventually found that selective hydrolysis and decarboxylation of the β-amino-acrylate ester to the carbinolamine (9) could be achieved with 1% hydrochloric acid in refluxing aqueous methanol for an hour. Under these conditions there was no appreciable hydrolysis of the C-5 ester, and most importantly of all, the crucial point of the synthesis was achieved in that Picter-Spengler cyclisation of C-9 to the aromatic ring did not occur. Previous work with model compounds had indicated that a para phenolic hydroxyl rather than methoxyl was re-

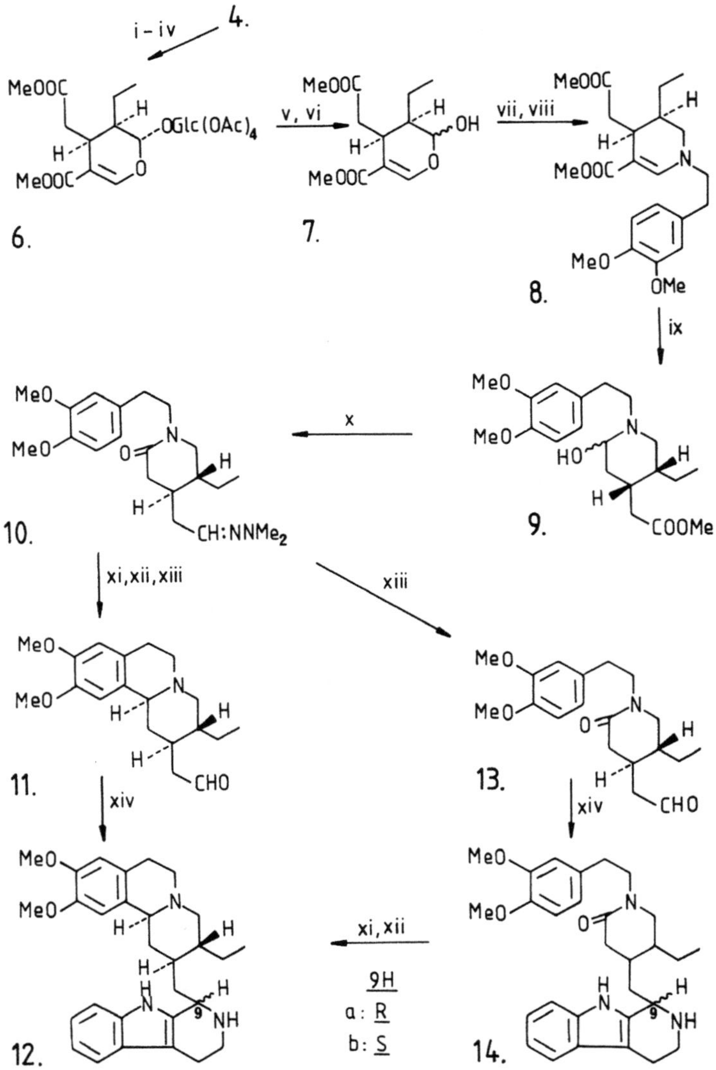

Reagents: i, Ac_2O/py; ii, CrO_3/Me_2CO; iii, CH_2N_2; iv, H_2/Pd; v, NaOMe; vi, β-glucosidase/pH5; vii, 3,4-dimethoxyphenethylamine; viii, $NaCNBH_3$; ix aq. HCl/MeOH/Δ; x, Me_2NNH_2; xi, $POCl_3$/Δ; xii, $NaBH_4$; xiii, aq. $Cu(OAc)_2$; xiv, tryptamine

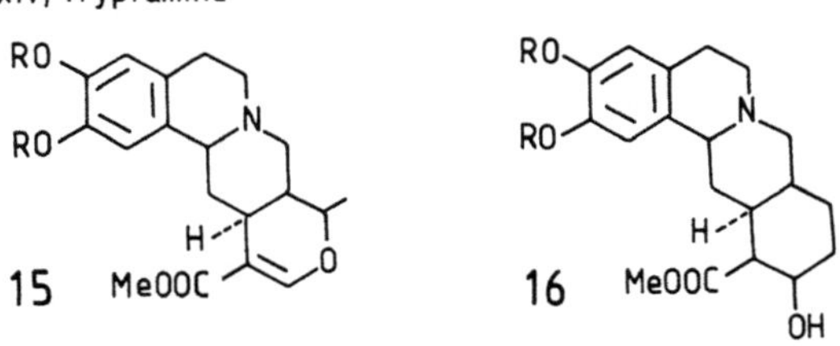

Fig. 2

quired for facile cyclisation, and it was gratifying to see this prediction realised. Subsequent heating with N,N-dimethylhydrazine opened the carbinolamine ring to form a hydrazone from the C-9 aldehyde, followed by lactamisation of the C-5 ester to afford the piperidone (10) in 85% yield.

From this key intermediate various alkaloids could be obtained by two separate routes. Firstly, Bischler-Napieralski cyclisation, reduction and hydrolysis of the hydrazone by the method of Corey and Knapp (1976) afforded protoemetine (11), which was converted via cephaeline into emetine (3) (Brossi et al. 1971). Condensation of (10) with tryptamine gave a 3:1 mixture of deoxytubulosine (12a) and deoxyisotubulosine (12b) identical with authentic samples. Alternatively, the hydrazone in (10) was hydrolysed first, and the aldehyde condensed with tryptamine to give two epimers (14a,b) which were then converted into (12a) and (12b).

3 Heteroyohimbine Analogues from Secologanin

It is a curious anomaly of Nature that dopamine-derived monoterpenoid alkaloids paralleling the common yohimbines and heteroyohimbines of the tryptamine series are unknown. There is no apparent reason why desacetylipecoside (5) should not be transformed in vivo into structures such as (15) and (16), but they have not as yet been isolated. Since we have had considerable success in synthesising both heteroyohimbines and yohimbines of the natural enantiomeric series from secologanin and tryptamine by biomimetic and other routes, we decided to explore the preparation of isoquinoline analogues as potential natural products (Sleigh 1978, Brown and Sleigh unpublished, Brown 1980, 1983).

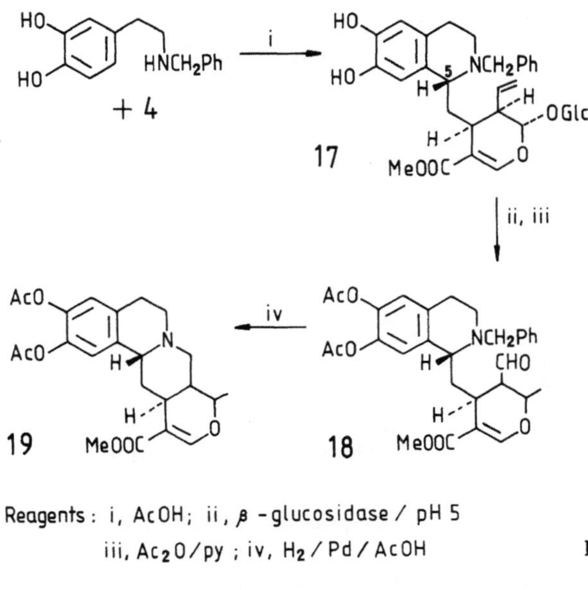

Fig. 3

In our initial approach condensation of N-benzyldopamine with secologanin afforded a single epimer – N-benzyldesacetylipecoside (17) – but unlike the indole series, glucolysis of (17) proceeded sluggishly and the rearranged aglucone (18) was only obtained in poor yield (5%). Nevertheless, catalytic hydrogenation of (18) did produce a mixture of compounds (19) with the expected spectroscopic properties (Fig. 3). Much better results were obtained by using 5-dehydrodesacetylipecoside (22) prepared by Bischler–Napieralski cyclisation of the phenethylamide (21) of secoxyloganin (Fig. 4). "One-pot" glucolysis of (22) in presence of sodium cyanoborohydride gave essentially two epimeric products (23) in over 50% yield with the same stereochemistry at the newly created chiral centres at C-2 and C-3. Rationalisation of this striking stereoselectivity can be made as in Fig. 5. Reduction of the C-5 imine gives a pair of epimers, whose aglucones both rearrange to a carbinolamine and thence to a dieneamine. Intramolecular protonation of the preferred transoid dieneamine affords the conjugated E-alkene, where steric constraints imposed by the C-4 methyl group and the β-orientation of the enolate group ensure that subsequent Michael addition can only occur from the upper (β) face and hence a β configuration for H-3. Protonation of the eneamine also occurs preferentially from the lower face to give 2,7 cis stereochemistry in the final product but the reason for this is less clear. By this procedure we had thus obtained the isoquinoline analogues of the heteroyohimbine indole alkaloids tetrahydroalstonine and akuammigine. Other stereoisomers could be obtained via the rearrangement of methyl secoxyloganin aglucone at pH5 to methyl elenolate (25) discovered in our laboratory (Brown et al. 1976). For example, rapid reductive amination of (25), Bischler-Napieralski cyclisation and reduction afforded (27), the analogue of ajmalicine, as the major product, whereas other C-2 and C-3 isomers predominated when equilibrating conditions were used (Fig. 6).

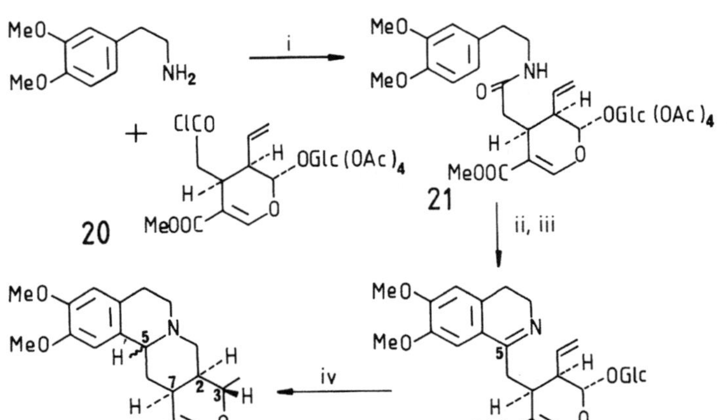

Reagents : i, Ph Me; ii, $POCl_3$; iii, NaOMe ; iv, β-glucosidase/pH 5/ $NaCNBH_3$ **Fig. 4**

Recently the interesting prospect of making isoquinoline analogues of yohimbine has appeared with our discovery that an alternative rearrangement of methyl secoxyloganin aglucone can be induced (Brown and Pratt 1980). At pH7 a highly stereose-

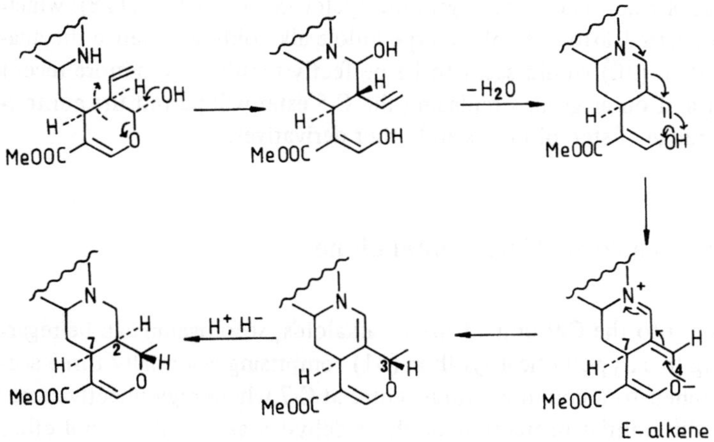

Fig. 5

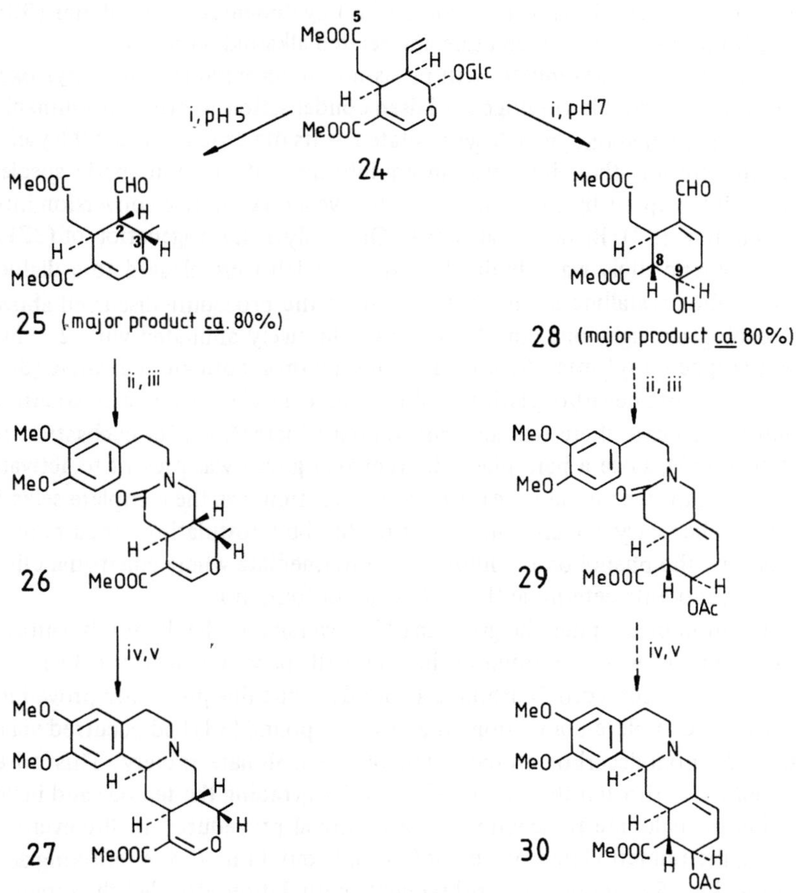

Reagents : i, β-glucosidase ; ii, 3, 4-dimethoxy-phenethylamine/ $NaCNBH_3$;
iii, Ac_2O/py ; iv, $POCl_3$; v, $NaBH_4$

Fig. 6

lective vinylogous aldol reaction occurs to give the cyclohexene aldehyde (28), which has already been converted into yohimbine type indole alkaloids, and hence the analogous sequence (28) to (30) would seem to be perfectly feasible. Even more recent variants with acetal and other groups replacing the C-5 ester will permit the controlled production of a range of stereoisomers and other derivatives.

4 Total Syntheses via a Novel Cyclopentenolone

In its transformation into the C-9 unit of ipecac alkaloids, secologanin can be regarded as corresponding to a hypothetical synthon (31) comprising essentially three acetaldehyde units attached to a common chiral centre at C-7 where regioselective reaction is ensured by differential protection of the aldehydes as acetal or enol ether groups. We envisaged that a lactol such as (35) would be an analogue of dihydrosecologanin aglucone and hence constitute a synthetic equivalent of the synthon (31). Very recently we have synthesised (35) via a novel cyclopentenedione dimer (32), and converted it into racemic protoemetine and related alkaloids (Fig. 7).

Treatment of methyl 2-hexenoate with potassium methoxide and dimethyl oxalate resulted in a novel double vinylogous Claisen condensation to afford presumably an unstable cyclopentenedione, which was isolated as its dimer (32) in ca. 50% yield. Importantly, in solution the dimer was in equilibrium with a monomeric species which could be intercepted by malonate anion to give a quantitative conversion into the cyclopentenolone (33) (Brown et al. 1984). On catalytic hydrogenation of (33) a pair of diastereomeric diols was obtained, both of which were cleaved to a dialdehyde isolated as the crystalline lactol (35). Following the procedure discussed above for dihydrosecologanin aglucone, the lactol was reductively aminated with 2-3'-hydroxy-4'-methoxyphenethylamine to afford the tetrahydronicotinate derivative (36). Selective hydrolysis and decarboxylation of the β-amino-acrylate ester again occurred as before, but in this case there was also concomitant Picter-Spengler cyclisation to the tricyclic base (37), since a para phenolic hydroxyl group was present to activate the aromatic ring. Another notable feature of this reaction was the complete selectivity for 5β stereochemistry, which could be rationalised by postulating a requirement for axial attack by the phenol on the immonium intermediate where equatorial ethyl and malonate substituents determine the half-chair conformation.

After methylation of the phenolic group in (37), inversion to the thermodynamically preferred 5α epimer was attempted by heating with propionic acid for 36 h – a standard method for tetrahydro-β-carboline alkaloids – but this procedure proved ineffectual. However, complete conversion to a new compound (38) had occurred via a slow but smooth monodecarbomethoxylation of the malonate moiety. This novel process enabled us to shorten the projected route by avoiding the tedious and inelegant conversion of malonate to acetate by conventional procedures. In the event, it was found that the unmethylated precursor (37) did equilibrate C-5 in refluxing acid to give majorly the H5-α epimer, and subsequent methylation afforded the required product (39). Yet further abbreviation of the route proved feasible by heating the tetrahydronicotinate (36) itself in propionic acid when both decarbomethoxylations,

Fig. 7

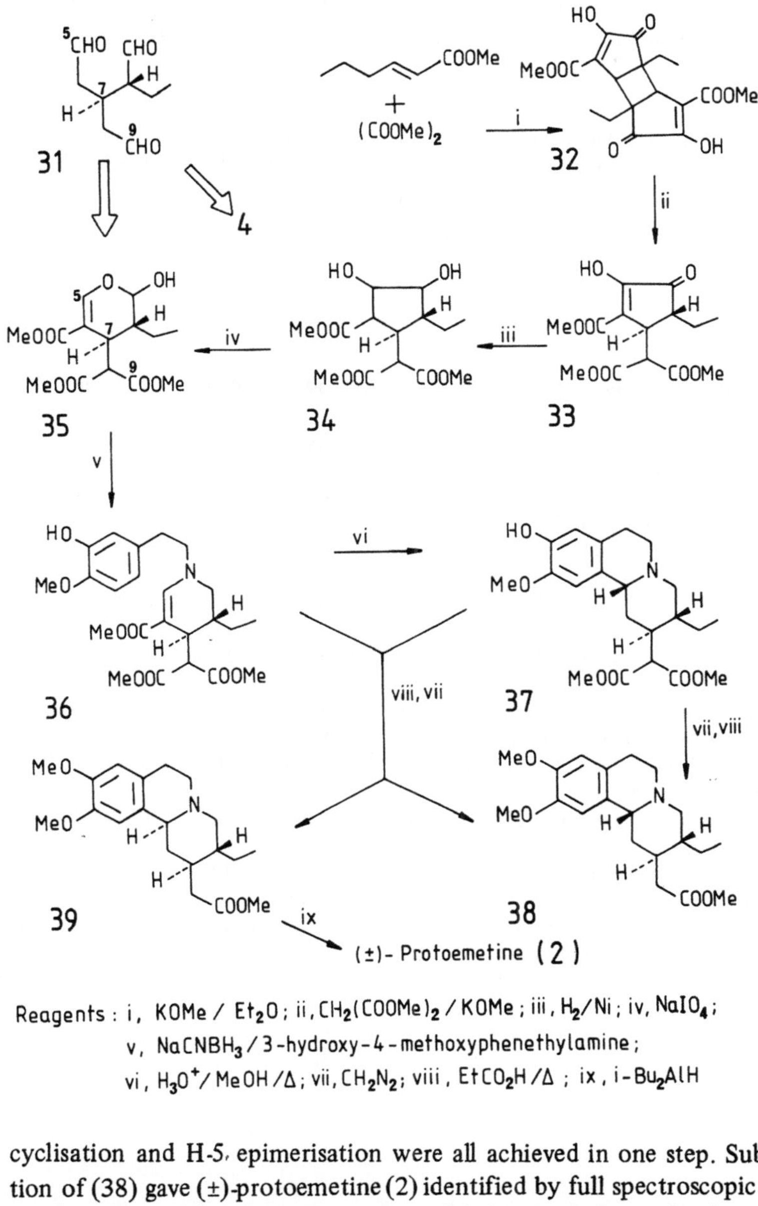

Reagents : i, KOMe / Et_2O; ii, $CH_2(COOMe)_2$ / KOMe ; iii, H_2/Ni ; iv, $NaIO_4$;
v, $NaCNBH_3$ / 3-hydroxy-4-methoxyphenethylamine ;
vi, H_3O^+/ MeOH /Δ; vii, CH_2N_2; viii, $EtCO_2H$ /Δ ; ix, i-Bu_2AlH

cyclisation and H-5, epimerisation were all achieved in one step. Subsequent reduction of (38) gave (±)-protoemetine (2) identified by full spectroscopic analysis and by condensation with tryptamine to form (±)-deoxytubulosine (12a) and its C-9 epimer (12b) (Brown and Jones, 1984). Since protoemetine has previously been converted into cephaeline and emetine, our sequence also constitutes a formal total synthesis of these alkaloids.

In addition to the above products with natural relative stereochemistry, this synthetic approach can be easily modified to produce the other diastereoisomers. Not only is the 5β epimer of protoemetine readily available from (38), but by using dimethoxyphenethylamine as in the synthesis of protoemetine from secologanin discussed above (cf. Fig. 2) the roles of C-5 and C-9 in the lactol (35) can be interchanged and yield 2,7 cis isomers. An even more interesting prospect lies in achieving enantioselection during the original Michael reaction of the dimer by the use of chiral substrates or bases.

Acknowledgements. Thanks are due to my dedicated colleagues, without whom none of this work would have been possible: Drs. W.P. Blackstock, S.B. Fraser, A.G. Lashford, S.K. Sleigh, J. Leonard, S.B. Pratt, Messrs. M. Wingfield and M.F. Jones.

References

Battersby AR, Parry RJ (1971) Biosynthesis of the ipecac alkaloids and of ipecoside. J Chem Soc Chem Commun:901–902
Brossi A, Teitel S, Parry GV (1971) In: Manske RHF (ed) The alkaloids, vol XIII, chap 3. Academic, London New York, and references therein
Brown RT (1980) In: Phillipson JD, Zenk MH (eds) Indole and biogenetically related alkaloids. Academic, London New York, pp 171–184
Brown RT (1983) In: Yoshida Z-I, Ise N (eds) Biomimetic chemistry. Kodansha/Elsevier, Amsterdam, p 21
Brown RT, Jones MF (1984) Total synthesis of monoterpenoid isoquinoline alkaloids. Tetrahedron Lett:3127–3130
Brown RT, Pratt SB (1980) Stereoselective synthesis of yohimbine alkaloids from secologanin. J Chem Soc Chem Commun:165–167
Brown RT, Chapple CL, Duckworth DM, Platt R (1976) Conversion of secologanin into elenolic acid and 18-oxayohimban alkaloids. J Chem Soc Perkin I:160–162
Brown RT, Lashford AG, Pratt SB (1979) Stereoconservative synthesis of ipecac alkaloids from secologanin. J Chem Soc Chem Commun:367–369
Brown RT, Blackstock WP, Wingfield M (1984) A simple synthetic route to substituted cyclopentenolones. Tetrahedron Lett:1831–1834
Corey EJ, Knapp S (1976) Facile conversion of N,N-dimethylhydrazones to carbonyl compounds by cupric ion-catalyzed hydrolysis. Tetrahedron Lett:3667–3668
Nagakura N, Höfle G, Coggiola D, Zenk MH (1978) The biosynthesis of the ipecac alkaloids and of ipecoside and alangiside. Planta Med 34:381–389
Sleigh SK (1978) Ph D Thesis, Univ Manchester

The Biosynthesis of Isoquinoline Alkaloids

R.B. HERBERT[1]

1 Introduction

The tetrahydroisoquinoline skeleton occurs in numerous plant alkaloids. The simplest bearers of this structural unit are exemplified by anhalonidine (12) (Sect. 2). Slightly more elaborate are the benzylisoquinoline alkaloids exemplified at their simplest by reticuline (25) and papaverine (27) (Sect. 3).

In many of the isoquinoline alkaloids the tetrahydroisoquinoline skeleton is clearly seen as part of the structure, in others it is somewhat obscured, e.g. chelidonine (76) (Sect. 3.4), and in still others, e.g., colchicine (77) (Sect. 4), which is a phenethylisoquinoline, and the *Erythrina* alkaloids (Sect. 3.3) severe modification in the course of biosynthesis has obliterated any suggestion of an isoquinoline moiety.

2 Simple Isoquinolines

The simple isoquinolines, anhalonidine (12) and anhalamine (13) are biosynthesized in the peyote cactus along with the hallucinogenic phenethylamine, mescaline (6). It is clear, by inspection, that the isoquinoline alkaloids, (12) and (13), are formed from an Ar-C_2-N unit like that seen in mescaline (6) with the inclusion of an additional C_1 or C_2 unit.

The pathways leading to these bases, which begin in primary metabolism with the α-amino-acid, tyrosine (1) are known in considerable detail. This is the result of quite extensive experiments particularly with various phenethylamine precursors (Lundström 1971, Herbert 1980). A key intermediate is 3,4-dihydroxy-5-methoxyphenethylamine (5) which stands at a fork leading separately to mescaline (6) and the isoquinoline alkaloids: methylation at C-3 leads to mescaline (6) whereas methylation at C-4 ultimately affords anhalonidine (12) and anhalamine (13) (Fig. 1). The pathways which have been deduced leading to mescaline (6) and to (7) are nicely supported by a study which explored the specificity of methylation by an O-methyltransferase isolated from peyote (Basmadjian and Paul 1971, Basmadjian et al. 1978).

All but one and two skeletal carbon atoms in anhalamine (13) and anhalonidine (12), respectively, derive then from tyrosine (1). The origin of these skeletal carbon

1 Department of Organic Chemistry, University of Leeds, Leeds LS2 9JT, United Kingdom

The Chemistry and Biology of Isoquinoline
Alkaloids, ed. by Phillipson et al.

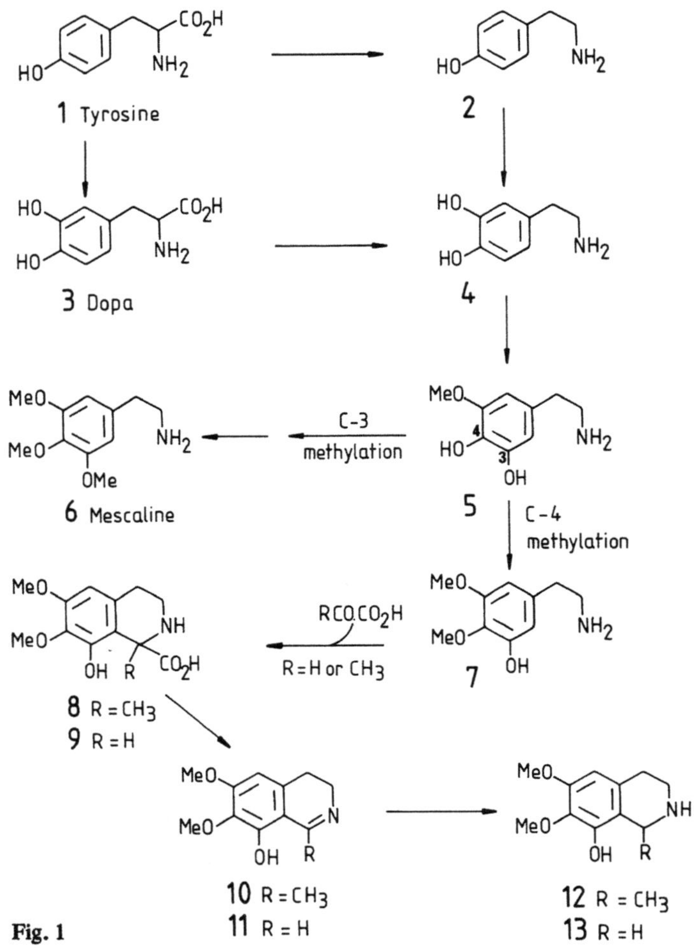

Fig. 1

atoms proved quite elusive until it was found that the amino acid (8) was implicated in anhalonidine biosynthesis. It was specifically incorporated into the alkaloid (12), is a natural peyote constituent, and was decarboxylated by fresh peyote slices to the intermediate imine (10) (Kapadia et al. 1970).

Thus the biosynthesis of anhalonidine (12) involves condensation of pyruvic acid with (7) to give (8) which then leads on to the alkaloid. A similar pathway is adduced for anhalamine (13), where glyoxylic acid replaces pyruvic acid (Kapadia et al. 1970). We have found in a case with some parallels that the β-carboline alkaloids, harman (17) and eleagnine (16) are biosynthesized via the naturally occuring amino-acid (14); (15) is also implicated as an intermediate (Herbert and Mann 1982) (Fig. 2). We are currently examining the chirality of precursor (14) as a final test of its intermediacy in β-carboline biosynthesis.

Lophocerine (19) derives from tyrosine and the C_5 unit, formed from C-1 and the pendant isobutyl group, derives from mevalonic acid. In contrast to the isoquinoline alkaloids just discussed the molecule which condenses with a phenethylamine inter-

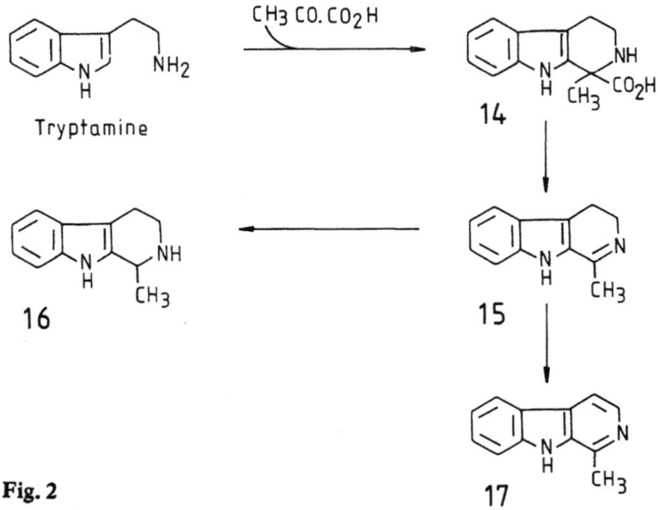

Fig. 2

Fig. 3

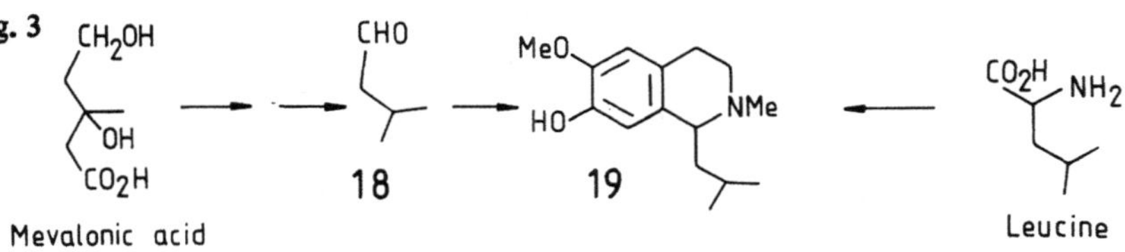

mediate is an aldehyde: (18). An alternative source of the C_5 unit is leucine, which is used apparently not via mevalonic acid (Fig. 3) (O'Donovan and Barry 1974).

3 Benzylisoquinoline Alkaloids

The 1-benzyltetrahydroisoquinoline skeleton is a very important one in the field of plant alkaloids and is the subject of various, interesting modifications. Indeed, with the exception of terpenoid indole alkaloids, no other skeleton is so diversely modified. In this family of benzylisoquinolines, reticuline (25) is preeminent, being a key intermediate in the biosynthesis of many alkaloids.

The benzylisoquinoline skeleton is formed from two molecules of tyrosine, and one of these molecules is utilized via dopamine (Fig. 4). From the work of Professor Zenk and his colleagues it is now clear that the other tyrosine molecule is used either via 4-hydroxyphenylacetaldehyde (20) or 3,4-dihydroxyphenylacetaldehyde (21) yielding, respectively, (22) or norlaudanosoline (23) (Schumacher et al. 1983). It is worth noting that labelled norlaudanosoline (23) was the first compound ever to be tested as an alkaloid precursor apart from α-amino acids and closely related compounds (Battersby et al. 1964). The successful study of plant alkaloid biosynthesis with advanced precursors was built upon that first successful experiment, which was

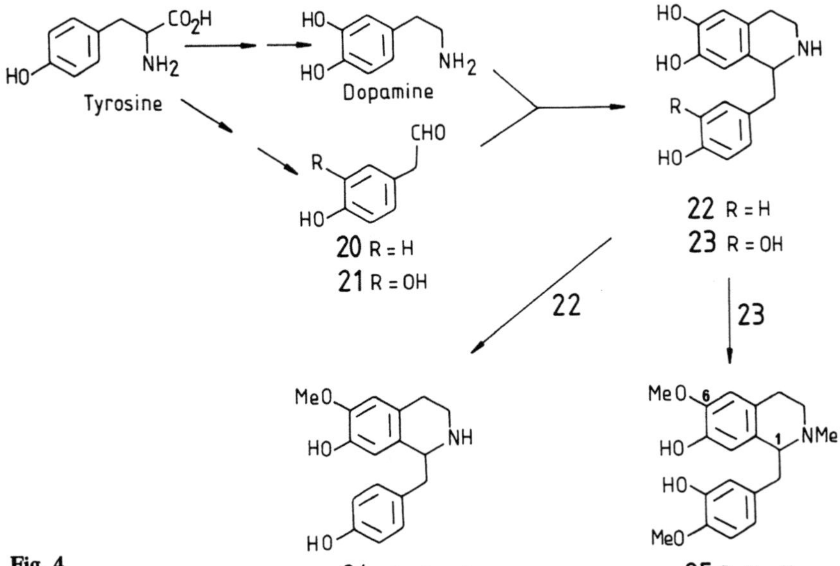

Fig. 4

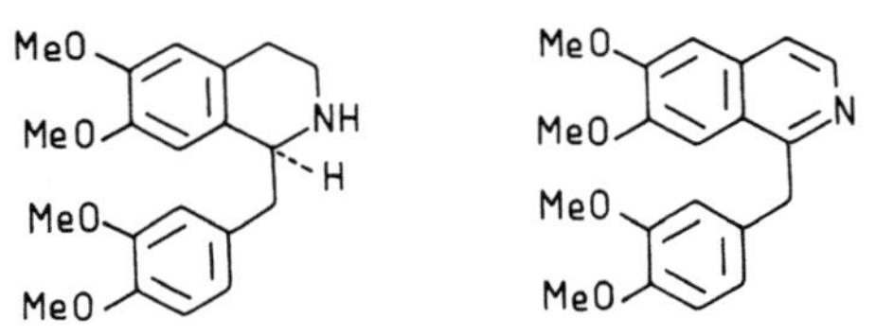

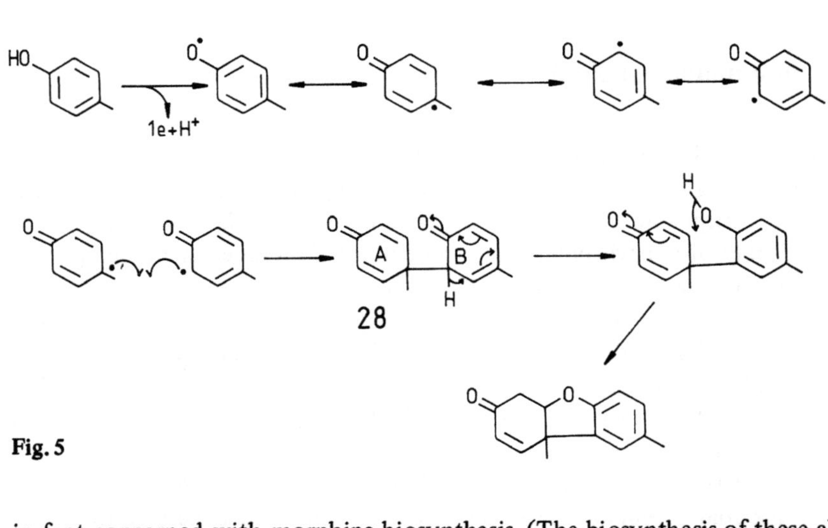

Fig. 5

in fact concerned with morphine biosynthesis. (The biosynthesis of these alkaloids is dealt with by Brochmann-Hanssen in Chap. 15).

Appropriate O- and N-methylation of (23) and (22) then affords alkaloids such as reticuline (25) and coclaurine (24). Papaverine (27) is formed specifically via the tetra-O-methylated isoquinoline, (S)-(–)-tetrahydropapaverine (26) (Brochmann-Hanssen et al. 1975, Uprety et al. 1975).

3.1 Aporphine Alkaloids

A single bond between two aromatic rings is all that distinguishes the group of aporphine alkaloids, e.g.. bulbocapnine (32), from alkaloids like reticuline (25). Hypothesis, supported by the results of numerous experiments, indicates that this bond is introduced by the coupling of two phenolic radicals (Herbert 1980). In this hypothesis, put forward some years ago now (Barton and Cohen 1957), it was proposed that one-electron oxidation of two phenolic nuclei takes place to give two mesomeric radicals. Coupling of the two aromatic nuclei then occurs ortho or para to each of the hydroxy-groups (Fig. 5). Rearomatization occurs after coupling provided that the site of coupling has a hydrogen atom present [as in ring B of (28)]; the presence instead of an alkyl group at this site secures the ring in the dienone form [as in ring A of (28)]. Subsequent "dienone-phenol" rearrangement, or reduction followed by "dienol-benzene" rearrangement rearomatizes this ring (Fig. 6).

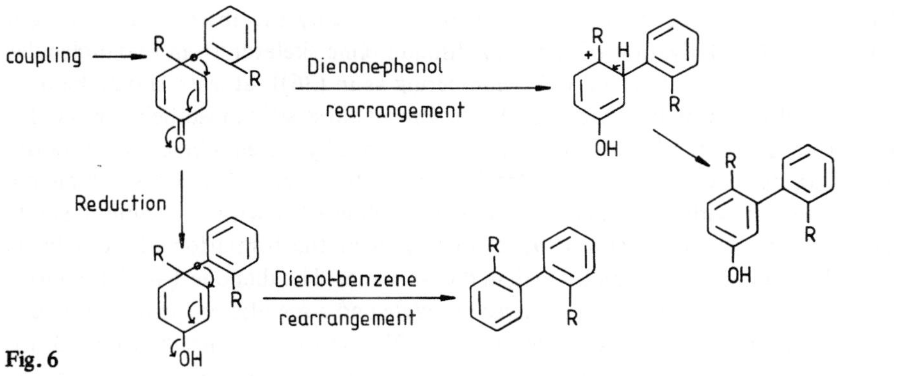

Fig. 6

Instead of C-C bond formation, coupling may also occur via one of the phenolic oxygen atoms, and this is exemplified many times over in the biosynthesis of bisbenzylisoquinoline alkaloids, e.g.. cocsulinin (30) which derives from two molecules of (S)-N-methylcoclaurine (29) (Bhakuni et al. 1978).

It is important to note that according to this hypothesis oxidative coupling can only occur if a phenolic hydroxy-group is present on each aromatic ring, and coupling can occur only ortho or para, never meta, to these hydroxy-groups. The correctness of these statements has been established by the results of numerous experiments with diverse living plants, although it is known that some deviations occur in Nature.

The biosynthesis of aporphine alkaloids illustrates splendidly this hypothesis of phenol oxidative coupling; most of the possible ways of joining two aromatic residues have been observed. The results obtained show that a pathway is in general not predictable by inspection, and is only elucidated by careful experimentation. Further, as we shall see, the methylation pattern in the alkaloid may be misleading in that it suggests a different pathway to that eventually established. As one example, the alkaloid, boldine (34), is formed by a different pathway in each of two different plants (Fig. 7).

The simplest sequence of biosynthesis is found in the formation of bulbocapnine (32). Reticuline (25) is the immediate benzylisoquinoline precursor. Oxidative coupling ortho to each of the phenolic hydroxy-groups occurs to give (31) from which bulbocapnine (32) is then derived (Blaschke et al. 1974).

Reticuline (25) is also the key precursor for isoboldine (33), which is a minor alkaloid of *Papaver somniferum* (Brochmann-Hanssen et al. 1973) (the hydrophenanthrene alkaloids of this plant, e.g. thebaine (35) also derive from reticuline). Orientaline (36) with a methylation pattern different to that in reticuline (25) was not incorporated into (33). Coupling of the two rings in reticuline (25) is thus ortho to a hydroxy-group on one ring and para on the other.

The importance of a particular O-methylation pattern in deciding the product of coupling is illustrated in an excellent way by the results of a study on the biosynthesis of thebaine (35) and isothebaine (39) in *Papaver orientale.* Thebaine (35) is formed via reticuline (25) but not via orientaline (36), whereas the reverse is true for isothebaine (39): it is derived exclusively via orientaline (36) (Battersby et al. 1965) (Fig. 7).

The course of isothebaine biosynthesis is hinted at by the absence of an oxygen function at C-10. Derivation of the benzylisoquinoline skeleton from two molecules of tyrosine means that C-7 and C-4′ [numbering as in (36)], at least, should be oxygenated in alkaloids such as the aporphines. One of these sites in isothebaine is C-10. This suggested that the loss of this oxygen function is by a dienol-benzene rearrangement following oxidative coupling. Detailed results have been obtained which are entirely consistent with this suggestion. (+)-Orientaline (36) with the same absolute stereochemistry as isothebaine (39) is implicated in the formation of isothebaine (39), and is a much better precursor than the (–)-isomer. The dienone, (–)-orientalinone (37), is a natural constituent of *Papaver orientale* and a specific precursor for isothebaine. One of the diastereoisomeric alcohols (38) derived chemically from (37) is identified as the next intermediate and the one on which dienol-benzene rearrangement occurs. It is clear that this step is, like the preceding ones, stereospecific and therefore subject to enzyme catalysis (Battersby et al. 1969).

The oxygenation-methylation pattern in the *Dicentra eximia* alkaloid, corydine (44), suggested a pathway proceding through reticuline (25) or orientaline (36), but neither of these bases was utilized for the biosynthesis of either corydine or two other co-occuring alkaloids, namely dicentrine (43) and glaucine (42). However, positive results were obtained with norlaudanosoline (23). Results of experiments with a number of possible O- and N-methylated norlaudanosoline derivatives pinpointed norprotosinomenine (40) as the key precursor. The likely following steps, which includes the so-far unique dienones (41), are indicated (Fig. 7) (the point, at which N-methylation occurs, is unknown). Boldine (34) is seen to be a likely intermediate on the way to glaucine (42) and dicentrine (43), and it was found to be efficiently incorporated into both of these bases (Battersby et al. 1971).

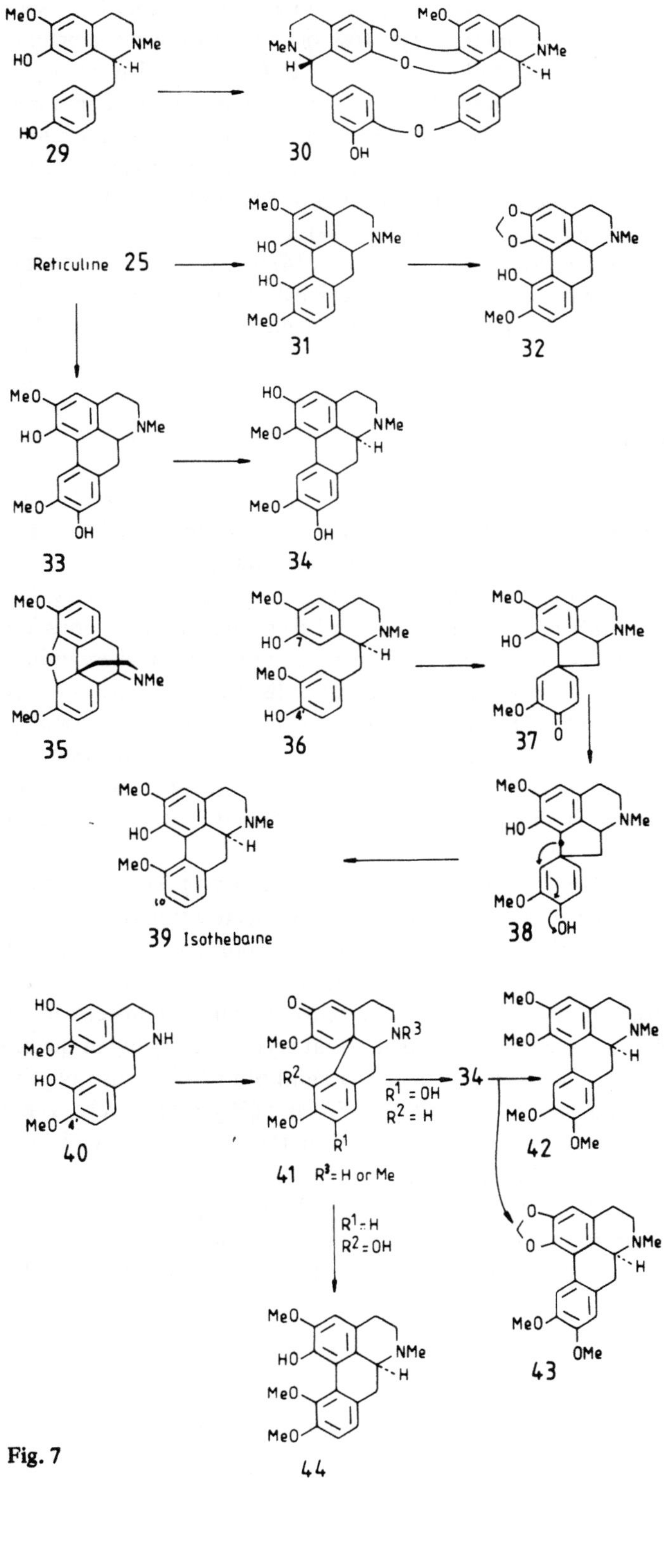
29
30
Reticuline 25
31
32
33
34
35
36
37
38
39 Isothebaine
40
41 R³= H or Me
R¹ = OH
R² = H
R¹ = H
R² = OH
42
43
44

Fig. 7

Rigorous distinction between possible benzylisoquinoline precursors has been made for boldine (34) in *Litsea glutinosa.* Norprotosinomenine (40) is not a precursor for (34) in this plant, in contrast to its biosynthesis in *D. eximia.* The methylation pattern in boldine (34) is misleading and the correct benzylisoquinoline intermediate here is (S)-reticuline [as (25)]. It follows that isoboldine (33) is the aporphine intermediate formed by oxidative coupling, and indeed it was found to be a precursor for boldine. It is also a natural constituent of the plant (Bhakuni et al. 1977).

The conversion of isoboldine (33) into boldine (34) involves demethylation and remethylation rather than methyl transfer since boldine (34) derived from [6-$O^{14}CH_3$, 1-^{3}H]reticuline [as (25)] showed a 64% loss of ^{14}C label. Parallel results were observed for the biosynthesis of the naturally occurring dienone, crotonosine (45), from (R)-coclaurine [as (24)] (Haynes et al. 1965).

3.2 Hasubanonine and Protostephanine

Consideration of the structures of protostephanine (51) and hasubanonine (52), co-occurring alkaloids of *Stephania japonica*, suggests that these alkaloids are benzylisoquinoline variants. But experimental support for this idea was a long time coming, and came only after an epic of experimentation with various possible precursors. The first conclusion was that both of these alkaloids, as befits benzylisoquinolines, *are* derived from two C_6–C_2 units originating in tyrosine, and that one of these units is (46). The second, and elusive, conclusion was that benzylisoquinolines are involved in the biosynthesis of (51) and (52) but quite uniquely they must all have two hydroxy-groups on one of the rings – ring B [as in (47) and (48)] – not just one as in all other examples. The first benzylisoquinoline precursor identified is (47) and biosynthesis takes the course shown in Fig. 8; the timing of N-methylation is apparently not critical (Battersby et al. 1981).

As one of the rings [see (48)] must have two hydroxy-groups present for oxidative coupling, a different mechanism may be suggested for this reaction to that which occurs in the biosynthesis of other benzylisoquinoline alkaloids (see, e.g. above). It seems possible that coupling here involves the formation of an ortho-quinone and bond-formation by substitution of the other ring onto this quinone instead of the coupling of radicals. We have preliminary evidence that aryl-aryl coupling in the biosynthesis of Lythraceae alkaloids, e.g., vertine (54) and lythrine (56), also requires dihydroxy-aryl precursors having so far identified (53) and (55), respectively, as precursors for (54) and (56); the corresponding mono-O-methyl derivatives failed to act as alkaloid precursors (Hedges et al. 1983), (Fig. 9).

To isolate and examine the enzymes involved in aryl-aryl coupling in these examples and to compare their properties with the enzymes involved in similar coupling in the biosynthesis of, e.g., aporphine alkaloids must surely be a most worthwhile project.

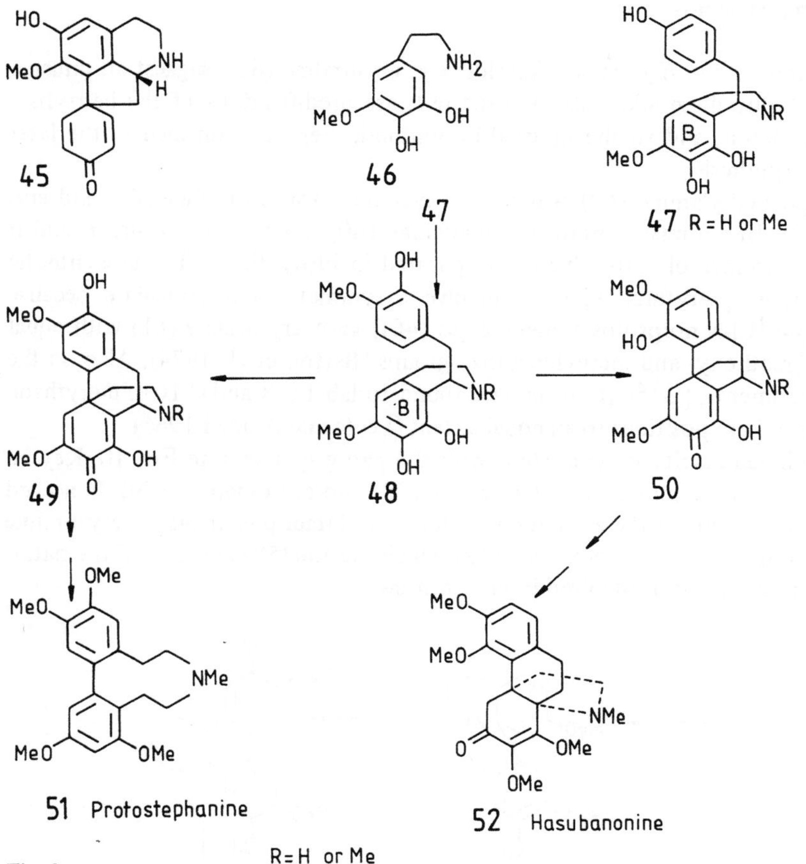
45
46
47
47 R = H or Me
48
49
50
51 Protostephanine
52 Hasubanonine
R = H or Me

Fig. 8

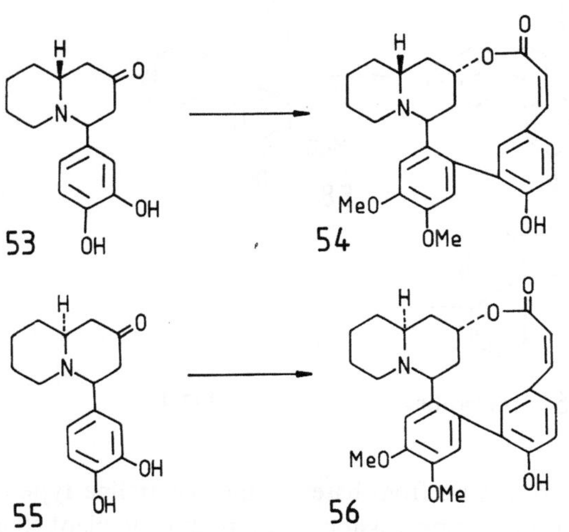
53
54
55
56

Fig. 9

3.3 Erythrina Alkaloids

The structures of the *Erythrina* alkaloids, e.g. erythraline (61), suggest an unusual biosynthesis. But these alkaloids, it turns out, are modifications of the benzylisoquinoline skeleton, and so the unusual biosynthetic steps are confined to the later stages of biosynthesis.

The bisphenethylamine (62) was a poor precursor for erythraline (61) and erythratine (60). In contrast N-norprotosinomenine (40) was a good precursor, and it is the (+)-(S)-isomer of (40) which is implicated in biosynthesis. The biosynthetic pathway to, e.g., erythraline, must involve a symmetrical intermediate because [4′-methoxy-^{14}C]-N-norprotosinomenine [as (40)] gave erythraline (61) with equal labelling of methoxy- and methylenedioxy-groups (Barton et al. 1974). Also, in the earliest experiments, [2-^{14}C]tyrosine was found to label C-8 and C-10 of β-erythroidine (63) and these positions to an equal extent (Leete and Ahmad 1966).

The combined results are consistent with the pathway shown in Fig. 10. Key intermediates are the symmetrical intermediate (58) and erysodienone (59). In accord with hypothesis both of these compounds served as intact precursors for erythraline (61). Moreover, only (–)-erysodienone (59) which has the (5S)-chirality of the naturally occurring alkaloids, is involved in biosynthesis.

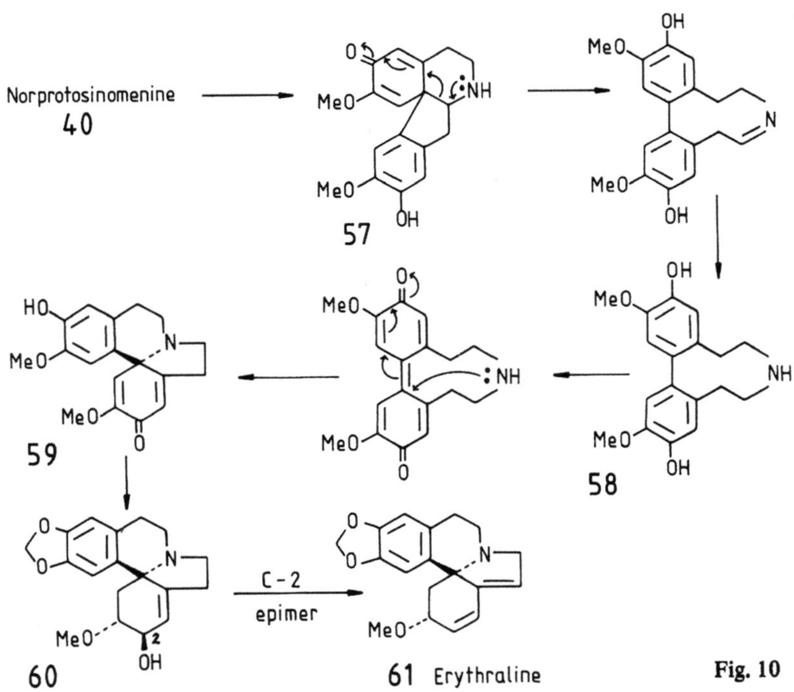

Fig. 10

The lactonic erythroidines, e.g., (63), arise from bases of the erythraline type in a sequence which clearly involves aromatic ring scission – a rare thing in plant alkaloid biosynthesis.

Coccuvine (64) is one of three *Erythrina* bases from *Cocculus laurifolius* to have its biosynthesis investigated. All three of these bases lack an oxygen function at C-16 and so the pathway just illustrated must be appropriately modified at some point to account for this. The incorporation of N-norprotosinomenine (40) was with loss of the O-methyl group at C-7 and complete retention of the 4′-O-methyl group, measured relative to a secure internal tritium marker. It follows that, unlike the biosynthesis of erythraline, no symmetrical intermediate is involved in coccuvine (64) biosynthesis (Bhakuni and Jain 1980).

It is interesting to note the use made of the dienone (41) = (57) in the biosynthesis of both *Erythrina* alkaloids of the Leguminosae and the *Dicentra* aporphine alkaloids of the Papaveraceae.

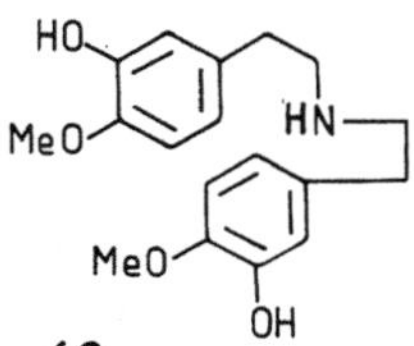

62

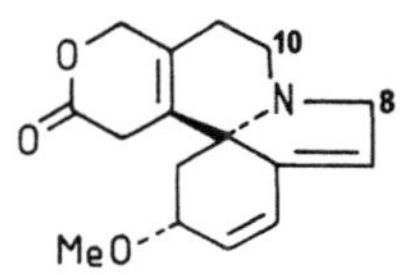

63

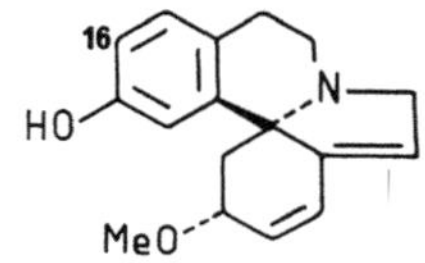

64

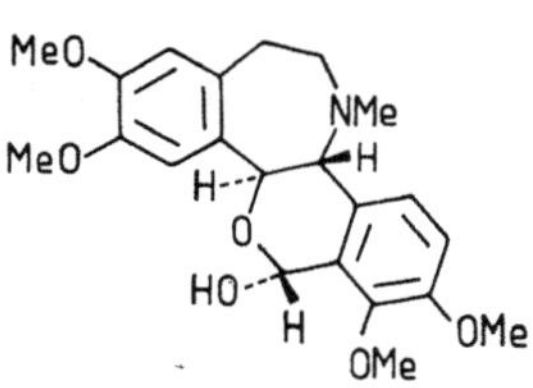

65

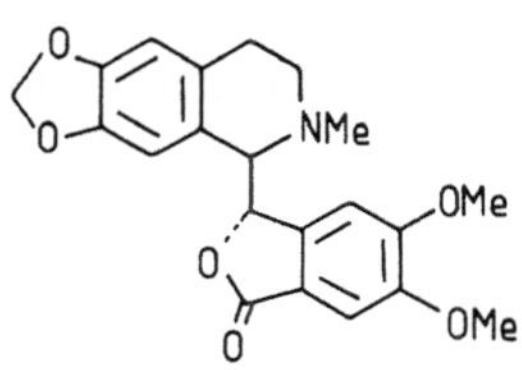

66

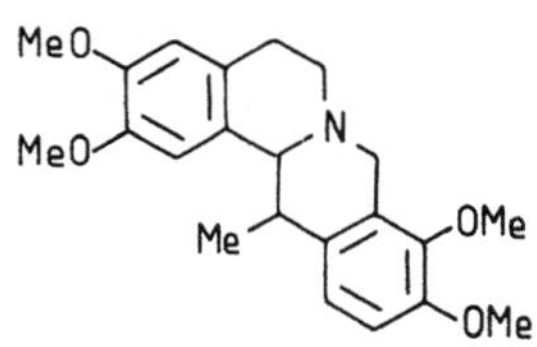

67

Fig. 11

3.4 Berberine and Related Alkaloids

The essential difference between the simple benzylisoquinoline skeleton, as in reticuline (25), and that of berberine (69) is the presence of an extra carbon atom (C-8): the so-called "berberine bridge". Less oxidized than berberine are the protoberberine alkaloids, e.g. scoulerine (73). This protoberberine skeleton serves as a substrate for further modification leading to alkaloids such as protopine (75), alpinigenine (65), chelidonine (76), corydaline (67) and hydrastine (66) (Herbert 1980). To do justice to all the various alkaloids would require a separate chapter and so we must be content with a brief outline of the biosynthetic routes to berberine (69) and to chelidonine (76), which are illustrative.

Berberine (69) is formed from (S)-reticuline [as (25)] via canadine (68), and the N-methyl group of reticuline provides the berberine-bridge carbon, plausibly as shown (Fig. 12) (Barton et al. 1965, Battersby et al. 1975). It may be noted that methylene-dioxy-groups, as in (68), originate through a similar oxidation of an *ortho*-methoxy-phenol, which proceeds plausibly as (70) → (71) → (72) (e.g. Barton et al. 1963).

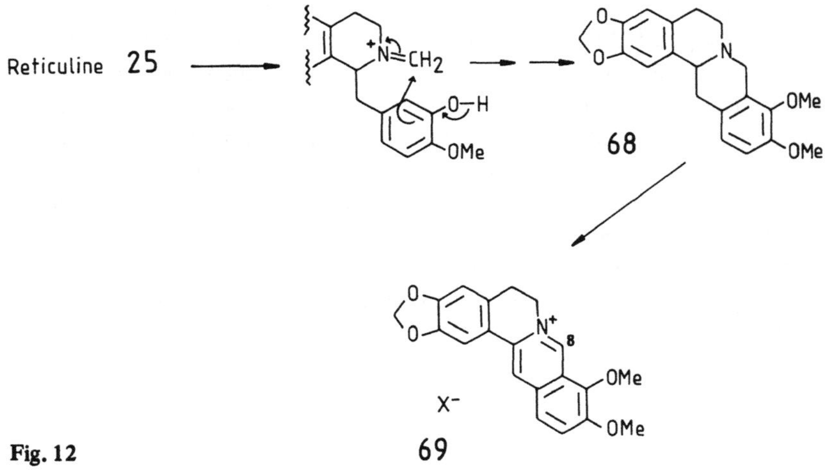

Fig. 12

(S)-Scoulerine (73) derived from (S)-reticuline [as (25)] is the source via (S)-stylopine (74) for chelidonine (76). Transformation of stylopine (74) to chelidonine (76) occurs with retention of tritium label at C-5 and C-8, but loss of tritium from C-14, and loss of the (pro-S)-hydrogen atoms from C-13 and C-6. In each case proton removal is from the α-face of the biosynthetic intermediates.

The suggested pathway to chelidonine (76) is illustrated in Fig. 13 (Battersby et al. 1979, Takao et al. 1976, 1983).

4 Colchicine

Most notable in an unusual structure for the non-basic alkaloid, colchicine (77), is the tropolone ring. Robinson thought that the alkaloid was "*sui generis* and no structural relation with other alkaloids [could] be recognised" (Robinson 1955). And so it seemed for a long time.

Phenylalanine, via cinnamic acid, serves as a C_6—C_3 unit which accounts for ring A and carbons 5, 6, and 7 (Herbert 1980). The tropolone moiety originates in tyrosine: label from [4'-^{14}C] and [3-^{14}C]-tyrosine were incorporated into colchicine (77) and appeared, respectively, at C-9 and C-12; label at C-1 and C-2 of tyrosine is lost (Herbert 1980). From the results with labelled tyrosine it is clear that the tropolone ring in colchicine (77) is formed from the benzene ring of this amino-acid by ring-expansion with the inclusion of a single carbon from the side chain, i.e. C-3. But how?

The answer lay with the structure of androcymbine, an alkaloid produced by *Androcymbium melanthioides*: this plant is a relative of *Colchicum autumnale* which is a rich source of colchicine. Androcymbine was assigned the structure (78) (Battersby et al. 1972a). Manifestly it arises by oxidative coupling within the phenethylisoquinoline (80). Phenethylisoquinolines were then quite unknown as naturally occurring alkaloids, but subsequently a number of alkaloids based on this skeleton have been identified, namely melanthioidine, some homoaporphines, schelhammeridine, and cephalotaxine.

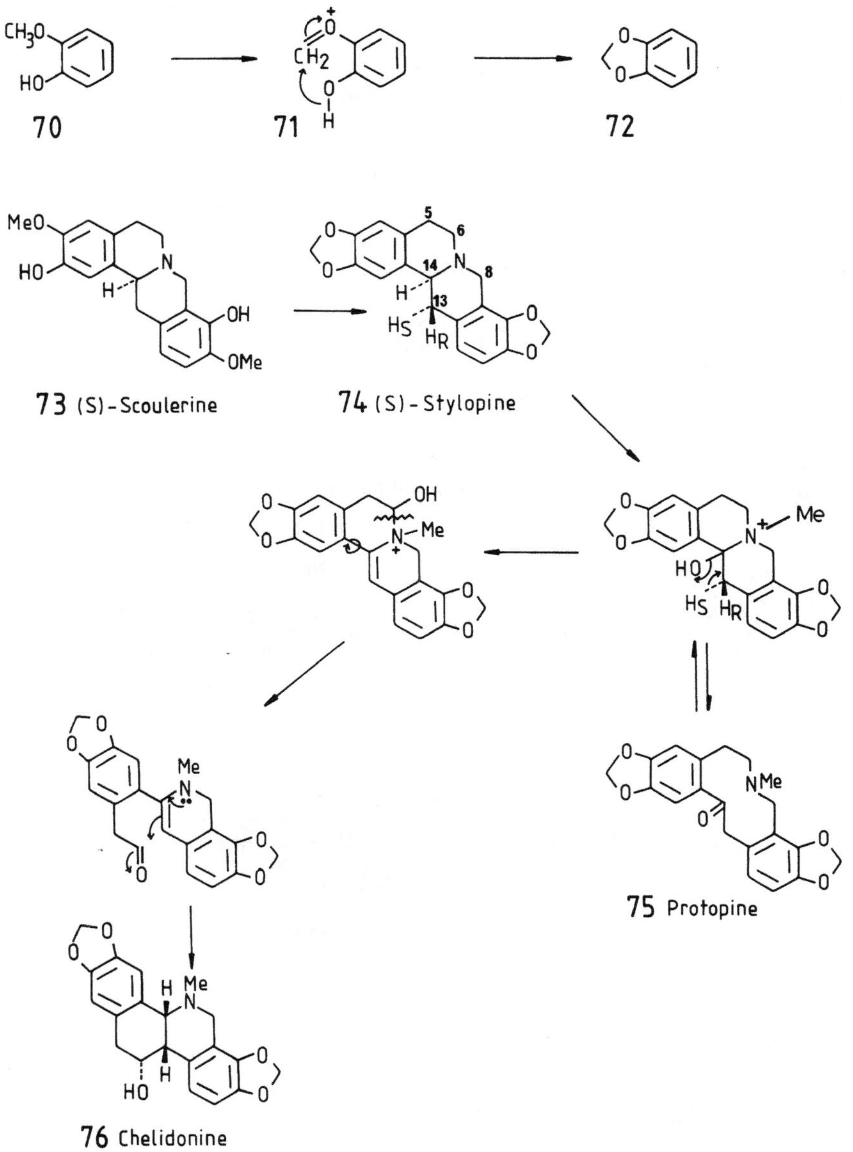

Fig.13

Of singular value in unravelling the biosynthesis of colchicine (77) was the idea that this alkaloid could be derived from the androcymbine skeleton and was therefore a modified phenethylisoquinoline. A reasonable way of transforming the androcymbine skeleton into that of colchicine would be to begin by forming the hydroxy-derivative (82). Homoallylic ring expansion and subsequent minor modification would yield colchicine (77) (Fig. 13).

The efficient and specific incorporations into (77) of labelled O-methylandrocymbine (79) and the phenethylisoquinoline (80), called autumnaline, clearly proved the correctness of these ideas. Further results established the pathway shown in shortened form in Fig. 14. It is to be noted that the (S)-isomer (80) of autumnaline, with the

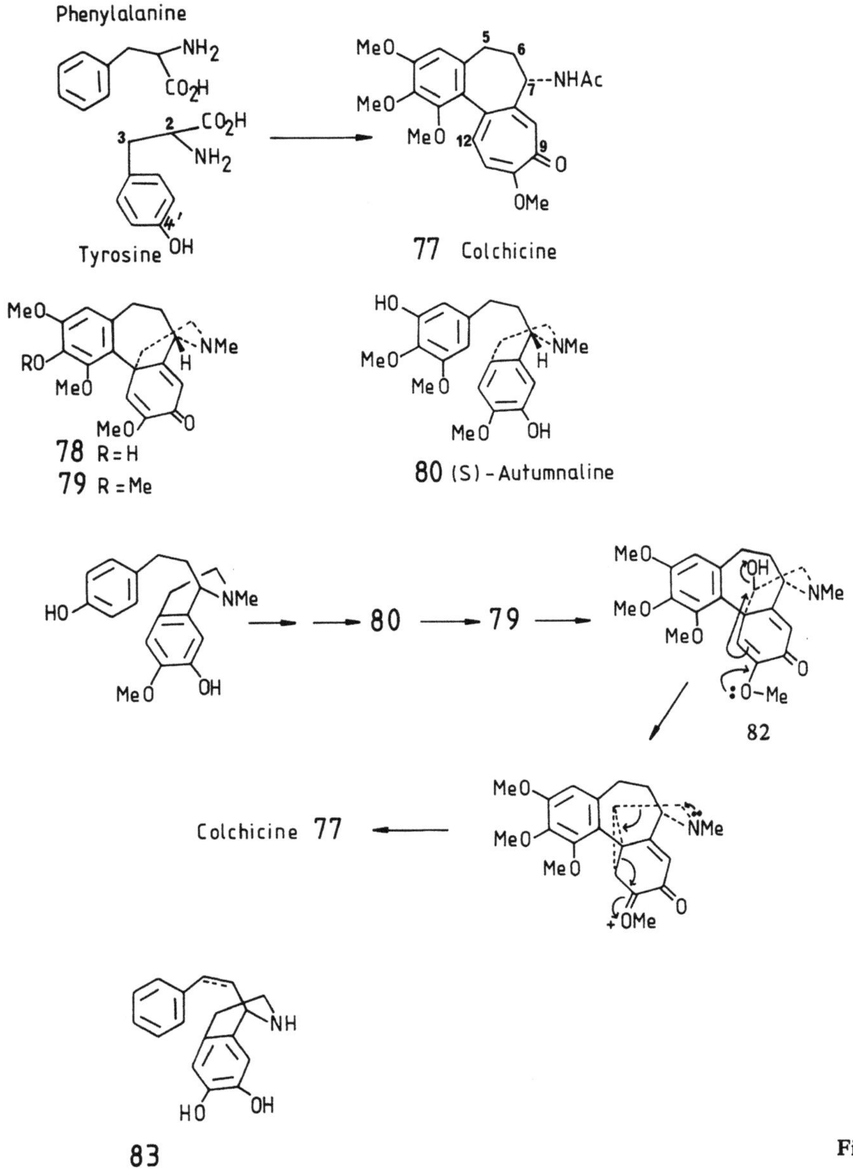

Fig. 14

same absolute configuration as colchicine (77), is involved in its biosynthesis; androcymbine and colchicine are also of the same absolute configuration. Further, tritium-labelling results demonstrate that coupling within (80) to give (79) occurs in a para-para sense rather than in the alternative ortho-para sense (Battersby et al. 1972b, 1974; Barker et al. 1967, see also Battersby et al. 1983).

The steps which lie between phenylalanine and tyrosine on the one hand and the first identified phenethylisoquinoline (81) are not completely understood but dopamine (4) is an intermediate derived from tyrosine, and cinnamic acid is an intermediate derived from phenylalanine (Battersby et al. 1972b). The next step after cinnamic acid is known not to be hydroxylation of the aromatic ring and we have some preliminary evidence that reduction of cinnamic acid to give cinnamaldehyde is the next step. Tentatively then the first phenethylisoquinoline could possibly be (83) formed by condensation of dopamine (4) with cinnamaldehyde or its dihydroderivative. Firm evidence on the structure of this first phenethylisoquinoline must await the results of further experiments.

References

Barker AC, Battersby AR, McDonald E, Ramage R, Clements JH (1967) Biosynthesis of colchicine: Ring expansion and later stages. Structure of speciosine. J Chem Soc Chem Commun: 390–392

Barton DHR, Cohen T (1957) In: Festschrift Dr A Stoll, Birkhäuser, Basel, p 117

Barton DHR, Kirby GW, Taylor JB, Thomas GM (1963) Phenol oxidation and biosynthesis, part VI. The biogenesis of amaryllidaceae alkaloids. J Chem Soc:4545–4558

Barton DHR, Hesse RH, Kirby GW (1965) Phenol oxidation and biosynthesis, part VIII. Investigations on the biosynthesis of berberine and protopine. J Chem Soc:6379–6389

Barton DHR, Bracho RD, Potter CJ, Widdowson DA (1974) Phenol oxidation and biosynthesis, part XXIV. Origin of chirality in the erythrinan system and derivation of the lactone rings of α- and β-erythroidine. J Chem Soc Perkin Trans 1:2278–2283

Basmadjian GP, Paul AG (1971) The isolation of an O-methyltransferase from peyote and its role in the biosynthesis of mescaline. Lloydia 34:91–93

Basmadjian GP, Hussain SF, Paul AG (1978) Biosynthetic relationships between phenethylamine and tetrahydroisoquinoline alkaloids in peyote. Lloydia 41:375–380

Battersby AR, Binks R, Francis RJ, McCaldin DJ, Ramuz H (1964) Alkaloid biosynthesis, part IV. 1-Benzylisoquinolines as precursors of thebaine, codeine and morphine. J Chem Soc: 3600–3610

Battersby AR, Brown RT, Clements JH, Iverach GG (1965) On the biosynthesis of isothebaine. J Chem Soc Chem Commun:230–232

Battersby AR, Brocksom TJ, Ramage R (1969) Further studies on the synthesis and biosynthesis of isothebaine. J Chem Soc Chem Commun:464–465

Battersby AR, McHugh JL, Staunton J, Todd M (1971) Biosynthesis of the apparently "directly coupled" aporphine alkaloids. J Chem Soc Chem Commun:985–986

Battersby AR, Herbert RB, Pijewska L, Santavy F, Sedmera P (1972a) Alkaloid biosynthesis, part XVII. The structure and chemistry of androcymbine. J Chem Soc Perkin Trans 1:1736–1740

Battersby AR, Herbert RB, McDonald E, Ramage R, Clements JH (1972b) Alkaloid biosynthesis, part XVIII. Biosynthesis of colchicine from the 1-phenethylisoquinoline system. J Chem Soc Perkin Trans 1:1741–1746

Battersby AR, Sheldrake PW, Milner JA (1974) Biosynthesis of colchicine: Incorporation of a ^{13}C-labelled precursor in a higher plant. Tetrahedron Lett:3315–3318

Battersby AR, Francis RJ, Hirst M, Ruveda EA, Staunton J (1975) Biosynthesis, part XXI. Investigations on the biosynthesis of stylopine in *Chelidonium majus.* J Chem Soc Perkin Trans 1:1140–1147

Battersby AR, Staunton J, Summers MC, Southgate R (1979) Studies of enzyme-mediated reactions, part 9. Stereochemistry of oxidative ring cleavage adjacent to nitrogen during the biosynthesis of chelidonine. J Chem Soc Perkin Trans 1:45–52

Battersby AR, Jones RCF, Minta A, Ottridge AP, Staunton J (1981) Biosynthesis, part 25. Proof that hasubanonine and protostephanine are biosynthesised from the 1-benzylisoquinoline system. J Chem Soc Perkin Trans 1:2030–2039

Battersby AR, McDonald E, Stachulski AB (1983) Biosynthesis, part 26. Synthetic studies on structural modification of late biosynthetic precursors for colchicine. J Chem Soc Perkin Trans 1:3053–3063

Bhakuni DS, Jain S (1980) Late stages in the biosynthesis of abnormal *Erythrina* alkaloids. Tetrahedron 36:2153–2156

Bhakuni DS, Tewari S, Kapil RS (1977) Biosynthesis of boldine (1,10-dimethoxy-6-aα-aporphine-2,9-diol). J Chem Soc Perkin Trans 1:706–709

Bhakuni DS, Labroo VM, Singh AN, Kapil RS (1978) Biosynthesis of the bisbenzylisoquinoline alkaloid cocsulin. J Chem Soc Perkin Trans 1:121–125

Blaschke G, Waldheim G, Schantz M von, Peura P (1974) Nachweis von Reticulin als Biosynthese Vorstufe in *Corydalis cava*. 4. Mitt. Untersuchungen zur Biosynthese von Alkaloiden. Arch Pharm 307:122–130

Brochmann-Hanssen E, Chen CH, Chiang H-C, Fu C-C, Nemoto H (1973) Opium alkaloids XIV: Biosynthesis of aporphines – detection of orientaline in opium poppy. J Pharm Sci 63:1291–1293

Brochmann-Hanssen E, Chen C, Chen CR, Chiang H, Leung AY, McMurtrey K (1975) Opium alkaloids, part XVI. The biosynthesis of 1-benzylisoquinolines in *Papaver somniferum*. Preferred and secondary pathways; stereo-chemical aspects. J Chem Soc Perkins Trans 1:1531–1537

Haynes LJ, Stuart KL, Barton DHR, Bhakuni DS, Kirby GW (1965) On the biosynthesis of crotonosine. J Chem Soc Chem Commun:141–142

Hedges SH, Herbert RB, Wormald PC (1983) Biosynthesis of lythraceae alkaloids: Incorporation of DL-[4,5-$^{13}C_2$,6-^{14}C] lysine and cis- and trans-4-(3,4-dihydroxyphenyl)-quinolizidin-2-one into vertine and lythrine. J Chem Soc Chem Commun:145–147

Herbert RB (1980) In: Coffey S (ed) Rodd's chemistry of carbon compounds, 2nd edn. Elsevier, Amsterdam, p 291

Herbert RB, Mann J (1982) The biosynthesis of the β-carboline alkaloids, harman and eleagnine. J Chem Soc Perkin Trans 1:1523–1525

Kapadia GJ, Rao GS, Leete E, Fayez MBE, Vaishnav YN, Fales HM (1970) On the origin of carbon 1 in tetrahydroisoquinoline alkaloids. J Am Chem Soc 92:6943–6951

Leete E, Ahmad A (1966) Biosynthesis of the *Erythrina* alkaloids. The incorporation of tyrosine-2-^{14}C into the erythroidines. J Am Chem Soc 88:4722–4725

Lundström J (1971) Biosynthetic studies on mescaline and related cactus alkaloids. Acta Pharm Suec 8:275–302

O'Donovan DG, Barry E (1974) Biosynthesis of lophocerine in *Lophocereus schottii*, part II. J Chem Soc Perkin Trans 1:2528–2529

Robinson R (1955) The structural relations of natural products. Oxford Univ Press, London

Schumacher H-M, Rüffer M, Nagakura N, Zenk MH (1983) Partial purification and properties of (S)-norlaudanosoline synthase from *Eschscholtzia tenuifolia* cell cultures. Planta Med 48: 212–22)

Takao N, Iwasa K, Kamigauchi M, Sugiura M (1976) Studies on the alkaloids of papaveraceous plants XXV. Biosynthesis of the alkaloids of *Corydalis incisa* Pers. and *Chelidonium majus* L. Incorporation of tetrahydroberberines, N-methosalts of tetrahydroprotoberberines and protopine. Chem Pharm Bull 24:2859–2868

Takao N, Kamigauchi M, Okada M (1983) Biosynthesis of benzo[c]phenanthridine alkaloids sanguinarine, chelirubine and macarpine. Helv Chim Acta 66:473–484

Uprety H, Bhakuni DS, Kapil RS (1975) Biosynthesis of papaverine. Phytochemistry 14:1535–1537

Biosynthesis of Morphinan Alkaloids

E. BROCHMANN–HANSSEN[1]

1 Introduction

By the turn of the century a large number of alkaloids had been isolated, and investigators began to turn their attention to the ways by which these natural products were constructed in the living plant. Many interesting ideas were proposed during the next decades. After World War II suitable radioisotopes became available for labelling and tracer studies, and the liquid scintillation counter was developed for accurate measurement of carbon-14 and tritium activities. Thus, by the late 1950's the time was ripe for experimental work to test the validity of the various hypotheses that had been presented. A burst of research activity focusing on the morphinan alkaloids emerged from several laboratories, and within less than a decade the major steps in the biosynthesis of morphine had been elucidated.

A guiding principle in this research was the relationship of morphine to the 1-benzylisoquinolines, as first pointed out by Gulland and Robinson (1925), who put forth the correct structure of morphine based on this biogenetic approach. They proposed that morphine (1) and related alkaloids are generated in the plant by oxidative coupling of a suitable benzyltetrahydroisoquinoline precursor, such as norlaudanosoline (2). Already 15 years earlier Winterstein and Trier (1910) had suggested that the benzylisoquinoline system in nature is built up from two units derived from 3,4-dihydroxyphenylalanine (dopa). Since dopa comes from tyrosine, the early investigations consisted of feeding experiments with radioactively labelled tyrosine, dopamine and norlaudanosoline and isolation of radioactive morphine, codeine and thebaine (Battersby and Harper 1958, Battersby and Francis 1964, Battersby et al. 1962, 1964a, Leete 1959, Leete and Murrill 1964, Kleinschmidt and Mothes 1959). When specifically labelled precursors were used, the label appeared in the morphine alkaloids in the positions predicted according to the hypothesis. It follows, therefore, that morphine is biosynthesized from two phenethyl residues, both being derived from tyrosine, and that norlaudanosoline is a likely intermediate. The biosynthesis of 1-benzyltetrahydroisoquinolines has been discussed by Herbert in the preceding chapter (Chap. 14).

Barton and Cohen (1957) carried Robinson's idea one step further, based on Pummerer's work on the oxidation of phenols to produce relatively stable radicals (Pum-

1 Department of Pharmaceutical Chemistry, University of California, San Francisco, California 94143, USA

The Chemistry and Biology of Isoquinoline Alkaloids, ed. by Phillipson et al.

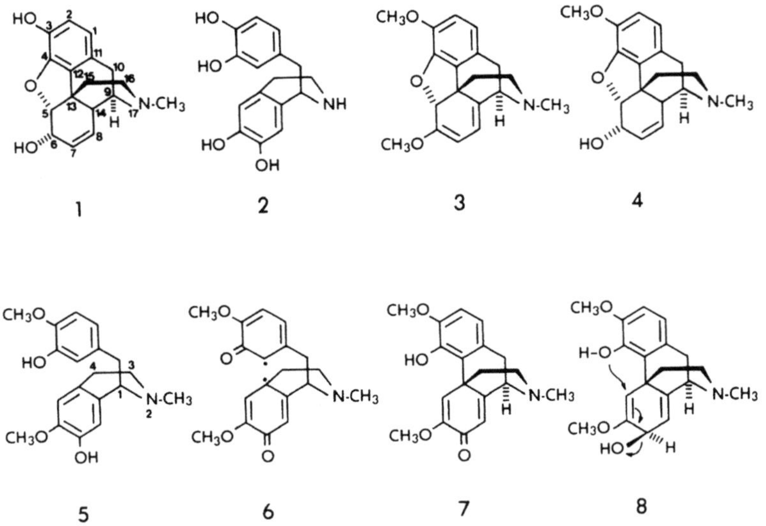

Fig. 1

merer et al. 1925). They expanded this principle to encompass the biogenesis of alkaloidal ring systems by intramolecular coupling of phenol radicals.

2 Role of Reticuline

Tracer experiments had clearly demonstrated that thebaine (3) is the first morphine-type alkaloid produced in the opium poppy and that it is converted to codeine (4) which, in turn, is demethylated to morphine (1) (Battersby and Harper 1960, Rapoport et al. 1960, Stermitz and Rapoport 1961). It was, therefore, reasonable to expect that a precursor closer to thebaine than norlaudanosoline should have an N-methyl group and two O-methyl groups in the molecule. A new benzylisoquinoline alkaloid isolated from *Annona reticulata* and named reticuline (5) filled this requirement (Gopinath et al. 1959). More importantly, reticuline had the phenolic characteristics and the substitution pattern suitable for intramolecular ortho-para coupling according to Barton and Cohen (1957) (6,7). Feeding experiments with labelled reticuline gave better incorporation into morphine, codeine and thebaine than had been achieved with either tyrosine or norlaudanosoline (Battersby et al. 1964a, Barton et al. 1965). Again, there had been no scrambling of the radioactive label. N-Norreticuline was also well incorporated but somewhat less efficiently than reticuline indicating that O-methylation precedes N-methylation. When all three methyl groups of reticuline (5) were labelled with ^{14}C in addition to a reference label in the carbon skeleton (C-3), all labels appeared in thebaine with about the same relative activities as in the precursor. Therefore, the biotransformations from reticuline to thebaine did not produce any change in the methylation pattern.

The idea of oxidative coupling of reticuline or a reticuline-like substance to a dienone proved to be correct. By an interesting coincidence, this morphinandienone

(7) had just been isolated from *Croton salutaris* and named salutaridine (Barnes 1964 in Barton et al. 1965). Labelled salutaridine showed good incorporation into morphine, codeine and thebaine, and its natural occurrence in the opium poppy was demonstrated by an isotope dilution experiment (Barton et al. 1965). Recently the oxidative coupling of reticuline to salutaridine was carried out in cell-free systems prepared from the opium poppy (Hodges and Rapoport 1982). Ginsberg (1962) and Battersby (1963) proposed that salutaridine was reduced to a dienol prior to forming the 4,5-epoxybridge. This reduction, when carried out in vitro with sodium borohydride, gave a mixture of two epimers named salutaridinol I and salutaridinol II. Although both epimers could easily be converted to thebaine in vitro, feeding experiments demonstrated that only salutaridinol I (8) had the correct stereochemistry at C-7 for in vivo conversion to thebaine by allylic elimination (Barton et al. 1965, 1967).

Until this time, all tracer experiments on *P. somniferum* had been carried out with racemic mixtures. When it was realized that natural S(+)-reticuline (9) isolated from *Annona reticulata* had the wrong stereochemistry relative to the morphine alkaloids, Battersby et al. (1965b) decided to work with the optical isomers. When (+) and (-)-reticuline, carrying a ^3H label at the asymmetric center (C-1) and ^{14}C labels at three other positions in the molecule, were fed to separate batches of opium poppies, both enantiomers were incorporated into the morphine alkaloids with about equal efficiency but with loss of ^3H. The ^3H loss was almost 100% when (+)-reticuline was the precursor but only 60–80% from (-)-reticuline feedings. These surprising results were accommodated by proposing a reversible oxidation-reduction system via 1,2-dehydroreticulinium ion (10) accounting for the loss of ^3H and permitting inversion of configuration at C-1 (Fig. 2). Indeed, the feeding of labelled 1,2-dehydroreticulinium chloride gave the best incorporation into morphinan alkaloids achieved up to that time. Later the natural occurrence of this substance was demonstrated in the opium poppy (Borkowski et al. 1978). Prior to the feeding experiments with optically active reticuline this alkaloid had been isolated from opium and found to be present in both enantiomeric forms with an excess of the S(+)-isomer over the R(-)-isomer (Brochmann-Hanssen and Furuya 1964, Brochmann-Hanssen and Nielsen 1965). This was confirmed by isotope dilution studies on the fresh plant. In the seedlings, where the biosynthesis of morphine is not active, both isomers were found to be present in about equal amounts (Battersby et al. 1965b, Martin et al. 1967). Since then, reticuline has been isolated from many plants belonging to several plant families and has been found to play an important role in the biosynthesis of a large number of alkaloids.

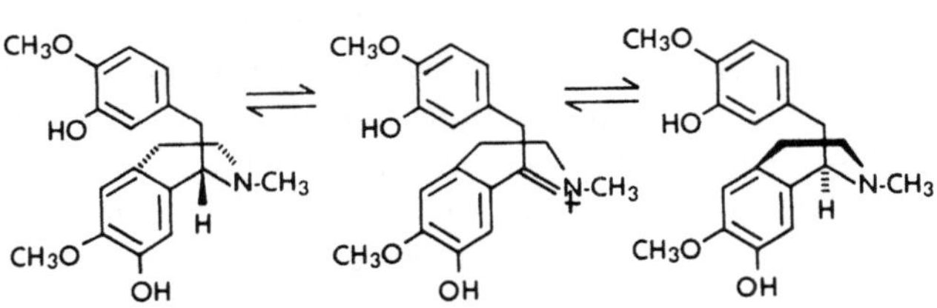

9 10

Fig. 2. Interconversion of S and R reticuline

3 Bioconversion of Thebaine to Morphine

The transformation of thebaine (3) had been considered on chemical grounds to involve hydrolysis of the enol ether to give neopinone (11) which, after isomerization to codeinone (12), could be reduced to codeine (4) (Barton and Cohen 1957) (Fig. 3). This hypothesis was tested independently by two research groups and found to be essentially correct (Battersby et al. 1967, Blaschke et al. 1967, Parker et al. 1972). Codeinone and neopinone were excellent precursors of codeine and morphine and could be detected in the plants by carrier dilution after exposure to $^{14}CO_2$. The intermediary neopinone also explained the presence of neopine (13) which had been isolated from opium (Hohmeyer and Shilling 1947).

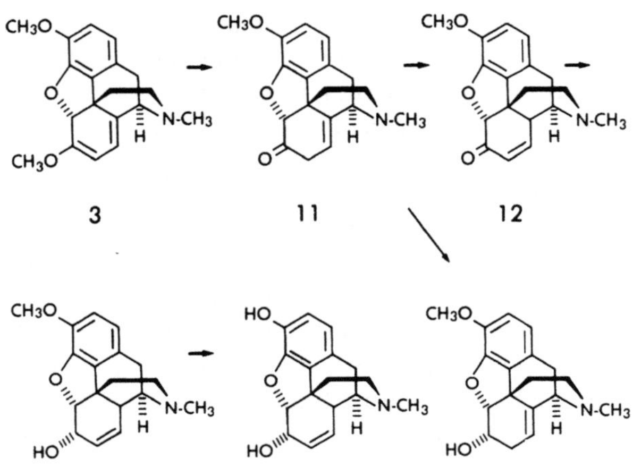

Fig. 3. Biotransformation of thebaine to morphine via codeinone and codeine

Many *Papaver* species are capable of producing thebaine, some in very high yield, but only two, *P. somniferum* and *P. setigerum*, appear to be able to generate morphine. Most other thebaine-producing plants lack the enzyme required for demethylation of the enol ether. Nonetheless, several *Papaver* species which contain thebaine are capable of 3-0-demethylation of the phenol ether to oripavine (14) which appears to represent the last step in the biosynthesis of morphinan alkaloids in these plants. In the opium poppy 6-0-demethylation of the enol ether of thebaine and 3-0-demethylation of the phenol ether of codeine proceed by the same oxidative mechanism (Horn et al. 1978) but appear to require different enzymes although these enzymes show relatively little substrate specificity. Major structural modifications of codeine and thebaine do not affect the efficiency of the dealkylations to any great extent (Kirby et al. 1972, Brochmann-Hanssen and Okamoto 1980, Brochmann-Hanssen and Cheng 1982). This lack of specificity led to speculations as to the presence of oripavine in *P. somniferum*. Recently oripavine was isolated from the dried capsules of opium poppies cultivated in Tasmania (Nielsen et al. 1983) and was later shown to be present in other chemical strains as well. This raised a question as to the metabolism of oripavine in the plant and its possible role in the biosynthesis of morphine.

When [2-^{3}H]oripavine was fed to two different strains of *P. somniferum*, very high incorporation was achieved into morphine (1) and also into morphinone (15) which was added as cold carrier during the extraction (Brochmann-Hanssen 1984). Radioactive oripavine was isolated from both chemical strains in larger amounts but of much lower specific activity than the precursor which had been fed, indicating dilution with a natural pool of oripavine. Thebaine and codeine were not radioactive. Therefore, demethylation of the phenolic ether of thebaine was not reversible. These results established the sequence: thebaine (3) → oripavine (14) → morphinone (15) → morphine (1) as a second pathway for the conversion of thebaine to morphine (Fig. 4). The close agreements and the high incorporation achieved with two chemical strains of *P. somniferum* as well as the presence of codeine and oripavine in both suggested that both pathways were operating simultaneously. The predominance of one pathway over the other may depend on the relative activities of 3-0-methyl oxidase and 6-0-methyl oxidase acting on thebaine.

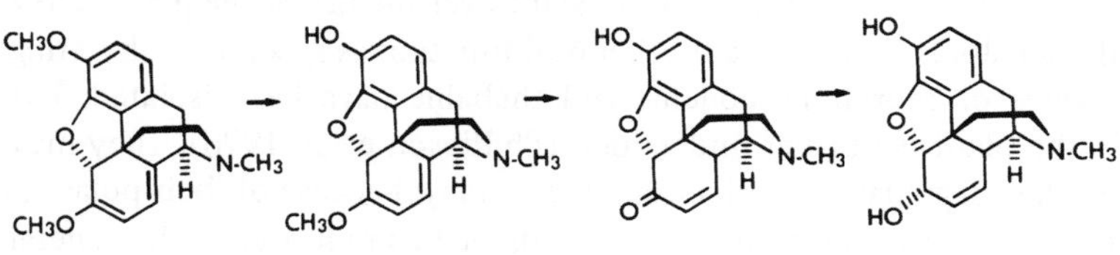

3 14 15 1

Fig. 4. Biotransformation of thebaine to morphine via oripavine and morphinone

4 Papaver bracteatum

The biosynthesis of thebaine in *P. bracteatum* follows the same pathway as in the opium poppy (Hodges et al. 1977, Brochmann-Hanssen and Wunderly 1978). In contrast to earlier studies, recent isolation work has revealed that *P. bracteatum* contains small amounts of codeine, neopine and oripavine as well as 14-β-hydroxycodeinone (16) and 14-β-hydroxycodeine (17) and both isomers of thebaine N-oxide (Küppers et al. 1976, Phillipson et al. 1976, Theuns et al. 1977, Meshulam and Lavie 1980). It appears that the O-methyl oxidases in *P. bracteatum* are extremely sluggish in comparison with the corresponding enzymes in *P. somniferum*. This leads to accumul-

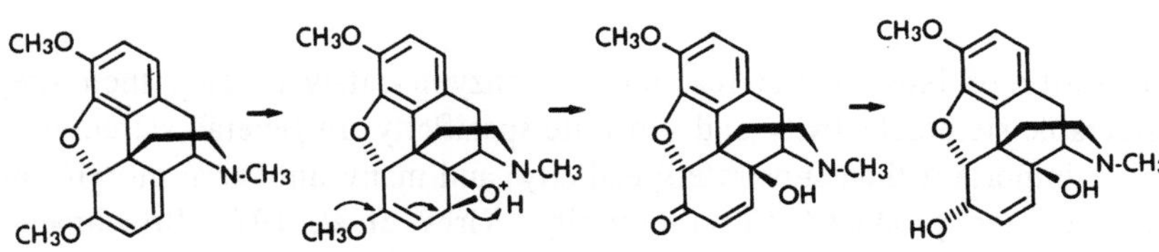

5 16 17

Fig. 5. Proposed biosynthesis of 14-β-hydroxycodeinone and 14-β-hydroxycodeine in *P. bracteatum*

ation of thebaine which is slowly and inefficiently converted to oripavine and to neopine and codeine. The unusual 14-β-hydroxyderivatives (16, 17) may be derived from thebaine via an oxygenated intermediate as illustrated in Fig. 5 (Meshulam and Lavie 1980).

5 Metabolism and in Vivo Degradation of Morphinans

The natural pools of alkaloids change and fluctuate not only as a result of the age and development of the plant but also on a daily and even hourly basis (Miram and Pfeifer 1960, Heydenreich and Pfeifer 1962, Fairbairn and Wassel 1964, Neubauer 1964). Miller et al. (1973) found that morphine was N-demethylated to normorphine in the opium poppy. This demethylation, like all other biosynthetic reactions preceding it, was irreversible and its rate was fairly constant over the life of the plant. They suggested that the alkaloids may serve a biochemical function as specific methylating agents. The N-oxides of morphine, codeine and thebaine have been isolated, but their function in the plant is not yet understood (Phillipson et al. 1976). They may represent just another degradative mechanism or they may, because of their polar nature, play a role in the translocation of the alkaloids or be implicated in the general oxidation-reduction processes of the plant.

Although morphine becomes the major alkaloid at the height of the plant's development, it is clearly not an end-product in this secondary nitrogen metabolism. Fairbairn and his coworkers demonstrated a fairly rapid metabolism of morphine to products which increased in complexity and molecular weight during the ripening of the plant. They suggested that these products may represent transitional forms in the metabolism and translocation of the alkaloids from the latex to the ovules and the seeds (Fairbairn and El-Masry 1967, 1968, Fairbairn et al. 1974, Fairbairn and Steele 1980). Morphine bound to water-soluble high molecular weight carbohydrates were found in the capsules of *P. somniferum* (Wold 1978). Whether these various products derived from morphine and other morphinan alkaloids play active roles with specific functions or whether they represent metabolic detoxification products which are channelled back into the plant's economy is still open to question.

6 Stereoselectivity and Substrate Specificity

The vast majority of biosynthetic reactions are enzyme-catalyzed and, therefore, show considerable stereoselectivity and substrate specificity. In general, stereoselectivity is more important than substrate specificity, and many unnatural morphinan derivatives have been produced biosynthetically (Kirby et al. 1972, Brochmann-Hanssen and Okamoto 1980, Brochmann-Hanssen and Cheng 1982, 1984, Brochmann-Hanssen et al. 1982). The oxidative coupling of reticuline to salutaridine as well as the succeeding biotransformations are not very substrate specific. Several reticuline analogues have been converted to the corresponding morphine and codeine analogues. On the other hand, a reaction which exhibits exceptional substrate specificity is the interconversion of the S and R isomers of reticuline, a reaction which is crucial for the production of morphine alkaloids in the opium poppy. No other natural or synthetic benzyltetrahydroisoquinoline has been found to undergo this reaction (Brochmann-Hanssen et al. 1975, 1982).

It is noteworthy that the early steps in the biosynthesis leading to reticuline involve a series of methylations which are stereoselective for the L(S)-configuration. The later steps from thebaine to normorphine consist of stepwise demethylation of compounds having the R-configuration at the equivalent asymmetric center.

The oxidative coupling of reticuline appears to be stereoselective for the R(-)-isomer in *P. somniferum.* However, there are plants which can perform this intramolecular ortho-para coupling with both isomers of reticuline, and one cannot rule out the possibility that this may also take place in the opium poppy although no product has yet been isolated to suggest this (Chambers et al. 1966, Chambers and Stuart 1968).

7 Non-Opiate Morphinans

The enantiomer of salutaridine was isolated from *Sinomenium acutum* and named sinoacutine (18) (Chu et al. 1964). It is derived from S(+)-reticuline (9) and is further converted to sinomenine (19) in the plant (Fig. 6). The steps whereby sinoacutine is converted to sinomenine are not yet known but several mechanisms have been proposed (Barton et al. 1968). Closely related morphinan derivatives which may be considered to be products of intramolecular para-para coupling of (-)- or (+)-reticuline include amurine (20), nudaurine (21), flavinantine (22), O-methylflavinantine (23) and pallidine (24). It is worth noting that the dienol nudaurine has the same stereochemistry at C-7 as salutaridinol I (8) (Döpke et al. 1968). In spite of the difference in substitution pattern tracer experiments have demonstrated that reticuline is a precursor of flavinantine (Stuart et al. 1969). The methylenedioxy group (20, 21) is derived from an o-methoxyphenol e.g., the enantiomer of pallidine (Barton et al. 1962). Biosynthetic relationships suggest that the alkaloids which have the same stereochemistry at C-9 as salutaridine (20, 21, 22, 23) are produced in plants which are capable of interconverting S- and R-reticuline (Fig. 7). *Corydalis pallida* (Papaveraceae) contains both sinoacutine and pallidine and is, therefore, capable of both ortho-para and para-para coupling of S(+)-reticuline (Kametani et al. 1970).

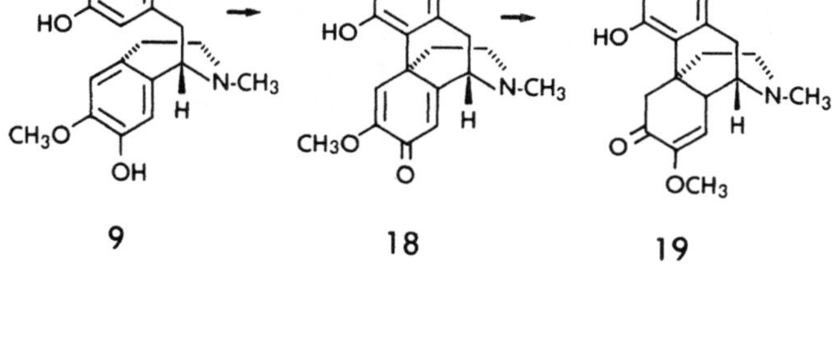

9 18 19

Fig. 6. Biosynthesis of sinomenine

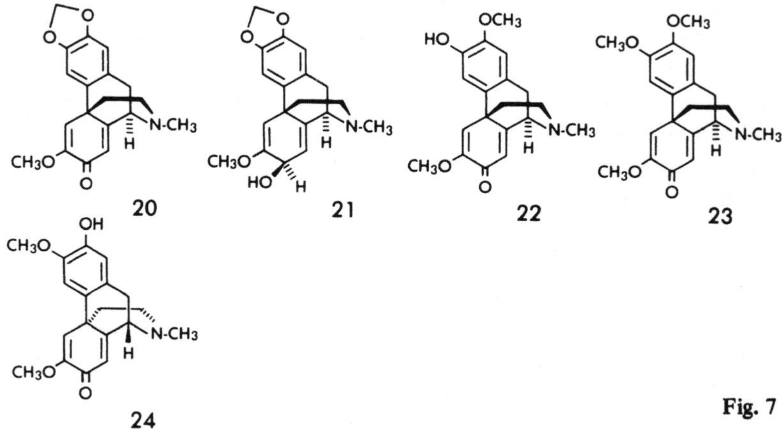

Fig. 7

8 Conclusion

Looking back over the last 25 years, we cannot help but be impressed with what has been accomplished with a combination of chemical mechanistic reasoning, tracer experiments and isolation work. At the same time we begin to realize that we have a long way to go before we will understand what transpires at the cellular level. The enzymology of biosynthetic transformations is still in its infancy, but from the growth and development of this branch of plant biochemistry will come major contributions in the future.

References

Barton DHR, Cohen T (1957) Some biogenetic aspects of phenol oxidation. Festschr Stoll. Birkhäuser, Basel, pp 117–143

Barton DHR, Kirby GW, Taylor JB (1962) On the origin of methylenedioxygroups in nature. Proc Chem Soc:340–341

Barton DHR, Kirby GW, Steglich W, Thomas GM, Battersby AR, Dobson TA, Ramuz H (1965) Investigations on the biosynthesis of morphine alkaloids. J Chem Soc 1965:2423–2438

Barton DHR, Bhakuni DS, James R, Kirby GW (1967) Phenol oxidation and biosynthesis. Part XII. Steréochemical studies related to the biosynthesis of the morphine alkaloids. J Chem Soc (C):128–132

Barton DHR, Kirby AJ, Kirby GW (1968) Phenol oxidation and biosynthesis, part XVII. Investigation on the biosynthesis of sinomenine. J Chem Soc (C):929–936

Battersby AR (1963) Tilden lecture. The biosynthesis of alkaloids. Proc Chem Soc:189–200

Battersby AR, Francis RJ (1964) Alkaloid biosynthesis, part V. Experiments on opium alkaloids using 3,4-dihydroxyphenethylamine. J Chem Soc:4078–4080

Battersby AR, Harper BJT (1958) Biogenesis of morphine. Chem Ind London 1958:364

Battersby AR, Harper BJT (1960) Rate studies on the incorporation of tyrosine into morphine, codeine and thebaine. Tetrahedron Lett 27:21–24

Battersby AR, Binks R, Harper BJT (1962) Alkaloid biosynthesis, part II. The biosynthesis of morphine. J Chem Soc:3534–3544

Battersby AR, Binks R, Francis RJ, McCaldin DJ, Ramuz H (1964a) Alkaloid biosynthesis, part IV. 1-Benzylisoquinolines as precursors of thebaine, codeine and morphine. J Chem Soc: 3600–3610

Battersby AR, Foulkes DM, Binks R (1965a) Alkaloid biosynthesis, part VIII. Use of optically active precursors for investigations on the biosynthesis of morphine alkaloids. J Chem Soc: 3323–3332

Battersby AR, Evans GW, Martin RO, Warren ME Jr, Rapoport H (1965b) Configuration of reticuline in the opium poppy. Tetrahedron Lett:1275–1278

Battersby AR, Martin JA, Brochmann-Hanssen E (1967) Alkaloid biosynthesis, part X. Terminal steps in the biosynthesis of the morphine alkaloids. J Chem Soc (C):1785–1788

Blaschke G, Parker HJ, Rapoport H (1967) Codeinone as the intermediate in the biosynthetic conversion of thebaine to codeine. J Am Chem Soc 89:1540–1541

Borkowski PR, Horn JS, Rapoport H (1978) Role of 1,2-dehydroreticulinium ion in the biosynthetic conversion of reticuline to thebaine. J Am Chem Soc 100:276–281

Brochmann-Hanssen E (1984) A second pathway for the terminal steps in the biosynthesis of morphine. Planta Med 51:343–345

Brochmann-Hanssen E, Cheng CY (1982) Aberrant biosynthesis of opium alkaloids: Biosynthetic conversion of the 6-ethyl analog of thebaine to codeine and morphine. J Nat Prod 45:437–439

Brochmann-Hanssen E, Cheng CY (1984) Biosynthesis of a narcotic antagonist. Conversion of N-allylnorreticuline to N-allylnormorphine in *Papaver somniferum.* J Nat Prod 47:175–176

Brochmann-Hanssen E, Furuya T (1964) A new opium alkaloid. Isolation and characterization of (±)-1-(3′-hydroxy-4′-methoxybenzyl)-2-methyl-6-methoxy-7-hydroxy-1,2,3,4-tetrahydroisoquinoline [(±)-reticuline]. Planta Med 12:328–333

Brochmann-Hanssen E, Nielsen B (1965) (+)-Reticuline – a new opium alkaloid. Tetrahedron Lett:1271–1274

Brochmann-Hanssen E, Okamoto Y (1980) Biosynthesis of opium alkaloids. Substrate specificity and aberrant biosynthesis. Attempted detection of oripavine in *Papaver somniferum.* J Nat Prod 43:731–735

Brochmann-Hanssen E, Wunderly S (1978) Biosynthesis of morphine alkaloids in *Papaver bracteatum* Lindl. J Pharm Sci 67:103–106

Brochmann-Hanssen E, Chen CH, Chen CR, Chiang HC (1975) Opium alkaloids, part XVI. The biosynthesis of 1-benzylisoquinolines in *Papaver somniferum.* Preferred and secondary pathways; stereochemical aspects. J Chem Soc Perkin Trans I:1931–1937

Brochmann–Hanssen E, Cheng CY, Chiang HC (1982) Biosynthesis of opium alkaloids: The effects of structural modification of reticuline on racemization and biotransformation. J Nat Prod 45:629–634

Chambers C, Stuart KL (1968) Flavinantine and flavinine. Novel morphinandione alkaloids from *Croton flavens.* Chem Commun:328–329

Chambers C, Haynes LJ, Stuart KL (1966) Norsinoacutine and salutaridine from *Croton balsamifera* Jacq. Chem Commun:449

Chu JH, Lo SJ, Chou YL (1964) The alkaloids of *Sinomenium acutum.* Hua Hsueh Pao 30:265–269; through Chem Abstr 61:12047

Döpke W, Flentje H, Jeffs PW (1968) The structures of amurine and nudaurine in *Papaver amurense.* Tetrahedron 24:4459–4476

Fairbairn JW, El-Masry S (1967) The alkaloids of *Papaver somniferum* L. V. Fate of the "end-product" alkaloid morphine. Phytochemistry 6:499–504

Fairbairn JW, El-Masry S (1968) The alkaloids of *Papaver somniferum* L. VI. "Bound" morphine and seed development. Phytochemistry 7:181–187

Fairbairn JW, Steele MJ (1980) Bound forms of alkaloids in *Papaver somniferum* and *P. bracteatum.* Phytochemistry 19:2317–2321

Fairbairn JW, Wassel G (1964) The alkaloids of *Papaver somniferum* L. I. Evidence for a rapid turnover of the major alkaloids. Phytochemistry 3:253–258

Fairbairn JW, Hakim F, El-Kheir Y (1974) Alkaloid storage, metabolism and translocation in the vesicles of *Papaver somniferum* latex. Phytochemistry 13:1133–1139

Ginsberg DC (1962) The opium alkaloids. Selected topics. Interscience, New York London, pp 90–91

Gopinath KW, Govindachari TR, Pai BR, Viswanathan N (1959) The structure of reticuline, a new alkaloid from *Annona reticulata.* Chem Ber 92:776–779

Gulland JM, Robinson R (1925) Constitution of codeine and thebaine. Mem Proc Manchester Lit Philos Soc 69:79–86

Heydenreich K, Pfeifer S (1962) On the metabolism in *Papaver somniferum* L. V. Hourly variation in alkaloid content. Sci Pharm 30:164–173

Hodges CC, Rapoport H (1982) Enzymic conversion of reticuline to salutaridine by cell-free systems. Biochemistry 21:3729–3734

Hodges CC, Horn JS, Rapoport H (1977) Morphinan alkaloids in *Papaver bracteatum*: Biosynthesis and fate. Phytochemistry 16:1939–1942

Hohmeyer AH, Shilling WL (1947) Isolation and purification of neopine. J Org Chem 12:356–358

Horn JS, Paul AG, Rapoport H (1978) Biosynthetic conversion of thebaine to codeinone. Mechanism of ketone formation from enol ether in vivo. J Am Chem Soc 100:1895–1898

Kametani T, Ihara M, Honda T (1970) The alkaloids of *Corydalis pallida* var. tenuis (Yatabe) and the structures of pallidine and kikemanine. J Chem Soc (C):1060–1064

Kirby GW, Massey SR, Steinreich P (1972) Biosynthesis of unnatural morphine derivatives in *Papaver somniferum.* J Chem Soc Perkin Trans I:1642–1647

Kleinschmidt G, Mothes K (1959) Physiology and biosynthesis of alkaloids in *Papaver somniferum.* Z Naturforsch 14b:52–56

Küppers FJEM, Salemink CA, Bastart M, Paris M (1976) Alkaloids of *Papaver bracteatum*: Presence of codeine, neopine and alpinine. Phytochemistry 15:444–445

Leete E (1959) The biogenesis of morphine. J Am Chem Soc 81:3948–3951

Leete E, Murrill JB (1964) The incorporation of dopamine into chelidonine and morphine. Tetrahedron Lett:147–151

Martin RO, Warren ME Jr, Rapoport H (1967) The biosynthesis of opium alkaloids. Reticuline as the benzyltetrahydroisoquinoline precursor of thebaine in biosynthesis with carbon-14 dioxide. Biochemistry 6:2355–2363

Meshulam H, Lavie D (1980) The alkaloidal constituents of *Papaver bracteatum* Arya II. Phytochemistry 19:2633–2635

Miller RJ, Jolles C, Rapoport H (1973) Morphine metabolism and normorphine in *Papaver somniferum.* Phytochemistry 12:597–603

Miram R, Pfeifer S (1960) On the changes in the alkaloid content of the opium poppy during a growing season. Sci Pharm 28:15–28

Neubauer D (1964) Distribution of the major alkaloids of the opium poppy in the various parts of the plant at different stages of development. Planta Med 12:43–50

Nielsen B, Röe J, Brochmann-Hanssen E (1983) Oripavine – A new opium alkaloid. Planta Med 48:205–206

Parker HI, Blaschke G, Rapoport H (1972) Biosynthetic conversion of thebaine to codeine. J Am Chem Soc 94:1276–1282

Pfeifer S, Heydenreich K (1962) The accumulation of opium alkaloids between flowering and biological ripening. Pharmazie 17:107–114

Phillipson JD, Handa SS, El-Dabbas SW (1976) N-Oxides of morphine, codeine and thebaine and their occurrence in *Papaver* species. Phytochemistry 19:1681–1684

Pummerer R, Puttfarcken H, Schopflocher P (1925) Oxidation of phenols. VIII. Dehydrogenation of p-cresol. Chem Ber 58B:1808–1820

Rapoport H, Stermitz FR, Baker DR (1960) The biosynthesis of opium alkaloids. I. The interrelationship among morphine, codeine and thebaine. J Am Chem Soc 82:2765–2772

Stermitz FR, Rapoport H (1961) The biosynthesis of opium alkaloids. Alkaloid interconversion in *Papaver somniferum* and *P. orientale.* J Am Chem Soc 83:4045–4050

Stuart KL, Teetz V, Franck B (1969) Alkaloid biosynthesis in *Croton flavens.* Chem Commun: 333

Theuns HG, Dam JEG van, Luteijn JM, Salemink CA (1977) Alkaloids of *Papaver bracteatum*: 14-β-hydroxycodeinone, 14-β-hydroxycodeine and N-methylcorydaldine. Phytochemistry 16: 753–755

Winterstein E, Trier G (1910) Die Alkaloide. Bornträger, Berlin, p 307

Wold JK (1978) Bound morphine and codeine in the capsule of *Papaver somniferum*. Phytochemistry 17:832–833

Enzymology of Benzylisoquinoline Alkaloid Formation

M.H. ZENK[1]

1 Introduction

The largest group of naturally occurring alkaloids comprises compounds based on the tetrahydroisoquinoline nucleus. A multitude of structural types derived from this nucleus is found in the plant kingdom (Shamma 1972, Shamma and Moniot 1978). A great number of biosynthetic hypotheses based on in vivo investigations has been proposed to explain this unprecedented diversity (I.O. Spenser 1968, Shamma and Moniot 1978). The biosynthesis of isoquinolines at the cell-free level, however, has hardly been touched. Most of our knowledge on alkaloid biosynthesis is based on hypothetical pathways inferred from experiments in which radioactively labelled precursors and intact whole plants were used. Low rates of incorporation of the precursors into the products of interest have usually been observed, and channelling of these intermediates into abnormal pathways may also occur (e.g. Stöckigt 1980). Although in vivo techniques are necessary to provide first ideas about a biosynthetic route, they do not reveal the actual sequence and mechanism of the biosynthetic process in detail. Complex pathways can only be fully elucidated by a determination of the structures of enzymatically formed intermediates and by a characterization of the single enzymes or enzyme systems involved in these reactions. Enzymes of the secondary pathways can usually not be obtained easily from differentiated plants since the stationary levels of these enzymes appear to be very low and also the high content of phenolic compounds and other undesirable cell constituents tends to inactivate these enzymes during the isolation procedure (Loomis and Battaile 1966). Plant cell suspension cultures capable of producing desired secondary products (Zenk 1978) are an excellent source of enzyme systems (Stöckigt 1980). Suspension cells can be grown under exactly defined conditions on a large scale, thus providing any desired amount of homogeneous plant material as enzyme source independent of seasons. Furthermore, levels of catalytically active proteins are considerably higher in suspension cells as compared with whole plants because the time required for product synthesis under cell culture conditions is much shorter than in field-grown plants. It was only the consistent application of cell culture systems that allowed the discovery of the main enzymes involved, for instance, in the formation of flavonoids (Hahlbrock and Grisebach 1975), heteroyohimbine (Stöckigt 1980, Zenk 1980), and quinoliz-

1 Lehrstuhl Pharmazeutische Biologie der Universität München, Karlstraße 29, 8000 München 2, FRG

The Chemistry and Biology of Isoquinoline Alkaloids, ed. by Phillipson et al.

idine alkaloids (Wink and Hartmann 1982). In order to extend our investigations on isoquinoline alkaloid biosynthesis (Nagakura et al. 1978) and to study enzymes involved in the synthesis of structurally diverse alkaloids, we concentrated on the group of benzylisoquinoline alkaloids. It has been demonstrated that cell-free extracts of *Berberis aggregata* callus cultures catalyse the formation of several protoberberine alkaloids from distant precursors, and therefore contain the appropriate enzymes involved in this pathway (Kelleher et al. 1980). It is our aim to clarify the major routes of the benzylisoquinoline pathways at the enzyme level and to isolate and characterize the biocatalysts involved. In this communication we present data related to the enzymatic formation of tetrahydrobenzylisoquinolines, especially reticuline, the central intermediate in isoquinoline biosynthesis (Shamma 1972), and protoberberine and morphinan alkaloids.

2 Enzymatic Formation of Tetrahydrobenzylisoquinoline Alkaloids

2.1 (S)-Norlaudanosoline Synthase

Already in 1910 Winterstein and Trier, in a far-sighted biosynthetic speculation, assumed that the initial reaction in isoquinoline biosynthesis might occur by a condensation of two aromatic units both derived from tyrosine, namely dopamine and 3,4-dihydroxyphenylacetaldehyde. As early as 1917, norlaudanosoline was postulated as the first isoquinoline intermediate in this pathway (Robinson 1917). This scheme was later modified as a result of biomimetic experiments (Hahn and Rumpf 1938) showing that the condensation of dopamine with phenylpyruvates led to benzylisoquinoline-1-carboxylic acids which in turn could be chemically decarboxylated to the dehydroderivative and possibly could be further hydrogenated to the desired tetrahydrobenzylisoquinolines. Indeed, synthetic (R,S)-norlaudanosoline-1-carboxylic acid derived from the condensation of dopamine with 3,4-dihydroxyphenylpyruvate was shown to be incorporated into morphine alkaloids by using the intact plant or latex of *Papaver* species (Wilson and Coscia 1975). These results were later claimed to be confirmed by additional in vivo (Battersby et al. 1975b) and in vitro techniques (Scott et al. 1978a,b). The stereochemical aspects of this reaction remained unsolved. On the other hand, recent work in our laboratory (Rueffer et al. 1981, Schumacher et al. 1983) has led to the discovery of a new enzyme named (S)-norlaudanosoline synthase, which catalyses the stereospecific condensation of dopamine with 3,4-dihydroxyphenylacetaldehyde, while 3,4-dihydroxyphenylpyruvate is inactive (Fig. 1). The assay procedure for this reaction exploits the fact that 50% of the label is lost and appears as water in the ambient solution during the condensation reaction with 2,6-^{3}H-dopamine. Crude extracts of 10 plant cell cultures repre-

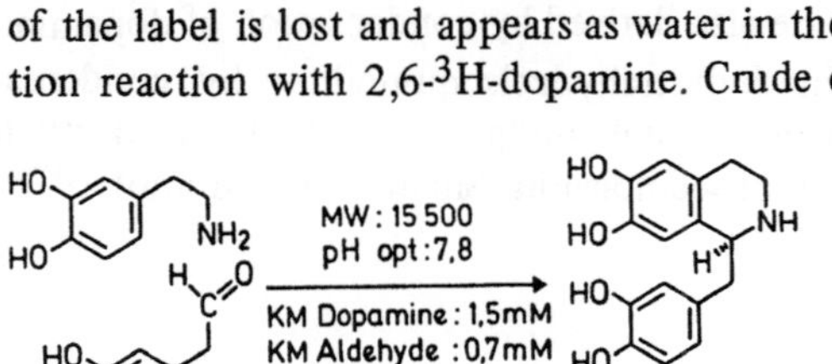

Fig. 1. Scheme of reaction catalysed by (S)-norlaudanosoline synthase indicating also the assay principle

senting 3 families contained this enzyme while it was absent in control tissues such as *Catharanthus, Daucus,* and *Nicotiana.* The enzyme was purified approximately 40-fold from cell cultures of *Eschscholtzia tenuifolia.* The molecular weight was determined to be 15,500, assuming a globular shape. The pH-optimum is 7.8 and the apparent K_M-values for dopamine and 3,4-dihydroxyphenylacetaldehyde are 1.5 mM and 0.7 mM, respectively. The synthase is highly specific in that only 3,4-dihydroxy-, 4-hydroxy-, and unsubstituted phenylacetaldehydes but none of the correspondingly substituted phenylpyruvates can serve as substrates. Isoelectric focusing and disc-gel electrophoresis revealed four norlaudanosoline synthase isoenzymes with isoelectric points of pH 4.6, 5.3, 6.7, 7.6. The isoenzyme with the IEP of 7.6 was purified. When the incubation of the crude extract was carried out on a large scale in the presence of dopamine and 3,4-dihydroxyphenylacetaldehyde, the isolated norlaudanosoline was chiral and possessed the S-configuration as measured by CD-spectroscopy. However, it was only about 25% optically pure. Since there was no evidence for the occurrence of a norlaudanosoline racemase in the enzyme preparation, one must assume that the racemic norlaudanosoline is formed by a chemical (non-enzymatic) reaction between the amine and the aldehyde. This chemical reaction was also observed in control assays lacking catalytically active protein (Schumacher et al. 1983) and occurred already within minutes after starting the reaction. Pyruvic acids also react chemically in the absence of enzyme. This is in sharp contrast to data presented by Scott et al. (1978a,b), who claimed that no norlaudanosoline-1-carboxylic acid was formed under their incubation conditions when "boiled enzyme" was used. The observed formation of norlaudanosoline-1-carboxylic acid under in vivo and in vitro conditions might be explained by the ease with which phenethylamines react non-enzymatically with phenylpyruvates. The original and the more recent (Prakash et al. 1979 and literature cited therein) results on the incorporation of norlaudanosoline-1-carboxylic acid into benzylisoquinoline alkaloids by different plant systems can be explained by the facile oxidative decarboxylation either in air under slightly basic conditions (Bobbitt et al. 1976) or by enzymes which occur widely in the plant kingdom such as peroxidases or laccases (Coutts et al. 1979). The 3,4-dihydroisoquinoline derivatives thus formed could be further reduced to the tetrahydroisoquinoline stage and could then be further metabolized under in vivo conditions.

Our results support the conclusion that the major pathway leading to benzylisoquinoline alkaloids involves (S)-norlaudanosoline synthase which condenses dopamine and phenylacetaldehydes in a stereospecific manner, thus forming (S)-norlaudanosoline, and (S)-desmethylcoclaurine as the first and central intermediates in the early stages of the pathway. It appears logical to assume a Pictet-Spengler type condensation of dopamine with phenylacetaldehyde on comparative biochemical grounds, because 1-phenylisoquinolines are also derived by condensation of dopamine with substituted benzaldehydes, phenethyltetrahydroisoquinolines by condensation with γ-phenylpropionaldehyde precursor and the ipecac alkaloids by an initial condensation with the aldehyde function of secologanin (Shamma and Moniot 1978) (Fig. 2).

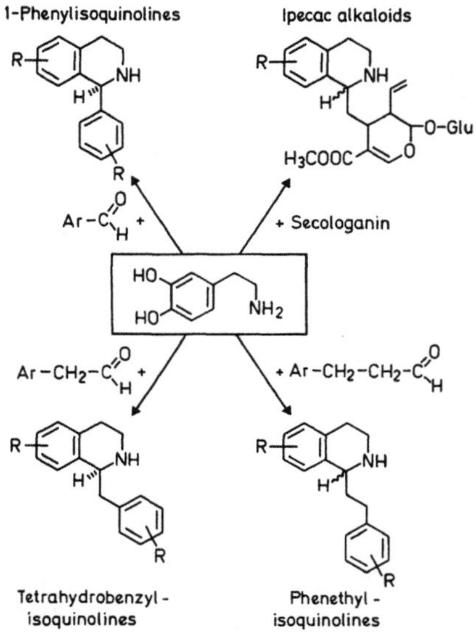

Fig. 2. Condensation of dopamine with differently substituted aldehydes as key reaction in the early stage of biosynthesis of isoquinoline alkaloids

2.2 S-Adenosylmethionine: (R), (S)-Norlaudanosoline-6-O-Methyltransferase

The most obvious metabolic alteration of norlaudanosoline is O-methylation. A series of elegant experiments were summarized by Davis et al. (1982), who showed that norlaudanosoline is methylated in the presence of SAM and rat liver enzymes (catechol-O-methyltransferase), yielding a mixture of monomethylated norlaudanosolines. Interestingly, (S)-norlaudanosoline yielded 79% 6-O-methylnorlaudanosoline and 14% of the 7-O-methyl derivative, while the (R)-isomer gave only 26% 7-O-methyl- and 68% of the 6-O-methylnorlaudanosoline (analysis by GC-MS and HPLC). In animals the site of enzymatic O-methylation is clearly dictated by the stereoisomeric form of the alkaloid substrate. In plants, it has been amply documented that reticuline is the branch point for a multitude of different isoquinoline alkaloid types. This means that norlaudanosoline must be transformed to reticuline by three methylation reactions, two O-methylations (position 6 and 4′) and one N-methylation. Since norreticuline has been undoubtedly demonstrated to be a precursor of reticuline in vivo, one has to assume that O-methylation precedes N-methylation. This is also in agreement with in vivo feeding experiments (Brochmann-Hanssen et al. 1975).

In our attempt to elucidate the enzymatic steps involved in reticuline biosynthesis in plants we have investigated the O-methylation of norlaudanosoline (Rueffer et al. 1983b). We screened cell-free extracts derived from cell cultures of four different plant families using (R,S)-norlaudanosoline (in order to prevent N-methylation) and ^{3}H-SAM as substrates and demonstrated that all species containing isoquinolines contained methylating activity. The highest amount of methyltransferase per volume of cell culture medium was observed in *Argemone platyceras* cultures. The target enzyme was isolated from this source and purified approximately 80-fold. This enzyme catalyses the formation of 6-O-methylnorlaudanosoline (80%), and to a minor extent

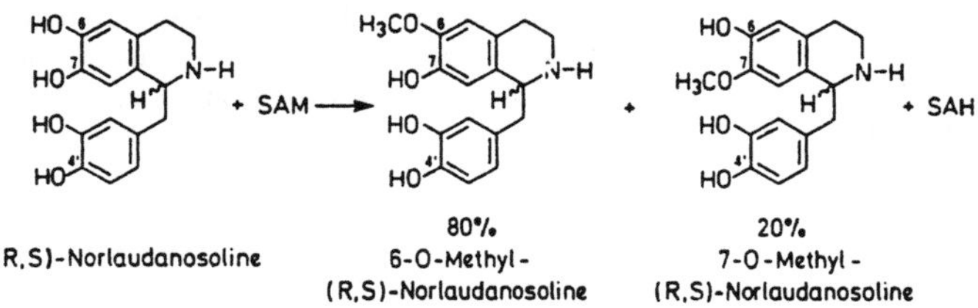

Fig. 3. Reaction sequence catalysed by 5-adenosylmethionine: (R),(S)-norlaudanosoline-6-O-methyltransferase

7-O-methylnorlaudanosoline (20%),from SAM and (S)- as well as (R)-norlaudanosoline as revealed by HPLC and MS (Fig. 3).

In contrast to the animal system, the 6-position is the preferred position for methylation in the plant system, regardless of the steric configuration of the norlaudanosoline substrate. Both stereoisomers of norlaudanosoline are about equally well accepted by the enzyme. The plant enzyme shows a considerably higher affinity for the substrates than the animal enzymes. This is demonstrated by the low K_M-values for norlaudanosoline (K_M 0.2 mM) and SAM (0.05 mM). The *Argemone* enzyme appears to prefer tetrahydrobenzylisoquinolines possessing unsubstituted vicinal OH-groups, (S)-norlaudanosoline being the best known substrate. Simple phenols like caffeic acid, adrenaline or quercetin do not serve as methyl-group acceptors. Interestingly, (R,S)-norlaudanosoline-1-carboxylic acid shows only about 5% of the activity of (S)-norlaudanosoline.

As determined by gel filtration, the apparent molecular weight of the enzyme is 47,000. Its pH optimum is 7.5. It should be pointed out that a study of the time course of the enzyme formation in suspension cultures of *Argemone* cells reveals that the enzyme is present only in very low amounts in the first 2 days of cultivation. One then observes a 30-fold increase in total activity as compared with only a 10-fold increase in dry cell material, when the culture is leaving the logarithmic growth phase. During the stationary phase one observes a drastic decrease of total enzyme activity.

2.3 S-Adenosylmethionine: (6-O-Methylnorlaudanosoline)-4′-O-Methyltransferase

Norlaudanosoline is metabolically methylated in the 6-position first. Hence it follows that a second O-methyl group has to be introduced at position 4′ of the tetrahydrobenzylisoquinoline molecule on the way to norreticuline. One must postulate the existence of an enzyme which specifically methylates 6-O-methyl-norlaudanosoline in the 4′-position with SAM serving as donor of the methyl group. Again,*Berberis* cell cultures (Hinz and Zenk 1981) proved to be an excellent plant source in the search for this new enzyme. Frenzel in our laboratory recently discovered this highly specific enzyme which catalyses the reaction depicted in Fig. 4. This enzyme should prove to be an interesting catalyst for specifically labelling norreticuline.

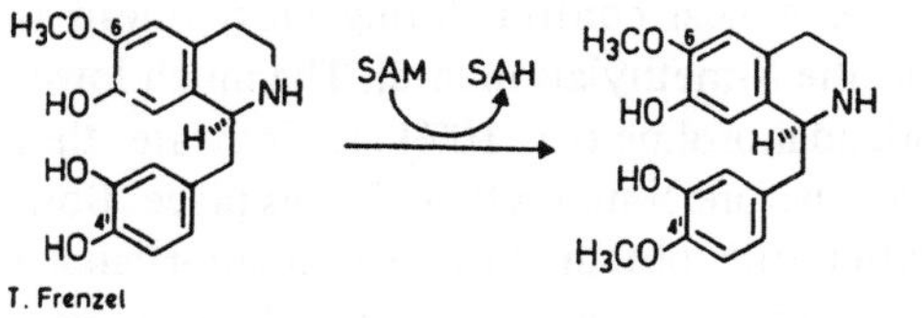

Fig. 4. Reaction sequence catalysed by S-adenosylmethionine: (6-O-methylnorlaudanosoline)-4′-O-methyltransferase (norreticuline synthase)

Purification and characterization of this enzyme is now in progress. During the course of purification of SAM-6-O-methylnorlaudanosoline methyltransferase from *Argemone* cell cultures, a second enzymatic activity was discovered which utilizes 6-O-methylnorlaudanosoline to introduce a second methyl group. However, methylation occurs in the 5′-position of the molecule, and the reaction yields nororientaline (Rueffer et al. 1983c). This new enzyme was named S-adenosylmethionine:(6-O-methylnorlaudanosoline)-5′-O-methyltransferase and catalyses the formation of nororientaline from 6-O-methylnorlaudanosoline and SAM. It was partially purified (40-fold), shows a molecular weight of 47,000, a pH optimum at 7.5 (K_M for the benzylisoquinoline 0.4 mM; for SAM 0.05 mM), and is highly specific in that it does not accept norlaudanosoline or laudanosoline as substrates. The enzyme certainly plays a role in the biosynthesis of orientaline by specifically determining the methylation pattern at the C-ring system.

2.4 S-Adenosylmethionine: (R),(S)-Norreticuline-N-Methyltransferase

The final step in the biosynthesis of reticuline, the branch point metabolite, consists in the N-methylation of norreticuline. The presence of N-methyltransferases had to be postulated, since it has previously been shown that isolated latex of *Papaver somniferum*, incubated with norlaudanosoline and ^{14}C-SAM, yielded labelled reticuline (Antoun and Roberts 1975). The same type of experiment with a cell-free extract from agar grown callus of *Berberis aggregata* yielded labelled protoberberine alkaloids (Kelleher et al. 1980). This latter study also demonstrated that (R,S)-tetrahydropapaverine is N-methylated to yield laudanosine. Using both (R)- and (S)-tetrahydropapaverine as substrates, more than 50 cell suspension cultures propagated from plants producing isoquinoline alkaloids were screened for N-methyltransferase activity (Wat et al. 1985). The enzyme was discovered in the following genera: *Berberis, Corydalis, Eschscholtzia, Fumaria,* and *Thalictrum.* Cell cultures of *Berberis vulgaris* and *B. wilsoniae* var. *subcaulialata* proved to be particularly rich sources of this protein. The protein occurs in the form of two isoenzymes and was purified 100-fold. The N-methyltransferase had a molecular weight of 68,000 and a pH optimum of 7.7. The K_M-value for the (S)-stereoisomer of tetrahydropapaverine (30 μM) was lower than that for the (R)-isomer (180 μM). However, in terms of relative activity, the enzyme is about five times more active with the (R)- than with the (S)-isomer of tetrahydropapaverine. By far the best substrates for this enzyme are norreticuline and nororientaline yielding reticuline and orientaline in the presence of SAM. The enzyme is about twice as active with the (S)-isomer of norreticuline as with the (R)-form. Nevertheless, the enzyme shows a remarkable lack of stereospecificity. This demonstrates that the stereochemical control during biosynthesis is exerted neither at the O-methylation nor at the N-methylation level. The much lower activity of this enzyme towards (R,S)-norlaudanosoline (ca. 10%) also indicates that a certain level of O-methylation is necessary before N-methylation takes place. Both observations are in absolute agreement with results obtained by Brochmann-Hanssen et al. (1975). The N-methyltransferase does not discriminate between the stereoisomers of norreticuline (and nororientaline) and it proves an easy way for the synthesis of the respective stereochemically pure isomers of reticuline (and orientaline) labelled in the N-methyl group with ^3H or ^{14}C at extremely high specific activities. This procedure is much superior to any chemical method and allows the resolution of the enantiomers of these important intermediates. The reaction catalysed by the N-methyltransferase using norreticuline as substrate is shown in Fig. 5.

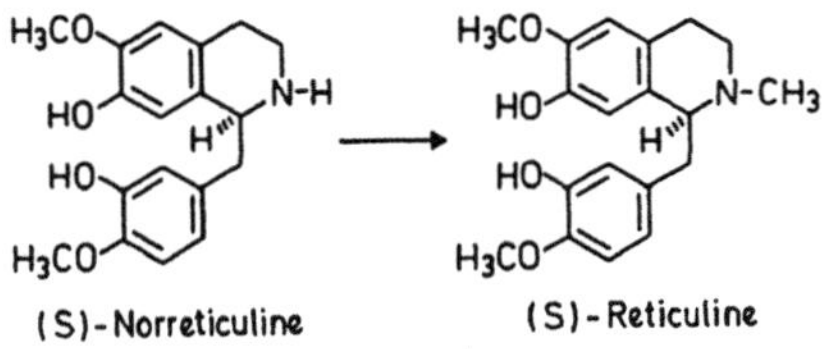

Fig. 5. Reaction sequence catalysed by S-adenosylmethionine: (R),(S)-norreticuline-N-methyltransferase

3 Enzymatic Formation of Protoberberine Alkaloids

The target molecule of this biosynthetic pathway is berberine itself, which is a useful antibacterial, gastric and antiinflammatory agent. Considerable efforts have been made in the past years to produce this compound in cell cultures (e.g. Sato and Yamada 1984). Berberine is the prototype of the berberine alkaloid family. Its biosynthesis has attracted considerable interest, and by in vivo precursor feeding experiments it has become quite clear that reticuline is transformed first into scoulerine, a tetrahydroberberine alkaloid, by conversion of the N-methyl group of reticuline into the so-called berberine bridge (D. Spenser and Gear 1962, Barton et al. 1963, Battersby et al. 1963). This initial cyclization reaction is followed by several steps involving methylation, another cyclization and oxidation, finally yielding the berberine molecule. Of all the enzymes taking part in these postulated reactions, only a single enzyme catalysing the first cyclization from reticuline to scoulerine has previously been discovered (Rink and Böhm 1975, Böhm and Rink 1975).

3.1 The Berberine Bridge-Forming Enzyme

While a differentiated plant (*Papaver bracteatum*) containing considerable quantities of protoberberine derivatives did not show any enzyme activity, cultured cells of *Macleaya microcarpa* yielded a cell-free system catalysing the conversion of (R,S)-reticuline into scoulerine of unknown stereochemistry (Rink and Böhm 1975). The enzyme was purified 7-fold and roughly characterized. Enzyme assays were carried out by using a N-$^{14}CH_3$-labelled substrate and subsequent TLC-chromatography of the product formed. The enzyme did not require a co-factor and acted only on reticuline and laudanidine. An essential prerequisite for the enzymatic reaction appeared to be the presence of an unsubstituted hydroxyl group at the C-5′ of the tetrahydrobenzylisoquinoline system (Böhm and Rink 1975). In our attempt to clarify the enzymatic steps involved in protoberberine synthesis, we (Steffens et al. 1984) devised a new assay system for this enzyme. The assay relies on the use of N-CT_3-(S)-reticuline as a substrate (synthesized enzymatically from (C-T_3)-SAM and (S)-norreticuline by catalysis with the N-methyltransferase described above). The transition to scoulerine results in the release of exactly 1/3 of the radioactivity contained within the reticuline molecule into the aqueous phase of the incubation mixture. Since (S)-reticuline was synthesized with a high specific activity (7.2 Ci $mmol^{-1}$), the sensitivity of the test is in the order of 10^{-15} mol s^{-1}. This is one of the most sensitive assays in secondary product biosynthesis developed thus far. The berberine bridge-forming enzyme was found in both the differentiated plants and the cell cultures of several species out of four protoberberine-containing plant families. The enzyme is absent from plants such as Solanaceae or Apocynaceae that do not contain isoquinolines. The enzyme was isolated from *Berberis beaniana* cell suspension cultures and purified to homogeneity (440-fold). Maximal catalytic activity was observed at pH 8.9 and 45°C. The enzyme has a molecular weight of approximately 49,000. Rink and Böhm (1975) already observed that the enzymatic reaction shows an absolute oxygen requirement. We could demonstrate that for each mole of (S)-reticuline, 1 mol of O_2 is taken up and 1 mol each of (S)-scoulerine and H_2O_2 is formed (Fig. 6). The K_m-value of reticuline is surprisingly low (0.14 μM). Only (S)-protosinomenine [145% of (S)-reticuline] and (R,S)-laudanosoline (20%) are accepted as substrates. The (R)-forms of these substrates are absolutely inactive. Neither (S)-orientaline nor (S)-isoorientaline are transformed. In addition, no transformation was observed with the racemates of laudanidine, laudanosine and the (R)- and (S)-reticuline N-oxides, although the latter compounds have been postulated as possible intermediates. Furthermore, the enzyme is exclusively contained in particles which can be isolated from the cells by sucrose density gradient centrifugation (ρ = 1.14 g ml^{-1}). The enzyme is strongly inhibited by EDTA and O-phenanthroline. However, the inhibition of enzyme activity by these chelating agents could not be reversed by the addition of various heavy metal salts.

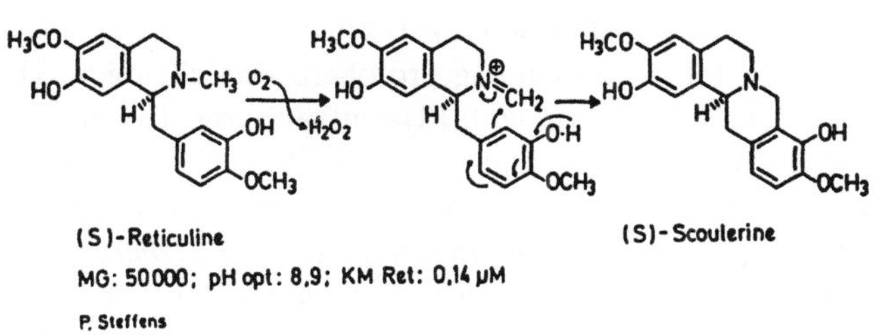

Fig. 6. Scheme of reactions catalysed by the berberine-bridge-forming enzyme

3.2 S-Adenosyl-Methionine: (S)-Scoulerine-9-O-Methyltransferase

When the substrate ^{14}C-labelled (S)-scoulerine was incubated with crude cell-free homogenates of various *Berberis* cell cultures in the presence of SAM, the formation of inter alia tetrahydrocolumbamine was observed. The crude homogenate was fractionated by $(NH_4)_2SO_4$ precipitation and DEAE as well as gel chromatography. After each step the presence of the methyltransferase was assayed by an enzymatic test involving unlabelled (S)-scoulerine and C^3H_3-SAM. The reaction product was extracted by isoamyl alcohol, followed by a determination of radioactivity in the organic phase. The enzyme was only slightly enriched but was free of any interfering enzymatic activities. This new enzyme (Mümmler, Rueffer, and Zenk unpublished) showed maximal catalytic activity at pH 8.9 and had a molecular weight of about 60,000. It accepts only a relatively limited number of substrates, and (S)-scoulerine (100%) and 2,3,9,10-tetrahydroxytetrahydroberberine (35%) are the best substrates for the methyltransferase. The reaction is stereochemically controlled since (R)-scoulerine and the quaternary protoberberine salt derived from scoulerine are only marginally active as substrates. The product of the methylation reaction using (S)-scoulerine is (S)-tetrahydrocolumbamine (Fig. 7).

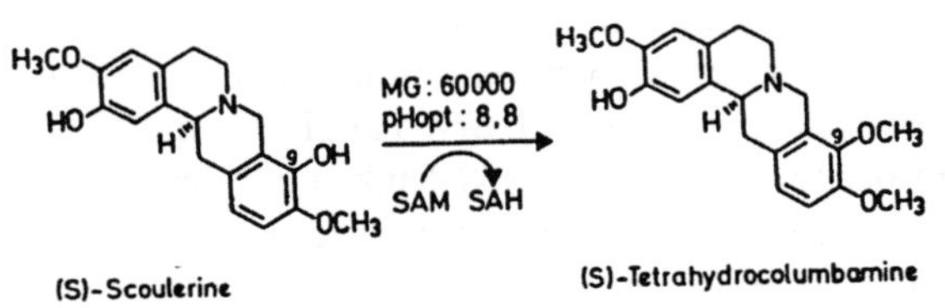

Fig. 7. Reaction sequence catalysed by S-adenosyl-l-methionine: (S)-scoulerine-9-O-methyltransferase

In summary, the substrate specificity shows the enzyme to act preferentially on (S)-scoulerine and not on the oxidized scoulerine product containing an aromatic C-ring. The methyltransferase is highly specific for the 9-hydroxy position of (S)-scoulerine. One can therefore conclude that methylation of the tetrahydroberberine molecule occurs prior to the oxidation of ring C.

3.3 (S)-Tetrahydroprotoberberine Oxidase

The final step in the biosynthesis of the protoberberine skeleton is the specific oxidation of ring C of the tetrahydroberberines. A new enzyme was discovered in our laboratory (Amann et al. 1984) which catalyses the dehydrogenation of more than 20 different tetrahydroberberines to the corresponding protoberberine alkaloids. The enzyme is absolutely specific for the substrate with (S)-configuration. The enzyme was assayed by using [8,14-^{3}H]-(R,S)-canadine and measuring the release of tritium into the aqueous medium during the oxidation reaction. The enzyme was found to occur in a number of plant cell cultures, especially in those strains of *Berberis* which produce considerable amounts of protoberberines. The enzyme was purified to homogeneity (purification factor 150-fold, 25% yield) from *Berberis wilsoniae* var. *subcaulialata*. Fluorescence spectroscopy revealed that the enzyme contains a flavine. It has a molecular weight of 100,000 and a pH optimum for maximal catalytic activity of 8.9. The K_m-values are 1.3 μM for (S)-tetrahydrojatrorrhizine and 26.7 μM for (S)-canadine. The enzyme displays a broad range of substrate specificity. The aromatic substitution patterns of rings A and D of the tetrahydroprotoberberine nucleus

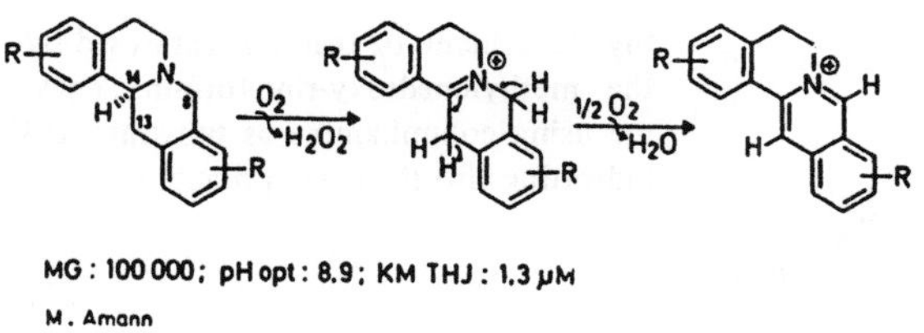

Fig. 8. Scheme of reaction catalysed by (S)-tetrahydroprotoberberine oxidase (STOX)

only quantitatively influence the catalytic activity. The reaction sequence catalysed by the enzyme is shown in Fig. 8. 1.5 mol of oxygen are consumed (during the oxidation), and one mole of H_2O_2 is produced by each mole of oxidized substrate. When catalase is added to the incubation medium, the oxygen consumption is reduced to 67% of the original rate. The enzyme does not oxidize canadine to berberine in the presence of $NaBD_4$, but 1 atom of deuterium is found to be incorporated at the C-14 carbon atom of re-isolated canadine. 13,14-Dehydrocanadine and canadine metho-salt are not oxidized by the enzyme. Methyl-substitution at C-8 or C-13 renders these tetrahydroprotoberberines unsuitable for enzymatic oxidation. It can therefore be concluded that the enzyme catalyses the dehydrogenation of the tetrahydroprotoberberine molecule at C-14 and C-7 and that the 7,14-dehydroberberinium intermediate aromatizes further in ring C to yield the protoberberine molecule (see also 4.1). The enzyme is exclusively found in a type of organelle which can be isolated from several cell cultures using sucrose density gradient centrifugation ($\rho = 1.14$ g/cm^3). Since the enzyme displays an absolute stereochemical specificity towards molecules with (S)-configuration and a broad substrate specificity, it is a useful reagent for the determination of the stereochemical configuration of tetrahydroberberines which may be present in minute amounts.

3.4 The Methylenedioxy-Ring-Forming Enzyme

Columbamine is formed as an intermediate in berberine biosynthesis. This is consistent with previous in vivo experiments (Battersby et al. 1975a) and suggests that the next step in the biosynthetic sequence is the formation of the methylenedioxy-ring in the A-ring of the protoberberine molecule, thus yielding berberine. Up to now, the formation of a methylenedioxy-ring has not been observed under cell-free conditions in plant biochemistry. Rueffer and Zenk (unpublished) recently isolated a protein from *Berberis stolonifera* cell cultures by gel-filtration prefractionation and DEAE cellulose chromatography, using 3-OC^3H_3-columbamine as a substrate. This enzyme preparation was active in catalyzing the reaction of columbamine to berberine (Fig. 9). The rather unstable enzyme has a pH-optimum of 8.9, and columbamine as substrate shows a K_m-value of 2 µM. Beecher and Kelleher (1983) recently suggested that the formation of the methylenedioxy-group, an oxidative process, requires the presence of a soluble oxidation-reduction co-factor. However, we observed no apparent need for such a low molecular dissociable organic co-factor in this reaction. The enzyme was found to occur in a number of cultures of *Berberis* and *Thalictrum* species. It is not found in plant cell cultures which do not produce isoquinolines such as *Nicotiana tabacum* or *Rauvolfia serpentina*.

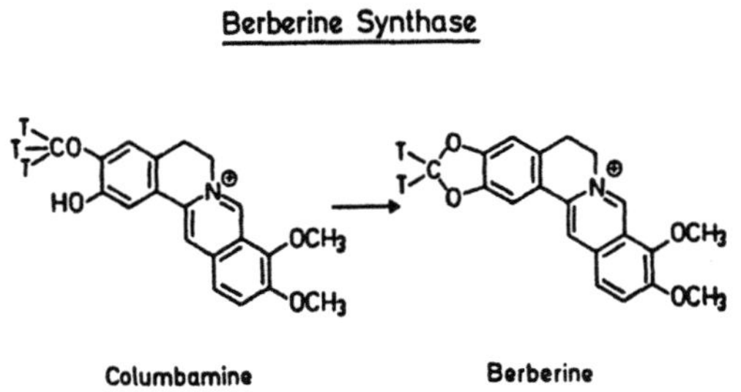

Fig. 9. Scheme of reaction catalysed by the methylenedioxy-ring-forming enzyme using columbamine as substrate and indicating also the assay principle

As shown above, the co-operation of eight enzymes is necessary in the biosynthesis of berberine from the metabolites delivered from primary metabolism, i.e. dopamine and 3,4-dihydroxyphenylacetaldehyde. In the course of this investigation seven of these enzymes have been discovered, isolated, and characterized. Berberine is the first alkaloid to be used medicinally whose biosynthesis is now completely understood at the enzyme level. As will be shown below, berberine itself is not the ultimate metabolite, since it can be transformed further to jatrorrhizine.

3.5 Jatrorrhizine Biosynthesis

Jatrorrhizine is the major alkaloid of *Berberis* cell cultures (Hinz and Zenk 1981). It contains an unusual 2-O-methylation pattern which makes it difficult to deduce its biosynthesis from reticuline. One would speculate rather that it is derived from protosinomenine which already possesses the methyl-group of the A-ring in the correct position. In a very elegant experiment Beecher and Kelleher (1983) proved that berberine is a biosynthetic precursor of jatrorrhizine by using chemically synthesized 9-^{14}C-berberine and callus cultures of *B. aggregata.* They also proposed a tempting mechanism for the one-sided opening of the methylenedioxy group (Fig. 10). Independently, our group (Rueffer et al. 1983a) could confirm these results. Feeding experiments with N-$^{14}CH_3$(S)-reticuline and N-$^{14}CH_3$(S)-protosinomenine showed that the incorporation of reticuline into jatrorrhizine is better by a factor of 6 than that of protosinomenine. Feeding experiments involving doubly-labelled 6-O-C^3H_3-laudanosoline N-$^{14}CH_3$ and *B. stolonifera* callus cultures ultimately proved that the

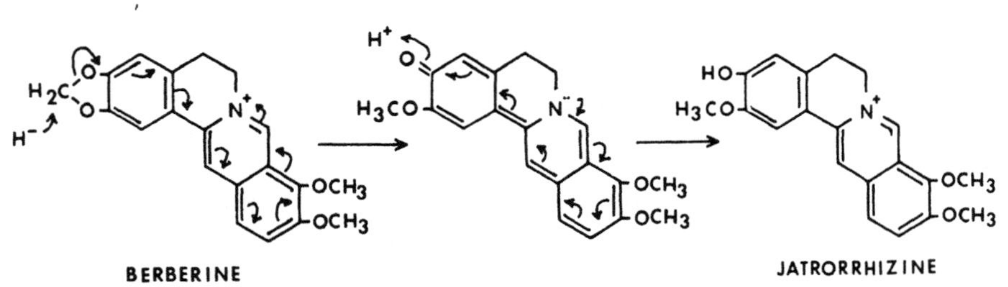

Fig. 10. Postulated mechanism for the biosynthetic conversion of berberine to jatrorrhizine. (After Beecher and Kelleher 1983)

incorporation of this compound into jatrorrhizine is accompanied by the loss of exactly 1/3 of the tritium located in the 6-O-C^3H_3-group during the transition of the labelled compound to jatrorrhizine. Berberine which was isolated from the same callus tissue showed the same isotope ratio as jatrorrhizine. These experiments proved that the major biosynthetic route to jatrorrhizine is via berberine (Fig. 10) and that a minor route proceeds through (S)-protosinomenine. The enzyme responsible for the formation of jatrorrhizine by opening the methylenedioxy group in berberine has not yet been discovered. Only the characterisation of this enzyme will allow us to decide whether the methylenedioxy-ring opening reaction takes place at the level of tetrahydroprotoberberine or the oxidized quaternary protoberberine.

4 Formation of Morphinan Alkaloids

Morphinan alkaloids are the pharmaceutically most important class of isoquinoline alkaloids. Codeine and morphine are found only in *Papaver somniferum* and closely related species. Both alkaloids are derived from (R)-reticuline, the absolute stereochemistry of which corresponds with the configuration of morphinan alkaloids. Reticuline is therefore an important branch point in isoquinoline biosynthesis. However, either enantiomer of reticuline, when fed separately, is incorporated into morphinan alkaloids essentially to the same extent. The question therefore arises by which mechanism (R)-reticuline is formed within the plant.

4.1 The 1,2-Dehydroreticulinium Ion

Feeding experiments with (S)-reticuline labelled with ^3H at C-1 and ^{14}C- at the N-methyl group showed that the incorporation into thebaine was accompanied by a considerable loss of tritium while ^{14}C was retained (Battersby et al. 1965a). In order to explain these results the 1,2-dehydroreticulinium ion was assumed to be an intermediate which allows both for the loss of ^3H and the inversion of configuration at C-1. Labelled 1,2-dehydroreticuline was synthesized and was indeed well incorporated into thebaine, codeine, and morphine (Battersby et al. 1965a, Borkowski et al. 1978). Furthermore, the labelled 1,2-dehydroreticulinium ion could be demonstrated to be a natural product by detecting it in *P. somniferum* (Borkowski et al. 1978). This finding established 1,2-dehydroreticuline as the branch point intermediate leading from (S)-reticuline to the morphinan pathway. In an attempt to clarify the mechanism of the inversion of configuration of (S)-reticuline we (Amann, Nagakura, and Zenk unpublished) prepared C-1 ^3H-labelled reticuline with high specific activity and tested for loss of tritium from this molecular species during the incubation with various crude protein fractions of plant cell cultures and differentiated plants. An enzyme activity which catalysed the removal of tritium from 1-^3H-reticuline was purified from *Berberis* and *Papaver somniferum.* The reaction product was shown unequivocally to be 1,2-dehydroreticuline by preparative isolation (MS, NMR, comparison with authentic sample) (Fig. 11). The enzyme catalyses the dehydrogenation of a considerable number of benzylisoquinolines, including tetrahydropapaverine. It is ab-

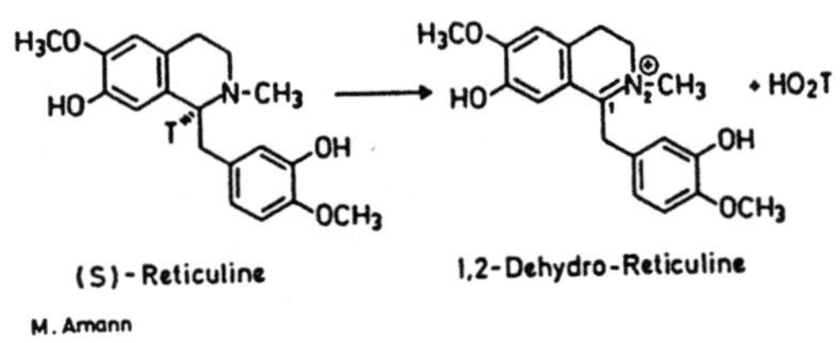

Fig. 11. The formation of 1,2-dehydroreticulinium ion from (S)-reticuline

solutely specific for the substrate with (S)-configuration. The substrates with the opposite (R)-configuration cannot be transformed. The enzyme has a pH-optimum of 8.9 and contains a covalently-bound flavine as prosthetic group. The hydrogen atom at C-1 is removed and appears as H_2O_2 in the incubation solution. Much to our surprise, this enzyme is identical with (S)-tetrahydroprotoberberine oxidase (STOX see Sect. 3.3). This finding is one of the rare examples of secondary metabolism where one and the same enzyme catalyses the transformation of two widely differing substrate classes, i.e., benzylisoquinolines and tetrahydroprotoberberines, yielding products which do not allow recognition of the underlying catalytic principle. While previously authentic 1,2-dehydroreticulinium ion had to be prepared (with low specific activity) in a 7-stage chemical synthesis, we can now prepare the desired dehydroderivatives of a number of benzylisoquinolines at extremely high specific activities with a minimum of effort and a maximum of yield by using the very stable STOX-enzyme as a reagent. If the proposed (Battersby et al. 1965b, Borkowski et al. 1978) biosynthetic pathway for the inversion of configuration of (S)-reticuline to (R)-reticuline via 1,2-dehydroreticuline holds true, an enzyme has to be postulated which stereospecifically hydrogenates 1,2-dehydroreticuline to (R)-reticuline. Various cell cultures and differentiated plants are currently being screened in order to detect this particular enzyme. It should be noted, however, that the reaction mechanism of the STOX-enzyme (H_2O_2-formation and subsequent degradation to H_2O by catalase) does not allow the tritium label at C-1 of (S)-reticuline to be retained substantially during its transition to morphine. The postulated transfer of tritium from the "oxidation-reduction" enzyme to 1,2-dehydroreticuline with the formation of a small (ca. 15%) amount of (R)-[1-^{3}H]-reticuline can not occur if the STOX-enzyme is involved. Since the STOX-enzyme does not dehydrogenate (R)-reticuline, the observed (Battersby et al. 1965b) loss of tritium from this enantiomer cannot be explained at present.

4.2 Formation of Salutaridine

The intramolecular condensation of (R)-reticuline to salutaridine has been established by tracer-feeding studies to be the key step in the formation of the morphinan alkaloids (Barton et al. 1963). The biochemical stereoselectivity of the reaction has no equivalent in synthetic organic chemistry. Chemical syntheses developed for this compound have only resulted in low yields (e.g. Szántay et al. 1982). Hodges and Rapoport (1982) reported recently that a cell-free extract from *Papaver somniferum* utilized hydrogen peroxide to convert (R,S)-[3-^{3}H]-reticuline to [^{3}H]-salutaridine with high yields (Fig. 12). It was necessary to separate hydrogen peroxide degrading acti-

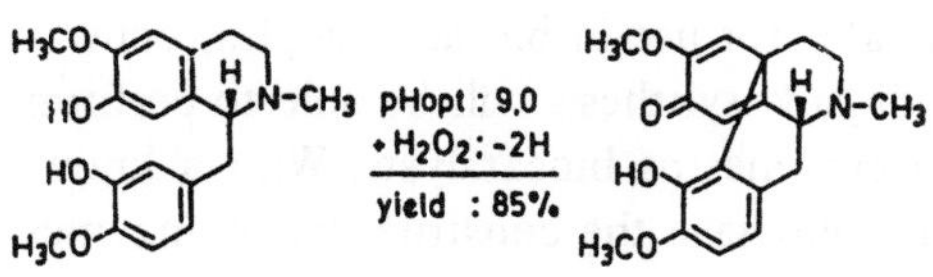

(R)-Reticuline Salutaridine

C.C.Hodges and H.Rapoport: Biochemistry 21,3729 (1982)

Fig. 12. Enzymatic conversion of reticuline to salutaridine by cell-free systems from *Papaver somniferum* as reported by Hodges and Rapoport (1982)

vities (catalase) from the active protein preparation catalysing this phenolic coupling by differential centrifugation. Without this step no or only little conversion of the precursor could be detected. This finding is a major breakthrough in the field of morphinan biosynthesis. It must be stated, however, that we were unable to repeat the reported experiments in our laboratory when we used our strain of opium poppy. We also used a specially devised radioimmunoassay for the quantitation of salutaridine (salutaridinole) (Wieczorek, Nagakura, Sund, and Zenk unpublished) in order to detect the conversion. Although the published experiments appear to have been carried out most carefully, we are now unable to confirm them. If this reaction could be confirmed in other laboratories and if the system could be purified to some extent, this most interesting enzyme system might provide important information with regard to the actual mechanism of the biochemical transformation.

Apart from the reduction of codeinone to codeine none of the other enzymatic steps between salutaridine and morphine are known.

4.3 Transformation of Codeinone to Codeine

Of all the enzymatic steps between the firmly established morphinan precursor salutaridine and morphine, only the conversion of (-)-codeinone to (-)-codeine is known. The system has been studied (Furuya et al. 1978, Hodges and Rapoport 1980) by using cell-free preparations of *Papaver somniferum* callus cultures and whole plants. The reaction was shown to depend on NADH as co-factor and is stereospecific. However, this reduction might be non-specific, since it has previously been shown that cell cultures of *Digitalis, Nicotiana,* and *Dioscorea* are also able to perform complicated reductions in vivo (Graves and Smith 1976). As far as a biotechnological application is concerned, this reaction is trivial since codeinone is stereospecifically and quantitatively reduced to codeine by borohydride (Gates 1953).

5 Conclusions

The combination of plant cell culture techniques with radioisotope and immunological techniques has allowed us to detect and characterize enzyme systems involved in isoquinoline and other alkaloid formation. In some cases, individual enzymes have been purified to homogeneity, which should make it possible to study their catalytic mechanisms in more depth and detail. Because of the tremendous variety of unusual and frequently complicated reactions catalysed by enzymes of the secondary pathways some of these enzymes may be candidates for use in biotechnological conversions. It can be envisaged that biomimetic organic synthesis will be able to perform complicated stereochemical reactions with enzymes as biocatalysts. With a knowledge of the natural enzymes one might also envisage the construction of enzymes with modified properties which might, e.g. extend their substrate ranges and one might even come up with new types of organic reactions which have not been realized in nature (Maugh 1984). In any case, the generations ahead of us will require sound knowledge of the biocatalysts involved in secondary metabolism, the procedures for their assay and isolation, their physical properties, and their reaction mechanisms. It will be our responsibility to provide them with detailed information on these enzymes so that they will be in a position to fully exploit gene and/or biotechnological processes in the production of known or new compounds for human welfare!

Acknowledgements. My thanks are due to the members of our laboratory for their enduring and patient work, especially to Dr. Martina Rueffer, who was involved in the discovery of five of the enzymes described here. I would also cordially thank Prof. Naotaka Nagakura, Kobe, Japan, who synthesized and supplied us with almost all of the reference alkaloids which we used in this study. Linguistic help of Dr. H. Ibelgaufts is gratefully acknowledged.

This work was supported by SFB 145 of Deutsche Forschungsgemeinschaft, Bonn, and by Fonds der Chemischen Industrie.

References

Amann M, Nagakura N, Zenk MH (1984) (S)-Tetrahydroprotoberberineoxidase, the final enzyme in protoberberine biosynthesis. Tetrahedron Lett 25:953–954

Antoun MD, Roberts MF (1975) Enzymic studies with *Papaver somniferum.* Planta Med 28:6–11

Barton DHR, Hesse HR, Kirby GW (1963) The origin of the berberine bridge carbon. Proc Chem Soc:267–268

Battersby AR, Francis RJ, Hirst M, Staunton J (1963) Biosynthesis of the berberine bridge. Proc Chem Soc:268

Battersby AR, Dobson TA, Ramuz H, Barton DHR, Kirby GW, Steglich W, Thomas GM (1965a) Investigations on the biosynthesis of morphine alkaloids. J Chem Soc:2423–2438

Battersby AR, Foulkes DM, Binks R (1965b) Use of optically active precursors for investigations on the biosynthesis of morphine alkaloids. J Chem Soc:3323–3332

Battersby AR, Staunton J, Wiltshire HR, Francis RJ, Southgate R (1975a) The origin of chelidonine and of other alkaloids derived from the tetrahydroprotoberberine skeleton. J Chem Soc Perkin Trans I:1147–1156

Battersby AR, Jones RCF, Kazlauskas R (1975b) Experiments on the early steps of morphine biosynthesis. Tetrahedron Lett:1873–1876

Beecher CWW, Kelleher WJ (1983) The incorporation of berberine into jatrorrhizine. Tetrahedron Lett 24:469–472

Bobbitt JM, Kulkarni CL, Wiriyachitra P (1976) Decarboxylation of phenolic tetrahydroisoquinoline-1-carboxylic acids in air. Heterocycles 4:1645–1648

Böhm H, Rink E (1975) Zur Rolle eines Reticulin-umwandelnden Enzyms (Berberin-Brücken-Enzym) im pflanzlichen Benzylisochinolin-Stoffwechsel. Biochem Physiol Pflanz 168:69–77

Borkowski PR, Horn JS, Rapoport H (1978) Role of 1,2-dehydroreticulinium ion in the biosynthetic conversion of reticuline to thebaine. J Am Chem Soc 100:276–281

Brochmann-Hanssen E, Chen CH, Chen CR, Chiang HC, Leung AY, McMurtrey KD (1975) The biosynthesis of 1-benzylisoquinolines in *Papaver somniferum*. J Chem Soc Perkin Trans I: 1531–1537

Coutts IGC, Hamblin MR, Tinley EJ (1979) The enzymatic oxidation of phenolic tetrahydroisoquinoline-1-carboxylic acids. J Chem Soc Perkin Trans I:2744–2750

Davis VE, Cashaw JL, McMurtrey KD, Ruchirawat S, Nimit Y (1982) Metabolism of tetrahydroisoquinolines and related alkaloids. In: Bloom F, Barchas J, Sandler M, Usdin E (eds) Beta-carbolines and tetrahydroisoquinolines. Liss, New York, p 99

Furuya T, Nakano M, Yoshikawa T (1978) Biotransformation of (RS)-reticuline and morphinan alkaloids by cell cultures of *Papaver somniferum*. Phytochemistry 17:891–893

Gates M (1953) Conversion of codeinone to codeine. J Am Chem Soc 75:4340–4341

Graves JMH, Smith WK (1967) Transformation of pregnenolone and progesterone by cultured plant cells. Nature (London) 214:1248–1249

Hahlbrock K, Grisebach H (1975) Biosynthesis of flavonoids. In: Mabry TJ, Mabry H (eds) The flavonoids. Chapman and Hall, London, p 866

Hahn G, Rumpf F (1938) Kondensation von Oxyphenyl-Äthylaminen mit α-Ketonsäuren. Ber dtsch chem Ges 71:2141–2153

Hinz H, Zenk MH (1981) Production of protoberberine alkaloids by cell suspension cultures of *Berberis* species. Naturwissenschaften 67:620

Hodges CC, Rapoport H (1980) Enzymic reduction of codeine in vitro cell-free systems from *Papaver somniferum* and *P. bracteatum*. Phytochemistry 19:1681–1684

Hodges CC, Rapoport H (1982) Enzymic conversion of reticuline to salutaridine by cell-free systems from *Papaver somniferum*. Biochemistry 21:3729–3734

Kelleher WJ, Rother A, Wellmann E, Grisebach H (1980) Enzymic formation of protoberberine alkaloids. Planta Med 40:127–136

Loomis WD, Battaile J (1966) Plant phenolic compounds and the isolation of plant enzymes. Phytochemistry 5:423–438

Maugh TH (1984) Semisynthetic enzymes are new catalysts. Science 223:154–156

Nagakura N, Höfle G, Zenk MH (1978) Desacetylipecoside: the key intermediate in the biosynthesis of the alkaloids cephaeline and emetine. J Chem Soc Chem Commun:896–898

Prakash O, Bhakuni DS, Kapil RS (1979) Biosynthesis of coclaurine. J Chem Soc Perkin Trans I: 1515–1518

Rink E, Böhm H (1975) Conversion of reticuline into scoulerine by a cell free preparation from *Macleaya microcarpa* cell suspension cultures. FEBS Lett 49:396–399

Robinson R (1917) A theory of the mechanism of the phytochemical synthesis of certain alkaloids. J Chem Soc:876–899

Rueffer M, El-Shagi H, Nagakura N, Zenk MH (1981) (S)-Norlaudanosoline synthase: the first enzyme in the benzylisoquinoline biosynthetic pathways. FEBS Lett 129:5–9

Rueffer M, Ekundayo O, Nagakura N, Zenk MH (1983a) Biosynthesis of the protoberberine alkaloid jatrorrhizine. Tetrahedron Lett 24:2643–2644

Rueffer M, Nagakura N, Zenk MH (1983b) Partial purification and properties of S-adenosylmethionine: (R),(S)-norlaudanosoline-6-O-methyltransferase from *Argemone platyceras* cell cultures. Planta Med 49:131–137

Rueffer M, Nagakura N, Zenk MH (1983c) A highly specific O-methyltransferase for nororientaline synthesis isolated from *Argemone platyceras* cell cultures. Planta Med 49:196–198

Sato F, Yamada Y (1984) High berberine producing cultures of *Coptis japonica* cells. Phytochemistry 23:281–285

Schumacher HM, Rueffer M, Nagakura N, Zenk MH (1983) Partial purification and properties of (S)-norlaudanosoline synthase from *Eschscholtzia tenuifolia* cell cultures. Planta Med 48: 212–220

Scott AI, Lee SL, Hirata T, Culver MG (1978a) Biosynthesis of plant products using cell free systems from tissue cultures. Rev. Latinoam Quím 9:131–138

Scott AI, Lee SL, Hirata T (1978b) The early stages of *Papaver* alkaloid biosynthesis. Cell-free synthesis of benzylisoquinolines. Heterocycles 11:159–163

Shamma M (1972) The isoquinoline alkaloids. Academic, London New York

Shamma M, Moniot JL (1978)) Isoquinoline alkaloid research 1972–1977. Plenum, London New York

Spenser D, Gear JR (1962) Biosynthesis of berberine. Proc Chem Soc:228–229

Spenser ID (1968) Biosynthesis of isoquinoline alkaloids. In: Florkin M, Stotz EH (eds) Comprehensive biochemistry, vol XX. Elsevier, Amsterdam, p 231

Steffens P, Nagakura N, Zenk MH (1984) The berberine bridge forming enzyme in tetrahydroberberine biosynthesis. Tetrahedron Lett 25:951–952

Stöckigt J (1980) The biosynthesis of heteroyohimbine-type alkaloids. In: Phillipson JD, Zenk MH (eds) Indole and biogenetically related alkaloids. Academic, London New York, p 113

Szántay C, Bárczai-Behe M, Péchy P, Blásko G, Dörnyei G (1982) Synthesis of (±)-salutaridine via phenolic oxidative coupling of (±)-reticuline. J Org Chem 47:594–596

Wat CK, Steffens P, Zenk MH (1985) Partial purification and characterization of norreticuline-N-methyltransferase from *Berberis* cultures (in press)

Wilson ML, Coscia CJ (1975) Studies on the early stages of *Papaver* alkaloid biogenesis. J Am Chem Soc 97:431–432

Wink M, Hartmann T (1982) Physiological and biochemical aspects of quinolizidine alkaloid formation in cell suspension cultures. In: Fujiwara A (ed) Plant tissue culture 1982. Maruzen, Tokyo, p 333

Winterstein E, Trier G (1910) In: Die Alkaloide. Bornträger, Berlin, p 307

Zenk MH (1978) The impact of plant cell culture on industry. In: Thorpe TA (ed) Publ Int Assoc Plant Tissue Culture, Calgary, p 1

Zenk MH (1980) Enzymatic synthesis of ajmalicine and related indole alkaloids. J Nat Prod 43: 438–451

Morphinan Alkaloids from Plant Cell Cultures

F. CONSTABEL[1]

1 Introduction

Morphinan alkaloids are an industrial commodity of plant origin. The market share of pharmaceuticals containing morphinans, predominantly codeine, is about 2.5% (Canada). The supply appears to be adequate, the price comparatively moderate. Restrictions, however, in the harvest and in the movement of raw and refined products due to socio-political implications has kindled interest in producing morphinan alkaloids by culturing poppy cells in vitro. The feasibility of developing such a process for industry would require fair accumulation of alkaloids in cells or their medium.

2 Alkaloid Accumulation

2.1 Occurrence

Analysis of extracts obtained with cell cultures derived from *Papaver bracteatum* and *Papaver somniferum* revealed the occurrence of thebaine in the former, of thebaine, codeine, and to a lesser extent of morphine in the latter material (Tables 1,2). Considering the many attempts to demonstrate morphinan alkaloids in cell cultures which remained unsuccessful and unreported, detection of occurrence would have been described until quite recently as a rare phenomenon. The availability of a radioimmunoassay and other improved analytical systems would explain the surge of reports of accumulation since 1980. According to recent analyses, concentrations of morphinan alkaloids are as low or lower than in leaf tissue (0.14% DW), and much lower than in dried latex, opium (ca. 20% DW).

2.2 Culture Conditions and Plant Regeneration

Plant cell culture is perceived as a method to study cell differentiation and morphogenesis by varying nutritional factors and hormonal regimes. (For review of poppy cell culture, see Nessler 1984). As any natural product formation, morphinan alka-

1 Plant Biotechnology Institute, National Research Council of Canada, Saskatoon, Saskatchewan, Canada S7N OW9

NRCC #23847

The Chemistry and Biology of Isoquinoline Alkaloids, ed. by Phillipson et al.

Table 1. Thebaine accumulation in *Papaver bracteatum* cell cultures

Material	Accumulation	Separation	Reference
Callus (~ 1-year-old strain)	60 μg g^{-1} DW 130 μg g^{-1} DW	TLC; GLC	Kamimura et al. (1976)
cv. Arya II Cell suspension	+	TLC	Shafiee et al. (1976)
cv. Arya II Callus with shoots (Several-years-old strain)	70 μg g^{-1} DW	TLC, HPLC	Staba et al. (1982)
cv. Arya II Callus with shoots and meristemoids (~2-year-old strain)	0.87 μg g^{-1} FW	HPLC	Kutchan et al. (1983)

Table 2. Morphinan alkaloid accumulation in *Papaver somniferum* cell cultures

Material	Accumulation Thebaine	 Codeine	 Morphine	Separation Method	Reference
Cell suspension (1-year-old strain)	1.49%	0.34%	1.31% DW	TLC	Khanna et al. (1978)
cv. Marianne Cell suspension (6-12-month-old strains)	-	0.15% DW	-	TLC; GLC	Tam et al. (1980)
Callus	+	+	+	TLC	Hsu (1981)
Callus with roots (several-years-old strain)	+	+?	-	TLC; HPLC	Staba et al. (1982)
Callus	12.7	34.3	1.4 μg g^{-1} FW	RIA; HPLC	Hodges and Rapoport (1982)
Callus	0.1 0.1	3.8 μg g^{-1} DW 33.0 μg g^{-1} DW	- -	TLC; GLC	Kamo et al. (1982)
Suspension culture with shoot buds	+	+	±	TLC; GLC	Yoshikawa and Furuya (1982)
Embryoids	0.2% DW			TLC	Schuchmann and Wellmann (1983)

Table 3. Media for *Papaver bracteatum* cell cultures

Medium	Auxin (mg l^{-1})	Cytokinin (mg l^{-1})	Supplements	Accumulation	Reference
MS	2,4-D (1.0)	KIN(0.1)	-	+	Kamimura
		KIN(1.0)	CM (10%)	+	et al. (1976)
RT	2,4-D (0.1)	-	-	+	Shafiee et al. (1976)
MS	IBA (1.0)	BA(3.0)	-	+	Staba et al. (1982)
MS	-	BA(5.0)	-	+	Kutchan
MS	2,4-D (2.0)	-	-	-	et al. (1983)

Table 4. Media used for *Papaver somniferum* cell cultures

Medium	Auxin (mg l^{-1})	Cytokinin (mg l^{-1})	Supplements	Morphinan alkaloids	Reference
W	a) 2,4-D(0.6)	-	CM (10%)	-	Ranganathan
	b) 2,4-D(0.6)	-	ME (0.1%)	-	et al. (1963)
	c) 2,4-D(0.6)	-	YE (0.1%)	-	
RT	a) 2,4-D(0.1)	-	TYR(125 mg l^{-1})	+	Khanna et al.
	b) 2,4-D(0.1)	-	ASC(50 mg l^{-1})	+	(1978)
MS	a) 2,4-D(0.1)	ZEA(0.025)	-	-	Morris and
	b)2,4-D(1.0)	KIN(0.1)	-	-	Fowler (1980)
	c) NAA(1.0)	KIN(0.1)	-	-	
B5	2,4-D(1.0)	KIN(0.1)	CH(1 g l^{-1})	+	Tam et al. (1980)
MS	2,4-D(0.5)	KIN(0.1)	-	+	Hsu (1981)
MS	a) 2,4-D(2.0)	KIN(0.25)	-	-	Nessler (1982)
	b) -	-	-	+	
MS	-	BA(3.0)	-	+	Staba et al. (1982)
MS	2,4-D(0.1)	KIN(0.1)	-	+	Kamo et al. (1982)
MS	a) 2,4-D(0.02)	-	CM(10%)	+	Hodges and
	b) 2,4-D(0.2) NAA (0.2)	2iP(0.03)	-	+	Rapoport (1982)
	c) NAA (0.2)	-	-	+	
	d) -	KIN(0.03)	-	+	

Media: B5 = Gamborg et al. (1968); MS = Murashige and Skoog (1962); W = White (1954).
Growth hormones: 2,4-D = 2,4-dichlorophenoxyacetic acid; NAA = α-naphthaleneacetic acid; IBA = indolylbutyric acid; KIN = kinetin; ZEA = zeatin; BA = 6-benzyladenine; 2iP = 2-isopentenyladenine
Supplements: CM = coconut water; ME = malt extract; YE = yeast extract; TYR = L-tyrosine; ASC = ascorbic acid; CH = casein hydrolyzate.

loid production by poppy cell cultures can be regarded as a cell differentiation and morphogenetic process. It might, therefore, be helpful to evaluate nutrient media which have been employed, in order to define optimum conditions. The result of such evaluation is, however, somewhat disappointing, as data (Tables 3, 4) do not yet clearly indicate superiority of specific media and hormone combinations. The medium after Murashige and Skoog (1962) has commonly been used and has permitted alkaloid accumulation in several instances. The medium after Gamborg et al. (1968) was found to be suitable as well. An alkaloid production medium as suggested for periwinkle cells (Zenk et al. 1977) appears not to have been employed successfully. 2,4-Dichlorophenoxyacetic acid in concentration of 1 mg l^{-1} and higher often prevented morphinan accumulation, it did not in the case of codeine accumulation as observed by Tam et al. (1980). The presence of cytokinins, i.e. benzyladenine and kinetin appears to be beneficial, it permitted codeine production according to Staba et al. (1982) and Hodges and Rapoport (1982). The addition of tyrosine and ascorbic acid as media supplements promoted morphinan accumulation (Kamimura et al. 1976, Khanna et al. 1978), confirmation of these observations would be desirable.

Transfer of *P. somniferum* callus on solid or in liquid medium to media without hormones led to the formation of embryoids resembling seed embryos, i.e. characterized by two stubby cotyledons and a short, thick hypocotyl-root axis. Contrary to seed embryos, such embryoids were found to accumulate up to 0.2% DW of thebaine (Schuchmann and Wellmann 1983). Plantlet formation through rooting of callus-borne shoots resulted in regenerants which also contained morphinan alkaloids. While, in general, regenerants showed a morphinan alkaloid pattern as in the source plants (Ikuta et al. 1974), in one case the main alkaloid was codeine, although the source plants as usual produced morphine as the major alkaloid (Yoshikawa and Furuya 1983).

2.3 Correlation With Laticifer Formation

In poppy plants morphinan alkaloids are primarily accumulated in latex which in turn is contained in structurally and physiologically specialized cells, laticifers. Roberts et al. (1983 and references therein) demonstrated that 1000 g and supernatant fractions of the latex are required for alkaloid biosynthesis and that a subpopulation of dense organelles of the 1000 g sediment have a function as a storage compartment for alkaloids. Morphinan alkaloid production with poppy cell cultures, therefore, would appear to require the concurrent differentiation of latex vessels. The high levels of alkaloid accumulation would thus appear to be intriguingly correlated with the formation of latex and latex vessels.

Callus and cell suspension cultures of opium poppy have, so far, not been seen to develop typical laticifers as displayed, for instance, in micrographs of shoots from seedlings and regenerants by Thureson-Klein (1970) and Nessler and Mahlberg (1977, 1979). Laticifer-like cells were demonstrated for *P. bracteatum* cell cultures by Kutchan et al. (1983). Absence of typical laticifers was often cited as the reason for failure to show occurrence of thebaine or codeine in poppy cell cultures. Today one may propose that accumulation of low amounts of morphinan alkaloids do not require latex and respective vessels. Indeed, other parts of the plant besides latex have been found metabolically active in alkaloid biosynthesis (Fairbairn et al. 1964). Incorporation of precursors was observed to be better with, but not restricted to (isolated) latex (Wilson and Coscia 1975). The situation may be likened to the occurrence of essential oils in mint cell cultures, which also had not developed specialized cells, glands, known to harbour and excrete these products in great amounts (Hudson and Goldsworthy 1981). The problem of concurrent cytodifferentiation and morphinan alkaloid accumulation appears compounded by the detection of cells which resemble early stages of laticifer formation (cf. Nessler and Mahlberg 1977), i.e. cells rich in vesiculating endoplasmic reticulum, which, however, do not accumulate morphinan alkaloids. The capacity of cultured cells, parenchyma, to accumulate morphinan alkaloids remains an open question.

Advances in cell technology may lift the level of accumulation in the absence of typical laticifers in cell cultures substantially. A first step in this direction would be to clearly determine the stage at which cell cultures show alkaloid accumulation. It would appear from observations by Hodges and Rapoport (1982) that highest alkaloid concentrations are displayed by calli subcultured only once or twice over periods of 2 weeks. With further subculturing and a total of only 9 weeks morphinan alkaloids were not detected any longer (see also Kamimura et al. 1976). With material derived from *P. somniferum* cv. Marianne, codeine was detected 4– ca. 8 months after callus initiation. Some time thereafter this capacity was lost, later the productive strains were lost as well. Other cell cultures have been described as about 1 year old. Poppy cell cultures, therefore, share the fate of similar cultures, i.e. repression of accumulation over time. Only a much more refined, better still a selection medium might help overcome the problem of loss of productivity. An experiment designed to define the environment for alkaloid producing cells may start with initiating protoplasts of poppy embryos and studying the developmental potential of young laticifers.

2.4 Genetic and Physiological Factors

Material of *P. bracteatum* employed in cell culture work usually originated from cv. Arya II.,*P. somniferum* material has generally been ill defined given the great number of cultivars and proveniences. It is the multitude of genotypes which would suggest variation in response of cells to in vitro culture conditions and screening for high-yielding material. Our efforts to identify morphinan producing cell cultures initiated with 46 different samples of seeds obtained through the Gene Resources Center/ Ottawa from a variety of Botanical Gardens failed completely. A repetition using a more sensitive morphinan alkaloid test might have provided a different result.

Alkaloid accumulation among different cell cultures was found to differ significantly. With young callus of *P. somniferum*, radioimmunoassays revealed 45% of an unreported number of cultures to have accumulated morphinan alkaloids (Hodges and Rapoport 1982). With cell suspensions derived from cultivar Marianne, 3 out of 32 cultures showed the presence of codeine (Tam et al. 1980). In 1982/83 we initiated new callus cultures with 1978 seed of cv. Marianne. None out of 77 cultures showed codeine accumulation. The singular occurrence of codeine as published earlier may today be explained by assuming an undetected regeneration or cytodifferentiation or, more likely, an unregistered temporary stress.

Stress has been identified as a factor to increase morphinan accumulation in poppy cell cultures. Lockwood (1984) reported exposure of cells to +5°C for a period of 3 days prior to harvesting as beneficial to thebaine accumulation. Temperature stress, then, would be a factor still to be explored and exploited.

Stress by infection with micro-organisms may be another route to follow. The suggestion is based on observations by Laughlin and Munro (1983) measuring a 75% increase in morphine concentration of leaves and stems subsequent to infection of plants with *Sclerotinia sclerotiorum.* Treatment of cells with elicitors derived from pathogenic micro-organisms may activate constitutive or secondary elicitors which in turn may enhance the accumulation of secondary metabolites. It may be added that host-specific parasites often show a lower effect on elicitor formation than non-specific ones. The elicitor substance, then, must not be obtained with poppy parasites, but could be obtained with still unspecified polysaccharides, peptides, glycoproteins or enzymes of certain fungi (Wolters and Eilert 1983).

3 Conclusions

Morphinan alkaloid production by poppy plant cell cultures is not wholly dependent on the simultaneous differentiation of laticifers. Whether cell cultures lacking typical laticifers can be manipulated to accumulate better than low amounts of morphinan alkaloids remains to be seen. The formation of latex vessels would substantially increase the capacity of cell cultures to accumulate morphinan alkaloids. Since Nessler (1982) reported the presence of laticifers in somatic embryos and since Schuchmann and Wellmann (1983) determined the thebaine concentration of such material to be 0.2% DW, poppy cell cultures may still be amenable to fermenter aided production of morphinan alkaloids. The key factor would be to induce mass-embryogenesis in the final stages of the culture process. In regard of the alkaloids formed, observations so far would indicate that in *Papaver somniferum* cell cultures, thebaine and in particular codeine prevail, while morphine accumulation is missing. In this respect in vitro cultures differ from source plants significantly. The accumulation of codeine suggests the possible biotechnological application of cell cultures, since it would obviate the present necessity of codeine synthesis from morphine (Hodges and Rapoport 1982). Technical advances, it is hoped, will lift the level of codeine production in cells and, thus, will render yields acceptable for industrial development.

References

Fairbairn JW, Paterson A, Wassel G (1964) The alkaloids of *Papaver somniferum* L. II. ^{14}C isotopic studies of the rapid changes in the major alkaloids. Phytochemistry 3:577–582

Gamborg OL, Miller RA, Ojima K (1968) Nutrient requirements of suspension cultures of soybean root cells. Exp Cell Res 50:151–158

Hodges CC, Rapoport H (1982) Morphinan alkaloids in callus cultures of *Papaver somniferum.* J Nat Prod 45:481–485

Hsu A-F (1981) Effect of protein synthesis inhibitors on cell growth and alkaloid production in cell cultures of *Papaver somniferum.* J Nat Prod 44:408–414

Hudson MJ, Goldsworthy A (1981) Microdetermination of essential oils in plant tissues. ASPP Newslett 8:25

Ikuta A, Syono K, Furuya T (1974) Alkaloids of callus tissues and redifferentiated plantlets in the Papaveraceae. Phytochemistry 13:2175–2179

Kamimura S, Akutsu M, Nishikawa M (1976) Formation of thebaine in the suspension culture of *Papaver bracteatum.* Agric Biol Chem 40:913–919

Kamo KK, Kimoto W, Hsu A-F, Mahlberg PG, Bills DD (1982) Morphinan alkaloids in cultured tissues and redifferentiated organs of *Papaver somniferum.* Phytochemistry 21:219–222

Khanna P, Khanna R, Sharma M (1978) Production of free ascorbic acid and effect of exogenous ascorbic acid and tyrosine on production of major opium alkaloids from in vitro tissue culture of *Papaver somniferum.* Indian J Exp Biol 16:110–112

Kutchan TM, Ayabe S, Krueger RJ, Coscia EM, Coscia CJ (1983) Cytodifferentiation and alkaloid accumulation in cultured cells of *Papaver bracteatum.* Plant Cell Rep 2:281–284

Laughlin JC, Munro D (1983) The effect of *Sclerotinia* stem infection on morphine production and distribution in poppy (*Papaver somniferum* L.). J Agric Sci 100:299–303

Lockwood B (1984) Alkaloids of cell suspensions derived from four Papaver species and the effect of temperature stress. Z Pflanzenphysiol 114:361–363

Morris P, Fowler MW (1980) Growth and alkaloid content of cell suspension cultures of *Papaver somniferum.* Planta Med 39:284–285

Murashige T, Skoog F (1962) A revised medium for rapid growth and bioassays with tobacco tissue cultures. Physiol Plant 15:473–497

Nessler CL (1982) Somatic embryogenesis in the opium poppy *Papaver somniferum.* Physiol Plant 55:453–458

Nessler CL (1984) Tissue culture and somatic embryogenesis in opium poppy. In: Bajaj VPS (ed) Tissue culture of economically important plants. Submitted

Nessler CL, Mahlberg PG (1977) Ontogeny and cytochemistry of alkaloid vesicles in laticifers of *Papaver somniferum* L. Am J Bot 64:541–551

Nessler CL, Mahlberg PG (1979) Laticifers in organs redifferentiated from *Papaver somniferum* callus. Can J Bot 57:675–685

Ranganathan B, Mascarenhas AF, Sayagaver BM, Jagannathan V (1963) Growth of *Papaver somniferum* in vitro. In: Maheshwari P (ed) Plant tissue and organ culture. Int Soc Plant Morphol, Univ Delhi, pp 108-110

Roberts MF, McCarthay D, Kutchan TM, Coscia CJ (1983) Localization of enzymes and alkaloidal metabolites in *Papaver* latex. Arch Biochem Biophys 222:599–609

Schuchmann R, Wellmann E (1983) Somatic embryogenesis of tissue cultures of *Papaver somniferum* and *Papaver orientale* and its relationship to alkaloid and lipid metabolism. Plant Cell Rep 2:88–91

Shafiee A, Lalezari I, Yassa N (1976) Thebaine in tissue culture of *Papaver bracteatum* Lindl, Population Arya II. Lloydia 39:380–381

Staba EJ, Zito S, Amin M (1982) Alkaloid production from *Papaver* tissue cultures. J Nat Prod 45:256–262

Tam WHJ, Constabel F, Kurz WGW (1980) Codeine from cell suspension cultures of *Papaver somniferum.* Phytochemistry 19:486–487

Thureson-Klein A (1970) Observations on the development and fine structure of the articulated laticifers of *Papaver somniferum.* Ann Bot (London) 34:751–759

White PR (1954) The cultivation of plant and animal cells. Ronald Press, New York

Wilson ML, Coscia CJ (1975) Studies on the early stages of *Papaver* alkaloid biogenesis. J Am Chem Soc 97:431–432

Wolters B, Eilert U (1983) Elicitoren-Auslöser der Akkumulation von Pflanzenstoffen. Dtsch Apoth Ztg 123:659–667

Yoshikawa T, Furuya T (1982) Morphinan alkaloid production by opium poppy suspension culture. In: Fujiwara A (ed) Proc 5th Int Congr Tissue Cell Culture. Maruzen, Tokyo, pp 307–309

Yoshikawa T, Furuya T (1983) Regeneration and in vitro flowering of plants derived from callus cultures of opium poppy (*Papaver somniferum*). Experientia 39:1031–1033

Zenk MH, El-Shagi H, Arens H, Stöckigt J, Weiler EW, Deus B (1977) Formation of the indole alkaloids serpentine and ajmalicine in cell suspension cultures of *Catharanthus roseus*. In: Barz W, Reinhard E, Zenk MH (eds) Plant tissue culture and its bio-technological application. Springer, Berlin Heidelberg New York, pp 27–43

The Production of Isoquinoline Alkaloids by Plant Cell Cultures

M. RUEFFER[1]

1 Introduction

Under optimal conditions plant cells are able to grow like microorganisms in suspension cultures without limitation or ageing. Moreover they are totipotent, in that one single cell or protoplast has the whole genetic information for the differentiated plant. Therefore one can expect that under appropriate conditions, plant cell suspension cultures have the ability to produce the whole range of natural products which are isolated from the differentiated plants. This could be of extreme importance for the production of substances especially for pharmaceutical use, since a survey indicates that 23% of all prescriptions contain natural compounds (Farnsworth and Morris 1976). Plant cell cultures would have several advantages in comparison to differentiated plants. They are independent for instance of geographical and climatical conditions, of plant diseases or animal destruction and therefore the price of the plant drugs could be stabilized. In addition, they could easily be cultivated under special state control, and thus would be of great importance for example for the production of morphine.

During recent years a variety of natural compounds among them a series of benzylisoquinolines have been isolated from plant cell cultures and the results of these researches are described in this chapter.

2 The Production of Isoquinoline Alkaloids

2.1 Benzylisoquinoline Alkaloids from Plant Cell Cultures

Since Reinhard (1967) described the production of protoberberine alkaloids by callus cultures of *Berberis vulgaris*, many different structure types of benzylisoquinoline alkaloids could be isolated from callus- and suspension cultures (Table 1). The greatest variety in the substitution pattern of the basic structure is found in the group of protoberberine alkaloids. Ikuta and Itokawa (1982b) for example described the isolation of 11 different protoberberine alkaloids from one plant cell culture (*Nandina domestica*, Berberidaceae).

1 Lehrstuhl Pharmazeutische Biologie der Universität München, Karlstraße 29, 8000 München 2, FRG

The Chemistry and Biology of Isoquinoline Alkaloids, ed. by Phillipson et al.

Table 1. Benzylisoquinoline alkaloids isolated from plant cell cultures

Structure type	Alkaloid name	Source (reference)
Benzylisoquinolines		
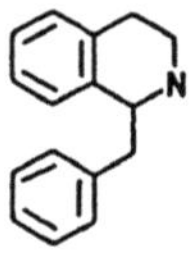 1	Coclaurine	*Fumaria capreolata* (Tanahashi et al. unpublished)
	Reticuline	*Fumaria capreolata* (Tanahashi et al. unpublished)
		Chasmanthera dependens (Rüffer et al. unpublished)
Bisbenzylisoquinolines 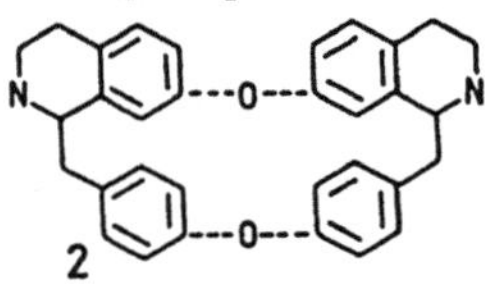 2	Berbamine Aromoline	*Stephania cepharantha* (Akasu et al. 1976)
Proaporphines 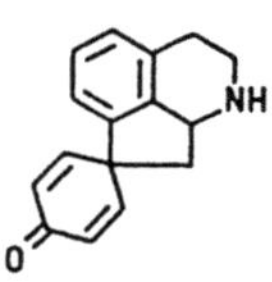 3	Stepharine	*Tinospora caffra, Chasmanthera dependens* (Rüffer et al. unpublished)
		Tinospora cordifolia (Arens unpublished)
Aporphines 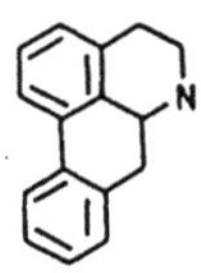 4	Cepharadione A	*Stephania cepharantha* (Akasu et al. 1975)
	Cepharadione B	(Akasu et al. 1975)
	Liriodenine	(Akasu et al. 1975)
	Lysicamine	(Akasu et al. 1975)
	Norcepharadione	(Akasu et al. 1975)
	Isoboldine	*Fumaria capreolata* (Tanahashi et al. unpublished)
	Magnoflorine	*Corydalis incisa, Corydalis pallida, Dicentra peregrina, Eschscholtzia californica, Papaver somniferum, P. setigerum, P. bracteatum, P. orientale, P. rhoeas* (Ikuta et al. 1974)
		Thalictrum minus, Coptis japonica, Mahonia japonica, Nandina domestica (Ikuta and Itokawa 1982a,b)
		Tinospora caffra, Chasmanthera dependens, Stephania japonica (Rüffer et al. unpublished) *Dioscoreophyllum cumminsii* (Furuya et al. 1983)

Table 1 (cont.)

Structure type	Alkaloid name	Source (reference)
	Stephanine	*Stephania glabra* (Gurova et al. 1980)
Protoberberines 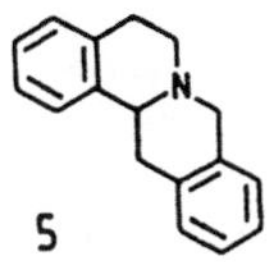 5	Berberine	*Argemone mexicana* (Khanna et al. 1982)
		Coptis japonica, Nandina domestica, Mahonia japonica, Thalictrum minus (Ikuta and Itokawa 1982a,b)
		Berberis vulgaris (Reinhard 1967)
		Berberis stolonifera (Hinz and Zenk 1981)
		Phellodendron amurense (Kuroda and Itokawa 1973)
	Columbamine	*Berberis stolonifera* (Hinz and Zenk 1981)
		Coptis japonica, Mahonia japonica, Nandina domestica, Thalictrum minus (Ikuta and Itokawa 1982a,b)
		Tinospora caffra Rüffer et al. unpublished)
	Coptisine	*Chelidonium majus* (Böhm et al. 1980)
		Coptis japonica, Mahonia japonica, Nandina domestica, Thalictrum minus (Ikuta and Itokawa 1982a,b)
		Fumaria capreolata (Tanahashi et al. unpublished)
	Dehydrocheilanthifoline	*Fumaria capreolata* (Tanahashi et al. unpublished)
	Dehydrocorydalmine	*Chasmanthera dependens* (Rüffer et al. unpublished)
	Desoxythalidastine	*Thalictrum minus* (Ikuta and Itokawa 1982a)
	Epiberberine	*Coptis japonica, Nandina domestica, Thalictrum minus* (Ikuta and Itokawa 1982a,b)
	Groenlandicine	*Coptis japonica, Nandina domestica* (Ikuta et al. 1974)

Table 1 (cont.)

Structure type	Alkaloid name	Source (reference)
	Jatrorrhizine	*Berberis stolonifera* (Hinz and Zenk 1981)
		Dioscoreophyllum cumminsii (Furuya et al. 1982)
		Coptis japonica, Nandina domestica, Mahonia japonica, Thalictrum minus (Ikuta and Itokawa 1982a,b)
		Chasmanthera dependens, Tinospora caffra, Stephania japonica (Rüffer et al. unpublished)
		Tinospora crispa (Sumimoto Chemicals 1982)
	Palmatine	*Coptis japonica, Thalictrum minus, Nandina domestica* (Ikuta and Itokawa 1982a,b)
		Dioscoreophyllum cumminsii Furuya et al. 1982)
		Phellodendron amurense (Kuroda and Itokawa 1973)
		Tinospora crispa (Sumimoto Chemicals 1982)
		Tinospora caffra, Chasmanthera dependens, Stephania japonica, Menispermum canadense (Rüffer unpublished)
	Scoulerine	*Fumaria capreolata* (Tanahashi et al. unpublished)
	Stylopine	*Papaver bracteatum* (Kamimura et al. 1976)
	Thalifendine	*Nandina domestica, Thalictrum minus* (Ikuta and Itokawa 1982a,b)
	Thalidastine	*Coptis japonica, Nandina domestica, Thalictrum minus* Ikuta and Itokawa 1982a,b)
Protopines 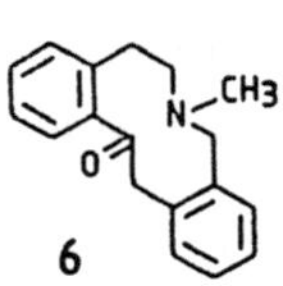 6	Allocryptopine	*Chelidonium majus* (Böhm et al. 1980)
		Macleaya microcarpa (Koblitz et al. 1975)
	Cryptopine	*Papaver somniferum* (Furuya et al. 1971)

Table 1 (cont.)

Structure type	Alkaloid name	Source (reference)
	Protopine	*Corydalis ophiocarpa* (Iwasa and Takao 1982)
		Fumaria capreolata (Tanahashi et al. unpublished)
		Macleaya microcarpa (Koblitz et al. 1975)
		Papaver somniferum (Furuya et al. 1971)
		Corydalis incisa, Corydalis pallida, Dicentra peregrina, Chelidonium japonica, Eschscholtzia californica, Papaver setigerum, P. bracteatum, P. orientale, P. rhoeas (Ikuta et al. 1974)
		Chelidonium majus (Böhm et al. 1980)
Benzophenanthridines 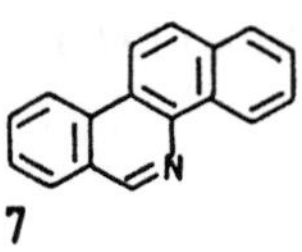 7	Chelidonine	*Chelidonium majus* (Böhm et al. 1980)
	Chelirubine	*Eschscholtzia californica, Macleaya cordata, Papaver bracteatum* (Ikuta et al. 1974)
	Dihydrochelerythrine	*Eschscholtzia californica* (Berlin et al. 1983)
	Dihydrochelirubine	*Eschscholtzia californica* (Berlin et al. 1983)
	Dihydromacarpine	*Eschscholtzia californica* (Berlin et al. 1983)
	Dihydrosanguinarine	*Chelidonium japonica, Eschscholtzia californica, Macleaya cordata, Papaver somniferum, P. setigerum, P. bracteatum, P. orientale, P. rhoeas* (Ikuta et al. 1974)
	Norsanguinarine	*Chelidonium japonica, Eschscholtzia californica, Macleaya cordata, Papaver somniferum, P. setigerum, P. bracteatum, P. rhoeas, P. orientale, Corydalis incisa, C. pallida, Dicentra peregrina* Ikuta et al. 1974)

Table 1 (cont.)

Structure type	Alkaloid name	Source (reference)
	Oxosanguinarine	*Eschscholtzia californica, Chelidonium japonica, Macleaya cordata, Papaver somniferum, P. setigerum, P. bracteatum, P. orientale, P. rhoeas* (Ikuta et al. 1974)
	Sanguinarine	*Chelidonium majus* (Böhm et al. 1980)
		Corydalis ophiocarpa (Iwasa and Takao 1982)
		Macleaya microcarpa (Koblitz et al. 1975)
		Eschscholtzia california, Macleaya cordata, Papaver somniferum, P. setigerum, P. bracteatum, P. orientale, P. rhoeas (Ikuta et al. 1974)
Phthalideisoquinolines 8 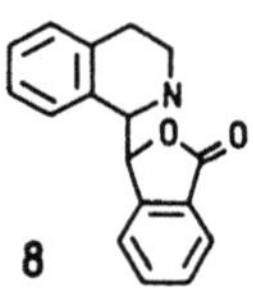	Narcotine	*Papaver rhoeas* (Khanna and Sharma 1977)
Morphinans 9 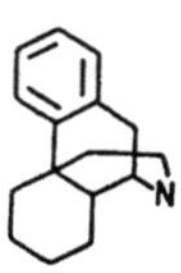	Codeine	*Papaver somniferum* (Tam et al. 1980, Hodges and Rapoport 1982, Furuya et al. 1971)
		Papaver setigerum (Cdn patent 1980)
		P. rhoeas (Khanna and Sharma 1977)
	Morphine	*Papaver somniferum* (Furuya 1981, Hsu 1981)
		Papaver rhoeas (Khanna and Sharma 1977)
	Thebaine	*Papaver bracteatum* Shafiee et al. 1978, Hsu 1981), Kamimura et al. 1976, Zito and Staba 1982)
		Papaver rhoeas (Khanna and Sharma 1977)
	Pallidine	*Fumaria capreolata* (Tanahashi et al. unpublished)

Table 1 (cont.)

Structure type	Alkaloid name	Source (reference)
Aristololactams 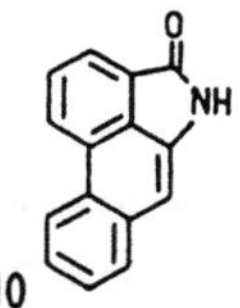 10	Cepharanone I Cepharanone II	*Stephania cepharantha* (Akasu et al. 1974)

In contrast only a few compounds from the large group of aporphines have been described so far. However, more complex structures like the benzophenanthridines and morphinans have been isolated not only from callus- but also from cell suspension cultures (see Table 1).

Most remarkable is the callus culture of *Stephania cepharantha* (Menispermaceae). Not only is it able to produce oxoaporphine alkaloids (Akasu et al. 1975) and bisbenzylisoquinolines (Akasu et al. 1976) but also two aristololactams which are only found in traces in the differentiated plants and are discussed as a hint for a possible systematic crosslink from the Menispermaceae to the Aristolochiaceae (Akasu et al. 1974). A further good example of the structure variability that can be produced by one culture is *Fumaria capreolata* (Fumariaceae) cultivated in our laboratory. Many structural types from a simple isoquinoline to a rather complicated structure like

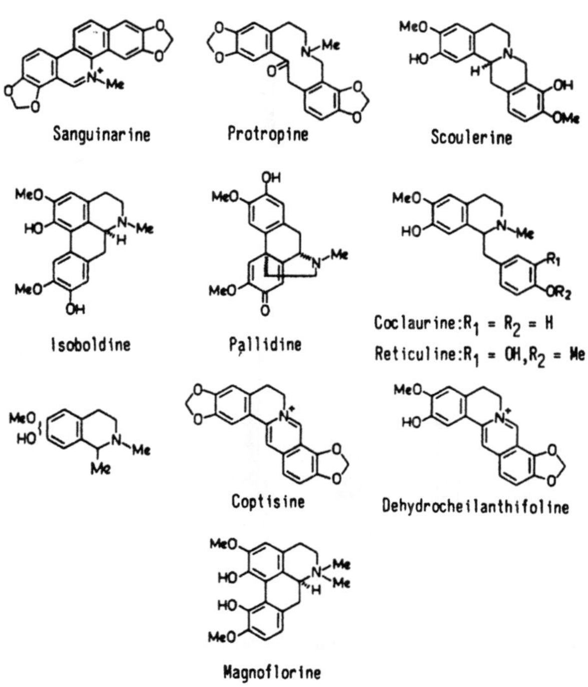

Fig. 1. Alkaloids isolated from a *Fumaria capreolata* (Fumariaceae) cell suspension culture

pallidine are found (Tanahashi unpublished) (Fig. 1). It is interesting to compare the alkaloids which have been produced by differentiated plants with their corresponding cell cultures.

2.2 Comparison Between Differentiated Plants and Cell Cultures

There are only a few detailed comparisons between plant cell cultures and differentiated plants to be found in the literature. Most striking is the lack of production of substantial amounts of morphinan alkaloids in cultures of *Papaver* species. Some examples are described where these alkaloids are detected in minor amounts (see Table 1), but the cell cultures do not reach the productivity of the plants. A possible reason for this may be the lack of specific enzymes in the biosynthetic pathway. On the other hand, one could postulate a correlation between the genetic expression of morphological differentiation and the production of these alkaloids since they are detected in higher amounts in redifferentiating calluses (Schuchmann and Wellmann 1983). The possible lack of specific enzymes is well illustrated by the bisbenzylisoquinoline alkaloids from *Stephania cepharantha* (Menispermaceae). The differentiated plant produces cepharanthine and isotetrandrine as major alkaloids, which are not detected in the cell culture, whereas the culture produces their possible precursors aromoline and berbamine in higher amounts than the plant, and this may be explained by the lack of specific methylating and the methylenedioxybridge forming enzymes (Akasu et al. 1976).

Sometimes the ratio between the amounts of several alkaloids in the plant differs markedly from that of the cell cultures. In the culture of *Berberis wilsoniae* var. *sub-*

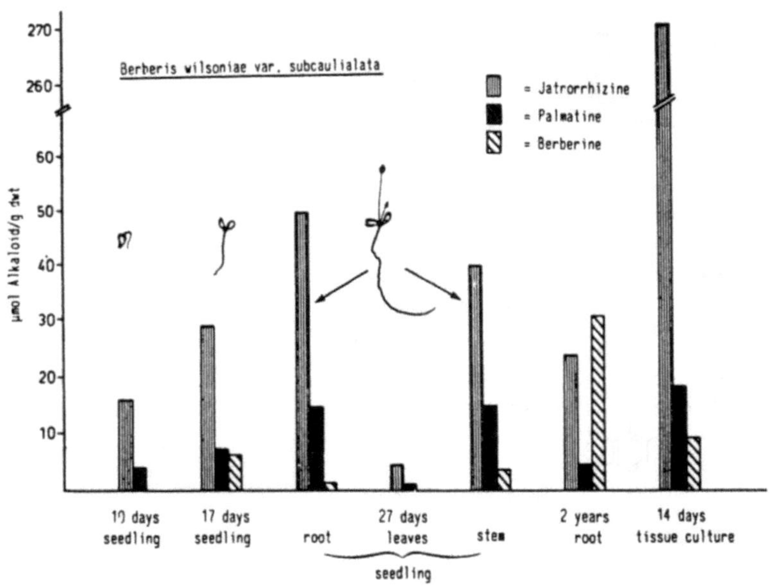

Fig. 2. Comparison between the alkaloid production of the cell culture, the seedling, and the intact root of *Berberis wilsoniae* var. *subcaulialata* (Berberidaceae)

caulialata we could observe that the major alkaloid was jatrorrhizine and the ratio of the three alkaloids jatrorrhizine: palmatine:berberine was 1:0.08:0.04 (Fig. 2). In contrast, a 2-year-old root shows a ratio of 1:0.2:1.4. The ratio of a seedling (1:0.2: 0.03), however, largely resembles that of the cell culture so that one may conclude that the enzyme pattern in the seedling is similar to that of the cell culture. The absolute amount in the cell culture exceeds by far that of the root of the seedling and this is therefore a good example of the capacity of cell cultures for the production of alkaloids in large amounts (Hinz and Zenk 1981).

From these results it can be seen that there may be a promising future for the production of alkaloids by cell cultures. However, it must be realised that there may not be any correlation between the amounts or the type of alkaloids which are produced by the plant or the cell culture.

2.3 Variant Cell Lines

An interesting but insufficiently explained phenomenon is the fact that in cell cultures which seem to be homogenous at the first glance, the amount of alkaloids may differ from cell to cell. Fluoromicroscopy of *Coptis japonica* (Ranunculaceae) cells for instance showed (Sato and Yamada 1984) that they produce varied amounts of protoberberine alkaloids as was shown for the indole alkaloid serpentine in *Catharanthus roseus* (Deus and Zenk 1982). The possibility for selecting cells or small colonies, plated on agar media, with a strong fluorescence to establish cultures with increased productivity was therefore considered practical (Sato and Yamada 1984). Further cultures of *Thalictrum dipterocarpum* (Ranunculaceae) were established by this method. One white cell line produced strongly blue fluorescent compounds in high amounts, whereas a second yellow coloured line gave good production of the protoberberines jatrorrhizine and berberine.

The selection procedure for these cultures producing protoberberine alkaloids is easy due to their yellow colour and their fluorescence in UV light. In the case of other benzylisoquinoline alkaloids, like the morphinans for example, immunological methods, especially the radioimmunoassay (Hodges and Rapoport 1982, Hsu et al. 1983, Wieczorek et al. unpublished) are useful for testing large numbers of small colonies (clones) to find possible cell lines with increased production of these alkaloids.

2.4 Stability

A prerequisite for the industrial use of cell lines selected by the method described above is their stability in producing alkaloids.

Concerning this aspect of cell selection differing results have been obtained so far for the production of benzylisoquinolines. The formation of berberine by cell cultures of *Coptis japonica* for example, produces a very stable culture which even increased its productivity during 3 years of continuous subcultivation and reached a maximal content of 1.67 g l^{-1} medium in one culture period (Fukui et al. 1982). However, Sato and Yamada (1984) could not stabilize the berberine production in their cell lines of *Coptis japonica.* A further example of instability of a culture is that of *Papaver somniferum* (Papaveraceae) which totally lost its morphinan forming capability after repeated subculturing (Hodges and Rapoport 1982), whereas in our laboratory a *Berberis stolonifera* (Berberidaceae) culture has continuously produced relatively large amounts of the protoberberine jatrorrhizine for the last 9 years (Zenk and Deus 1982, Deus-Neumann and Zenk 1984a). In every case a permanent control of the productivity seems to be necessary and repeated subcloning of the cultures may be necessary for the stabilization of their alkaloid production.

2.5 Localization of Alkaloids in the Cultures

2.5.1 Storage of Alkaloids in the Cells

An important aspect for the commercial production of benzylisoquinoline alkaloids by plant cell cultures is the question whether they are stored in the cells and therefore have to be extracted or excreted into the medium. So far various results have been obtained with different cell cultures.

Neumann and Müller (Neumann and Müller 1967) described the ability of *Macleaya microcarpa* (Papaveraceae) callus cultures to differentiate alkaloid accumulating cells. This also could be demonstrated by Kohlenbach (Lang and Kohlenbach 1983) with photographs of differentiation of "alkaloidal cells" from mesophyll protoplasts of *Macleaya*. In some cases these alkaloids even precipitate as crystals in the cells (Böhm and Franke 1982, Lang and Kohlenbach 1983). This phenomenon in *Macleaya* callus cultures suggested some specialized cells with the ability to accumulate alkaloids produced by the surrounding cells. In cell suspension cultures of *Macleaya microcarpa* Franke and Böhm (1982) observed that the smaller a cell aggregate, the lower was its relative amount of "alkaloidal cells". Indeed they could no longer be detected in aggregates consisting of 10 or fewer cells. Instead of being stored the alkaloids, mainly protopine, were excreted into the medium. Similar results were obtained for the cultures of *Chelidonium majus* (Papaveraceae) (Böhm et al. 1980) where the release of the alkaloids chelidonine, protopine, allocryptopine, coptisine, and sanguinarine into the medium could be increased to a factor 1:6.5 (alkaloids in the cell : alkaloids in the medium) by decreasing the cell aggregation and augmenting the auxin content in the medium (2–20 ppm).

These experiments are in contrast to our experiences with *Berberis* and *Thalictrum* cell cultures, where we found no substantial amounts of protoberberine alkaloids in the medium. Even very small cell aggregates in the 2-cell range showed intensive fluorescence, indicating the presence of protoberberine alkaloids within the cells. Furthermore we could isolate fluorescing protoplasts and vacuoles from *Thalictrum glaucum* (Ranunculaceae) suspension cultures, indicating that these alkaloids are stored within the vacuoles. Additionally we tested the uptake of the ^{3}H-labelled protoberberine alkaloids canadine and berberine into the isolated vacuoles of *Thalictrum glaucum* according to methods described by Deus-Neumann and Zenk 1984b) (Fig. 3).

(R)-Canadine was taken up in only small amounts whereas (R,S)-canadine showed a considerably enhanced uptake, indicating that the probably natural form (S)-canadine (Steffens et al. 1984) is selectively accumulated. In contrast the oxidized form of canadine namely berberine stays in the surrounding medium.

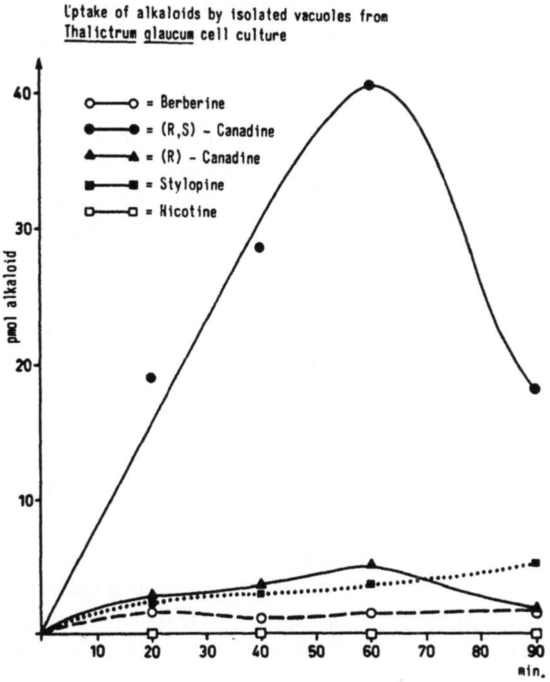

Fig. 3. Uptake measurement of ^{3}H-labelled alkaloids into the isolated vacuoles of *Thalictrum glaucum* (Ranunculaceae)

This phenomenon is in good agreement with data of Deus-Neumann and Zenk 1984b) with isolated vacuoles of *Catharanthus roseus* cultures, where they could observe that ajmalicine and not its oxidized form serpentine is transported into the vacuole.

A hypothetical explanation for this could be, that berberine with its positive charge is too polar to be accepted. Furthermore enzymatic experiments showed (Steffens et al. 1984, Amann et al. 1984) that the last steps in the biosynthesis of berberine do not take place in the cytosol and therefore probably no permeation processes would be necessary for the transfer into the vacuole. Alkaloids like nicotine, morphine or vindoline, which do not occur naturally in *Thalictrum*, are not accumulated in the vacuoles, whereas stylopine, a tetrahydroprotoberberine like canadine, could penetrate the tonoplast, even though only to a small extent, which supports a very specific uptake system located in the vacuole membrane.

Summarizing the data in this chapter one cannot conclude that generally benzylisoquinoline alkaloids are stored within the cells or excreted into the medium. It may be that specific types like the protopines or the benzophenanthridines are released into the medium whereas others like the protoberberines are deposited in the vacuole.

2.5.2 Stimulation of Releasing the Benzylisoquinolines into the Medium

Addition of Adsorbents. A possible method of enhancing the production of alkaloids by plant cell cultures could be to stimulate the release of alkaloids into the medium. It has been considered that low production of alkaloids in plant cell cultures may result from minor amounts of alkaloid released into the surrounding medium as a result of the lysis of cells and that this inhibits their further synthesis. To test this hypothesis one could add adsorbent materials like granulated charcoal or perhaps XAD (a non-ionic adsorbent resin based on polystyrol) to the medium to remove the alkaloids and thereby stimulating their further production. Good results were achieved with this method for the formation of natural compounds in *Matricaria chamomilla* cell suspension cultures (Knoop and Beiderbeck 1983). We tried to apply this procedure to the production of protoberberine alkaloids in *Berberis stolonifera* (Berberidaceae). The results were not encouraging, since no enhancement of alkaloid formation was observed, and only 0.1% of the alkaloid amount produced by the cells could be determined as adsorbed on the charcoal. Similar results were obtained by addition of XAD to the medium. However, this method should be reinvestigated with the *Macleaya microcarpa* system, where protopine is released into the medium (Franke and Böhm 1982).

Permeabilization of Cell Membranes. Besides a stimulation of excretion, another approach would be to permeabilize the cell membranes without killing the cells. Brodelius and Nilsson (1983) proposed such a process for the production of natural compounds using DMSO for the intermittant permeabilization of cell membranes of immobilized cells. We tried to apply this procedure to *Berberis stolonifera* cell cultures but only 0.07% of the alkaloids stored within the cells could be released by DMSO treatment. Furthermore microscopy of the treated cells showed that even after a regeneration period of 4 days the cells had not recovered. They were plasmolyzed irreversibly, and therefore it may be concluded that further application of this process is not promising.

2.6 Biotransformation

Besides the possibility of producing the alkaloids in situ the synthetic capacity of the plant cells could be used for biotransformation. So far only few experiments have been described. Furuya et al. (1978) tried to produce morphinan alkaloids by the addition of reticuline, known to be an important intermediate in their biosynthesis. (S)-reticuline was converted to (S)-scoulerine (14.7%) and further on to the protoberberine cheilanthifoline (0.5%). (R)-reticuline, the isomer needed for the morphinan pathway, remained unchanged indicating that probably the enzymes catalyzing the steps to salutaridine and further on to codeine and morphine are missing. The only reaction that could be performed in this pathway was the reduction reaction from codeinone to codeine (67%) but no further metabolization to morphine could be observed. More promising results were obtained in *Corydalis ophiocarpa* (Papaveraceae) (Iwasa and Takao 1982), where alkaloids normally not occurring in the cell culture could be produced by the addition of tetrahydroprotoberberine-α-N-methochloride. 34% of the administered precursor could be transformed to the protopine α-allocryptopine and 1% further on to the benzophenanthridine chelerythrine and therefore one may consider this as a realistic method for the future production of benzylisoquinolines.

2.7 Optimization of Media

Besides the selection of high yielding cell lines and specialized methods for increasing alkaloid formation, the right choice of culture media is indispensable for an optimal production of benzylisoquinolines.

Looking for example at the use of so far described basal media the production of jatrorrhizine in *Berberis stolonifera* cell cultures (Hinz and Zenk 1981) was optimal in LS-medium (Linsmaier and Skoog 1965), the cell suspension cultures of *Coptis japonica* (Sato and Yamada 1984) gave the best results in White medium (White 1963), and *Eschscholtzia californica* (Papaveraceae) (Berlin et al. 1983) was cultivated with the best results in B5 medium (Gamborg et al. 1968).

2.7.1 Carbon Source

Nearly all cultures described so far grow best with sucrose as carbon source, but there are exceptions. Yamamoto (1981) for instance observed that his culture of *Coptis japonica* showed the highest growth and production rate on maltose. The optimal concentration of sucrose may differ from culture to culture: 8% sucrose was best for *Eschscholtzia californica* (Berlin et al. 1983), 3% for *Coptis japonica* (Sato and Yamada 1984), 4% for *Tinospora caffra* (Menispermaceae) (Rueffer unpublished), and 3.5% for *Berberis stolonifera* (Hinz and Zenk 1981).

2.7.2 Inorganic Medium Constituents

Variation of the PO_4^{3-}, NO_3^-, and NH_4^+ concentrations of the basal medium did not effect markedly the production of jatrorrhizine in *Berberis stolonifera* (Hinz and Zenk 1981). Various PO_4^{3-} concentrations also did not influence the production of jatrorrhizine in *Tinospora caffra* (Rueffer unpublished), whereas an inhibition of alkaloid production was observed with higher concentrations of phosphate in *Eschscholtzia californica* (Berlin et al. 1983).

2.7.3 Plant Growth Regulators

Plant growth regulators have a remarkable influence on the production of benzylisoquinoline alkaloids in some cultures. Fukui et al. (1982) described an increase of alkaloid yields by addition of kinetin or benzylaminopurine (0.1–1.0 μM). He also could observe a modification in the alkaloid pattern with higher concentrations (10 μM) of these phytohormones: The production of palmatine and jatrorrhizine significantly decreased whereas berberine and coptisine formation remained largely uneffected.

No increase in the alkaloid yields were observed by varying the growth regulators in *Berberis stolonifera* (Hinz and Zenk 1981). In a systematic test of 35 different phenoxyacetic acids and 88 other synthetic growth regulators no better results could be achieved than with the initial LS-medium.

Regarding the growth regulator 2,4-dichlorphenoxyacetic acid (2,4-D) the so far published data show a decrease in alkaloid production as the concentration of this growth regulator is increased in the basal medium. In *Thalictrum minus* (Ranunculaceae) cultures (Ikuta and Itokawa 1982a) 7 different protoberberine alkaloids were produced in the medium containing 0.1 ppm 2,4-D (berberine, palmatine, jatrorrhizine, columbamine, thalifendine, thalidastine, and desoxythalidastine). A rise to 1.0 ppm 2,4-D resulted in a loss of desoxythalidastine formation and with 5.0 ppm 2,4-D in the medium no columbamine, thalifendine, and thalidastine were synthesized. A similar decrease of productivity was observed in *Eschscholtzia californica* (Berlin et al. 1983) and *Chasmanthera dependens* (Menispermaceae) (Rüffer et al. unpublished).

2.7.4 Biosynthetic Precursors

Another essential factor that might induce the production of benzylisoquinolines could be the addition of biosynthetic precursors to the medium. Kamimura et al. (1976), for instance, described a stimulation of thebaine formation with the addition of tyrosine and dopa to the medium of *Papaver bracteatum* cultures. An increase from 0.05% of dry weight to 0.07% in the production of berberine in *Argemone mexicana* (Menispermaceae) suspension cultures (Khanna et al. 1982) also was due to the addition of tyrosine, whereas phenylalanine in the medium only stimulated the growth index and the berberine yield remained constant.

In summary, one has to conclude that every culture has to be optimized individually for alkaloid production and no general composition of one optimal medium for all cultures can be given.

3 Conclusions

On surveying the accomplishments with plant cell cultures and alkaloid production, one may be optimistic for the future of producing benzylisoquinoline alkaloids by this technique. Disappointing is the fact that so far no good production of morphinans has been achieved, but perhaps this problem can be overcome by employing the proposed selection methods and intensifying the search for an optimal medium. Promising is the production of berberine, a useful antibacterial and stomachic, in *Coptis japonica.* The root of the intact plant synthesizes 200 mg (10% of dry weight) within 5 years whereas the cell suspension culture produces 1.67 g l^{-1} medium (11.4% of dry weight) within 3 weeks (Fukui et al. 1982). The best results so far have been the formation of the protoberberine alkaloid jatrorrhizine in *Berberis stolonifera* cell cultures (Hinz and Zenk 1981, Zenk and Deus 1982), where the second highest ever reported production of a natural compound could be observed with 2.7 g l^{-1} medium. Similar results were reported, also for the synthesis of jatrorrhizine, in *Berberis wilsoniae* where the first experiments for the cultivation in a fermentor were carried out (Rothenberger 1982).

These last examples suggest the future development of plant cell suspension cultures for a biotechnological use is a real possibility.

References

Akasu M, Itokawa H, Fujita M (1974) Four new fluorescent components isolated from the callus tissue of *Stephania cepharantha*. Tetrahedron Lett 41:3609–3612

Akasu M, Itokawa H, Fujita M (1975) Oxoaporphine alkaloids from callus tissues of *Stephania cepharantha*. Phytochemistry 14:1673–1674

Akasu M, Itokawa H, Fujita M (1976) Biscoclaurine alkaloids in callus tissues of *Stephania cepharantha*. Phytochemistry 15:471–473

Amann M, Nagakura N, Zenk MH (1984) (S)-Tetrahydroprotoberberineoxidase, the final enzyme in protoberberine biosynthesis. Tetrahedron Lett 25:953–954

Berlin J, Forche E, Wray V, Hammer J, Hösel W (1983) Formation of benzophenanthridine alkaloids by suspension cultures of *Eschscholtzia californica*. Z Naturforsch 38C:346–352

Böhm H, Franke J (1982) Accumulation and excretion of alkaloids by *Macleaya microcarpa* cell cultures, I. Experiments on solid medium. Biochem Physiol Pflanz 177:345–356

Böhm H, Franke J, Lammel L (1980) *Chelidonium majus* alkaloids. GDR Pat 143, 270 (Cl C12 P17/00)

Brodelius P, Nilsson K (1983) Permeabilization of immobilized plant cells, resulting in release of intracellular stored products with preserved cell viability. Eur J Appl Microbiol Biotechnol 17:275–280

Deus B, Zenk MH (1982) Exploitation of plant cells for the production of natural compounds. Biotechnol Bioeng XXIV:1965–1974

Deus-Neumann B, Zenk MH (1984a) Instability of indole alkaloid production in *Catharanthus roseus* cell suspension cultures. Planta Med 50:365–458

Deus-Neumann B, Zenk MH (1984b) A highly selective alkaloid uptake system in vacuoles of higher plants. Planta 162:250–260

Farnsworth NR, Morris RW (1976) Importance of secondary plant constituents as drugs. Am J Pharm 46:148

Franke J, Böhm H (1982) Accumulation and excretion of alkaloids by *Macleaya microcarpa* cell cultures, II. Experiments in liquid medium. Biochem Physiol Pflanz 177:501–507

Fukui H, Nakagawa K, Tsuda S, Tabata M (1982) Production of isoquinoline alkaloids by cell suspension cultures of *Coptis japonica*. Plant tissue culture 1982. Proc 5th Congr Plant Tissue Cell Culture, pp 313–314

Furuya T (1981) Production of opium alkaloids by tissue cultures of *Papaver*. Jpn Kokai Tokkyo Koho 8161,994 (Cl 12P17/18)

Furuya T, Ikuta A, Syono K (1971) Plant tissue culture XV. Alkaloids from callus tissues of *Papaver somniferum*. Phytochemistry 11:3041

Furuya T, Nakano T, Yoshikawa T (1978) Biotransformation of (R,S)-reticuline and morphinan alkaloids by cell cultures of Papaveraceae. Phytochemistry 17:891–893

Furuya T, Yoshikawa T, Kiyohara H (1983) Alkaloid production in cultured cells of *Dioscoreophyllum cumminsii*. Phytochemistry 22:1671–1673

Gamborg OL, Miller RA, Ojima K (1968) Nutrient requirement of suspension cultures of soybean root cells. J Exp Cell Res 50:151–158

Gurova TF, Popov YG, Fadeeva II, Shain SS, Tokachev O (1980) Callus formation and accumulation of alkaloids in the tissue culture of *Stephania glabra*. Rastit Resur 16:412–425

Hinz H, Zenk MH (1981) Production of protoberberine alkaloids by cell suspension cultures of *Berberis* species. Naturwissenschaften 67:620

Hodges CC, Rapoport H (1982) Morphinan alkaloids in callus cultures of *Papaver somniferum*. J Nat Prod 45:481–485

Hsu AF (1981) Effect of protein synthesis inhibitors on cell growth and alkaloid production in cell cultures of *Papaver somniferum*. J Nat Prod 44:408–414

Hsu AF, Brown D, Etskovitz RB, Chen PK, Bills DD (1983) Radioimmunoassay for the determination of morphine in capsules of *Papaver somniferum*. Phytochemistry 22:1665–1669

Ikuta A, Itokawa H (1982a) Berberine and other protoberberine alkaloids in callus tissue of *Thalictrum minus*. Phytochemistry 21:1419–1421

Ikuta A, Itokawa H (1982b) Studies on the alkaloids from tissue culture of *Nandina domestica.* Plant tissue culture 1982. 5th Int Congr Plant Cell Culture, pp 315–316

Ikuta A, Syona K, Furuya T (1974) Alkaloids of callus tissues and redifferentiated plantlets in the Papaveraceae. Phytochemistry 13:2175–2179

Iwasa K. Takao N (1982) Formation of alkaloids in *Corydalis ophiocarpa* callus culture. Phytochemistry 21:611–614

Kamimura S, Akutsu M, Nishikawa M (1976) Tissue Culture of *P. bracteatum.* III Formation of thebaine in the suspension culture of *P. bracteatum.* Agric Biol Chem 40:913–919

Khanna P, Sharma GL (1977) Production of opium alkaloids from in vitro tissue cultures of *Papaver rhoeas.* Ind J Exp Biol 15:951–952

Khanna P, Sharma OP, Saluka M (1982) Berberine from *Argemone mexicana.* Plant tissue culture 1982. Proc 5th Int Congr Tissue Cell Culture, pp 311–312

Knoop B, Beiderbeck R (1983) Adsorbenskultur – ein Weg zur Steigerung der Sekundärstoffproduktion in pflanzlichen Suspensionskulturen. Z Naturforsch 38C:484–486

Koblitz H, Schumann U, Boehm H, Franke J (1975) Tissue cultures of alkaloidal plants IV *Macleaya microcarpa.* Experientia 31:768–769

Kuroda H, Itokawa T (1973) Production of berberine and palmatine. Jpn Pat Kokai Tokkyo Koho 73–6153

Lang H, Kohlenbach HW (1983) Differentiation of Alkaloid cells in cultures of *Macleaya* mesophyll protoplasts. Planta Med 46:78–118

Linsmaier EM, Skoog F (1965) Organic growth factor requirements of Tobacco tissue cultures. Physiol Plant 18:100–127

Neumann D, Müller E (1967) Intrazellulärer Nachweis von Alkaloiden in Pflanzenzellen im licht- und elektronenmikroskopischen Nachweis. Flora (Jena) Abt A 158:479–491

Reinhard E (1967) Probleme der Produktion von Arzneistoffen durch pflanzliche Gewebekulturen. Dtsch Apoth Ztg 107:1201–1207

Rothenberger S (1982) Untersuchungen zur Bildung von Protoberberinalkaloiden in Zellkulturen von *Berberis wilsoniae.* PhD thesis

Sato F, Yamada Y (1984) High berberine producing cultures of *Coptis japonica* cells. Phytochemistry 23:281–285

Schuchmann R, Wellmann E (1983) Somatic embryogenesis of tissue cultures of *Papaver somniferum* and *Papaver orientale* and its relationship to alkaloid and lipid metabolism. Plant Cell Rep 2:88–91

Shafiee A, Lalezari I, Narges Y (1978) Production of thebaine in tissue cultures of *Papaver bracteatum.* US Pat 4,114,314 (Cl 47–85)

Steffens P, Nagakura N, Zenk MH (1984) The berberine bridge forming enzyme in tetrahydroprotoberberine biosynthesis. Tetrahedron Lett 25:951–952

Sumimoto Chemicals Co Ltd (1982) Berberine alkaloid production by tissue culture. Jpn Kokai Tokkyo Koho 57,144,992 (Cl C 12p17/18)

Tam WHJ, Constabel F, Kurz WGW (1980) Codeine from cell suspension cultures of *Papaver somniferum.* Phytochemistry 18:486–487

White PR (1963) The cultivation of animal and plant cells. Ronald Press, New York

Yamamoto H, Ishida M, Tomimoro T (1981) Studies on the fundamental cultures of *Coptis japonica* var. *japonica.* Shoyakugaku Zasshi 35:1–8

Zenk MH, Deus B (1982) Natural product synthesis by plant cell cultures. Plant tissue culture 1982. Proc 5th Int Congr Plant Tissue Cell Culture, pp 391–392

Zito SW, Staba EJ (1982) Thebaine from root cultures of *Papaver bracteatum.* Planta Med 45: 53–54

Cytodifferentiation and Papaver Alkaloid Accumulation

T.M. KUTCHAN, S. AYABE, and C.J. COSCIA [1]

1 Introduction

The latex found in many species of the Papaveraceae family is a remarkable biological fluid, not only because of its contents, the morphinan alkaloids, but also because of its unusual storage capacity. Latex arises upon articulation and anastamoses of the highly specialized laticifer cells present in these plants (Felklova and Babkova 1958, Fairbairn and Kapoor 1960). Laticifer cells appear as early as a few days after germination of *P. bracteatum* seedlings and these initial cells contain many vesicles with electron-dense caps (Nessler and Mahlberg 1978). Dissolution of transverse cell walls may be observed within a week after germination ultimately giving rise to networks of vessels. Latex in the ducts originates from the cell protoplasm.

In collaboration with Dr. Margaret Roberts of the University of London, we recently learned that latex also contains millimolar concentrations of dopamine, a fraction of which occurs in the same, relatively dense vacuoles that are rich in alkaloids (Roberts et al. 1983). Dopamine is an essential intermediate in the biogenesis of the alkaloids of the Papaveraceae (Shamma and Moniot 1978, Spenser 1969). Dopamine, which is formed from tyrosine, condenses with different carbonyl derivatives in Pictet-Spengler reactions to afford simple tetrahydroisoquinolines that can then give rise to a diverse array of alkaloids. Considerable evidence exists to suggest that the chirality of the tetrahydroisoquinoline alkaloids influences their transformation. It appears that morphinan alkaloids are formed from (-)-R-reticuline, whereas most other alkaloids including sanguinarine are derived from (+)-S-reticuline (Battersby et al 1965, 1975, Wilson and Coscia 1975, Borkowski et al. 1978, Brossi 1982).

Whether the laticifer cell is the site of morphinan alkaloid biosynthesis remains controversial. Fairbairn and coworkers (Fairbairn et al. 1968, Fairbairn and Steele 1981) have consistently demonstrated that latex isolated from *P. somniferum* transformed labeled tyrosine and L-dopa to morphine. In addition, *P. orientale* latex exhibited biosynthetic capacity extending over several early steps in the alkaloid biogenetic pathway (Wilson and Coscia 1975). Enzymes that can rapidly convert tyrosine to dopamine were also detected in *P. somniferum* latex (Roberts 1974, Roberts and Antoun 1978). Nevertheless, arguments have been put forward that latex biosynthetic activity is relatively low (Böhm et al. 1972) and most synthesis occurs in cells sur-

[1] E.A. Doisy Department of Biochemistry, St. Louis University School of Medicine, St. Louis, Missouri 63104, USA

The Chemistry and Biology of Isoquinoline Alkaloids, ed. by Phillipson et al.

rounding laticifers that serve as storage sites (Böhm and Franke 1982, Franke and Böhm 1982).

2 Laticifer Cell Development and Morphinan Accumulation in *P. bracteatum* Cell Cultures

Since we were able to promote cytodifferentiation and monitor attendant changes in levels of dopamine, thebaine and the benzophenanthridine alkaloid, sanguinarine, this appeared to be an attractive system to address such questions as the site of their accumulation and the biosynthetic role of the laticifer cell. Furthermore, it seemed reasonable that cultured *P. bracteatum* cells undergoing organogenesis could be readily used in cellular and subcellular localization studies to support previously obtained correlative evidence. In this chapter we describe methods of preparation and isolation of protoplasts utilizing hydrolytic enzymes and density centrifugation that permit differential enrichment of laticifer cells and their vacuolar contents. By analyses of the distribution of dopmaine, sanguinarine and thebaine on Renografin gradients, their cellular and subcellular compartmentation was compared.

2.1 Detection and Quantitation of Alkaloids

In our original experiments we discovered that the pool of dopamine in cultured *P. bracteatum* cells was substantial and under many growth conditions compared favorably with that in latex from the intact plant (1–3 mg ml^{-1}) (Roberts et al. 1983). Upon prefractionation of acid extracts of tissue on alumina columns prior to high pressure liquid chromatography (HPLC), dopamine was eluted as the major catecholic constituent (Fig. 1B).

Conversely, alkaloid extracts were quite complex mixtures with sanguinarine as a major component. The predominance of benzophenanthridine, protopine and aporphine alkaloids is common in cultured tissues of the Papaveraceae, despite their restricted occurrence in intact plants of this family (Ikuta et al. 1974, Kettenes-van den Bosch 1981, Phillipson 1983). HPLC was also used for the quantitation of these alkaloids and if preceded by fractionation on Extrelut columns (Roberts et al. 1983), the quantitation of thebaine was facilitated as well (Fig. 1A). While sanguinarine was found in readily detectable amounts under a variety of growth conditions (Table 1), thebaine was discovered to be a trace constituent (Kutchan et al. 1983). Sanguinarine levels were highest in black aggregates that appeared by electron microscopic inspection to be necrotic tissue. These results suggest this benzophenanthridine alkaloid accumulates in mature, unidentified cells.

2.2 Effect of Removal of Hormone From Culture Media

The transfer of cultures to a hormone-free media had a profound influence on alkaloidal metabolite levels. Thebaine content was elevated from 3- to 20-fold in these

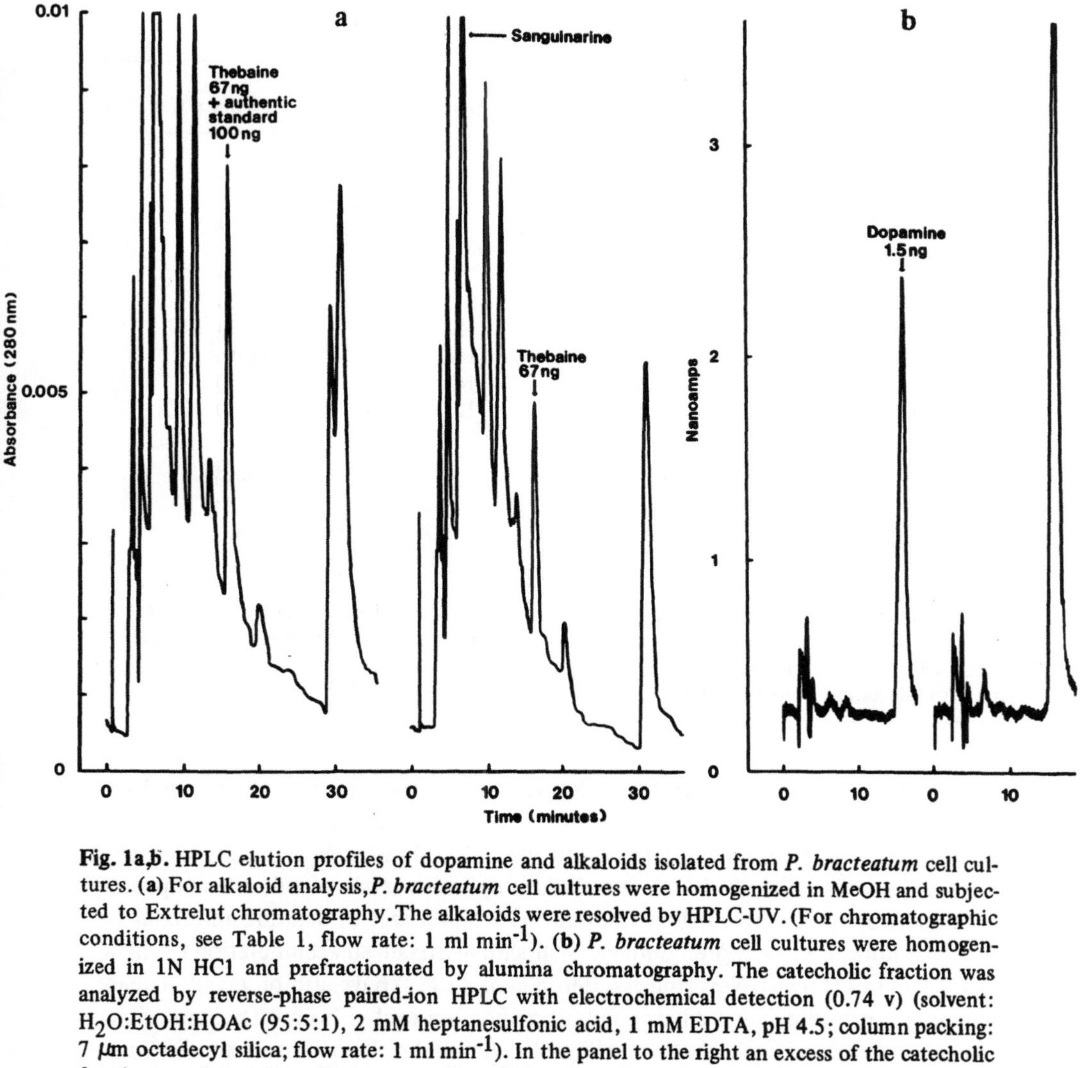

Fig. 1a,b. HPLC elution profiles of dopamine and alkaloids isolated from *P. bracteatum* cell cultures. (**a**) For alkaloid analysis, *P. bracteatum* cell cultures were homogenized in MeOH and subjected to Extrelut chromatography. The alkaloids were resolved by HPLC-UV. (For chromatographic conditions, see Table 1, flow rate: 1 ml min^{-1}). (**b**) *P. bracteatum* cell cultures were homogenized in 1N HC1 and prefractionated by alumina chromatography. The catecholic fraction was analyzed by reverse-phase paired-ion HPLC with electrochemical detection (0.74 v) (solvent: H_2O:EtOH:HOAc (95:5:1), 2 mM heptanesulfonic acid, 1 mM EDTA, pH 4.5; column packing: 7 μm octadecyl silica; flow rate: 1 ml min^{-1}). In the panel to the right an excess of the catecholic fraction was injected to display an unidentified more polar metabolite with a retention time of less than 10 min

P. bracteatum cells, depending upon the extent of differentiation and the initial hormonal supplement (Kutchan et al. 1983). Similar findings have been reported in several independent investigations (Kamimura and Nishikawa 1976, Kamimura et al. 1976, Yoshikawa and Furuya 1982, Schuchmann and Wellmann 1983). As seen in Fig. 2 dopamine levels diminished and then slowly returned to original values when static cultures were transferred to hormone-free cell supension cultures. This occurred for both dark-grown *P. somniferum* (Fig. 2B) as well as *P. bracteatum* (Fig. 2A). Note that dopamine levels are an order of magnitude greater in *P. bracteatum*. Growth was not adversely affected under these conditions as both light and dark grown *P. somniferum* had doubling times of 10 days, whereas *P. bracteatum* cells doubled in

Table 1. Benzophenanthridine alkaloid levels in *P. bracteatum* static cultures

Tissue Type	Age days	Illumination	Sanguinarine µg/g fresh wt.	(% dry wt. x 10^3)
		5 mg ml^{-1}		
Shoot cultures	21	dark	3.1	(9)
Shoot cultures	21	light	11	(31)
Shoot cultures	26	dark	13	(37)
Shoot cultures	26	light	33	(81)
Beige callus	21	dark	5.7	(13)
Beige callus	21	light	22	(43)
Black aggregates	21	dark	184	(250)
Black aggregates	21	light	167	(290)
Meristemoids	26	dark	121	(140)
		2 mg ml^{-1}		
Callus	21	dark	26	(49)
Callus	21	light	2.9	(5)
Callus	35	dark	6.8	(12)
Callus	35	light	10	(13)

Alkaloid fractions were obtained by Extrelut chromatography of methanolic extracts (Roberts et al. 1983). Concentrations were estimated by peak height analysis of alkaloids separated by HPLC with UV detection at 280 nm. Resolution was achieved on a 5 µm silica gel column eluting with n-hexane-chloroform-methanol-diethylamine (88.5:8.6:2.9:0.03).

19 days. Parent callus lines (grown on hormone-supplemented media) displayed similar growth rates (*P. somniferum*: 326% in 17 days; *P. bracteatum*: 160% in 32 days). In some experiments, sanguinarine levels were also elevated in the hormone-free *P. bracteatum* cells (Fig. 2), but inter-relationships with levels of other benzophenanthridine alkaloids in the alkaloid fraction make the effect of deleting hormonal supplements appear to be more complex than simple enhanced accumulation of metabolite (data not shown). Nevertheless, Berlin et al. (1983) have also observed increased benzophenanthridine alkaloid accumulation upon removal of hormones from cell suspension cultures of *Eschscholtzia californica.* Similarly, Schuchmann and Wellmann (1983) reported that upon transfer to hormone-free media, cultures of *P. somniferum* accumulated greater amounts of sanguinarine while *P. orientale* exhibited decreases in the content of this alkaloid.

2.3 Subcellular Localisation Studies

Cytological investigation of *P. bracteatum* callus revealed the presence of parenchymal cells, with lesser numbers of meristematic and even fewer laticifer cells (Kutchan et al. 1983). Since the hormone-free media induced differentiation and an increase in the number of laticifer cells, an attempt to correlate morphinan alkaloid occurrence with that of laticifers was made. Seventeen different samples, from tissue containing only traces of thebaine, were inspected by light or electron microscopy and found to be devoid of such laticifer cells. Conversely, laticifer cells were encountered in elec-

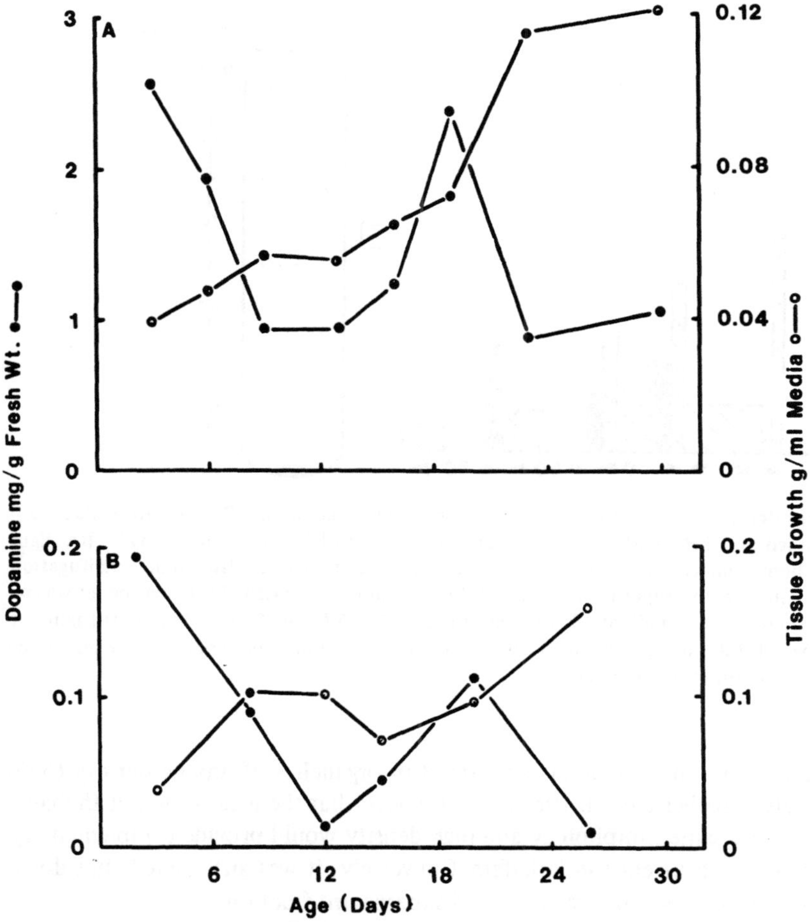

Fig. 2A,B. Changes in dopamine levels upon transfer of *P. bracteatum* cultures to hormone-free media. **(A)** *P. bracteatum* callus culture was transferred from Murashige and Skoog's revised to tobacco medium (MSRT) supplemented with 0.1 mg l^{-1} 2,4-dichlorophenoxyacetic acid (2,4-D) and 0.5 mg l^{-1} 6-benzylaminopurine (BA) (dark-grown) to suspension medium without hormones (dark grown). Aliquots (3 ml) were removed every 3–4 days and analyzed for growth and dopamine content (see Fig. 1). Values represent the average from three separate flasks analyzed in duplicate. **(B)** *P. somniferum* callus was maintained in the dark on MSRT without glycine and supplemented with 0.1% casein enzymatic hydrolysate and 0.2 mg l^{-1} naphthaleneacetic acid. After transfer to hormone-free suspension culture, aliquots were removed at 4–6 day intervals and analyzed as in **A**. Values represent the average of two separate experiments assayed in duplicate

tron microscopic grids from all six independent samples from tissues demonstrated to have 1–6 μg g^{-1} fresh wt. of thebaine (Kutchan et al. 1983). Since high levels of dopamine (1–4 mg g^{-1} fresh wt.) were found in tissue that by light and electron microscopic examination appeared to consist predominantly of parenchymal cells, a different cytological origin was suggested for this catecholamine. To corroborate these results and to investigate the site of sanguinarine accumulation as well, subcellular localization studies were undertaken. Our initial premise was based on the ob-

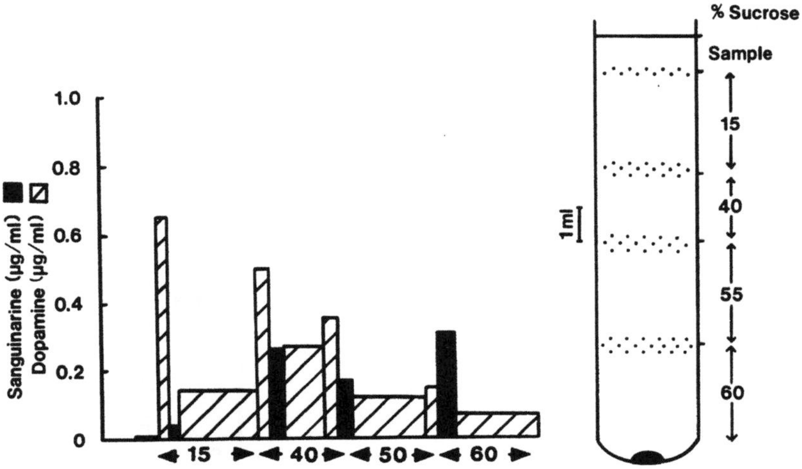

Fig. 3. Sucrose density gradient profiles of dopamine and sanguinarine, *P. bracteatum* static cultures (maintained on MSRT with hormonal supplements 0.1 mgl^{-1} 2,4-D and 0.5 mgl^{-1} BA, light grown) were homogenized with sand using a mortar and pestle. After differential centrifugation at 500 *g* x 5 min and the supernatant at 20,000 *g* x 20 min, the enriched vacuolar pellet was resolved on a sucrose step gradient (3 ml each 15%, 40%, 55% and 2 ml 60%) centrifuging at 125,000 *g* for 2 h (Matile and Wiemken 1974). Gradient fractions were analyzed for dopamine and sanguinarine content (see Fig. 1)

servation that morphinan alkaloids are stored in organelles of unique density (55% sucrose) in latex (Roberts et al. 1983). We reasoned that the localization of thebaine in vacuoles of the same morphology and high density would provide complementary evidence for their occurrence in laticifers. Conversely, it was anticipated that dopamine would occur in less dense vacuoles and supernatant fractions.

The first method adopted to resolve subcellular organelles was a modification of a sucrose step gradient developed by Matile and Wiemken (1974) to isolate meristematic vacuoles. They reported that large vacuoles remain on top of a 15% sucrose layer, whereas smaller meristematic vesicles migrate to the 15/40% interface under the conditions of this experiment. We introduced 55% and 60% sucrose layers to trap more dense vacuoles of the type found in latex (Fig. 3). The density profile revealed that most of the compartmentalized dopamine was in the larger and meristematic vacuoles (15–40% sucrose fractions), whereas most of the sanguinarine occurred in more dense vesicles (55–60% sucrose). Inspection of the various fractions by electron microscopy revealed that they were heterogeneous and vacuoles were present but not greatly enriched in any particular band. Thebaine could not be detected, but this was not surprising considering the low yields of other metabolites due to the relatively inefficient cell disruption accomplished by gentle hand homogenization. Since yeast vacuoles were recoverable after blending with very fine glass beads ("Bead Beater") (Cramer et al. 1983), we attempted to improve yields by this approach. Although the efficiency of cell disruption increased, the major proportion of dopamine, sanguinarine and thebaine were recovered in the 20,000 *g* supernatant rather than the pellet layered onto the sucrose gradient in this protocol. We attributed this to the destruction of vacuoles by the vigorous mechanical agitation.

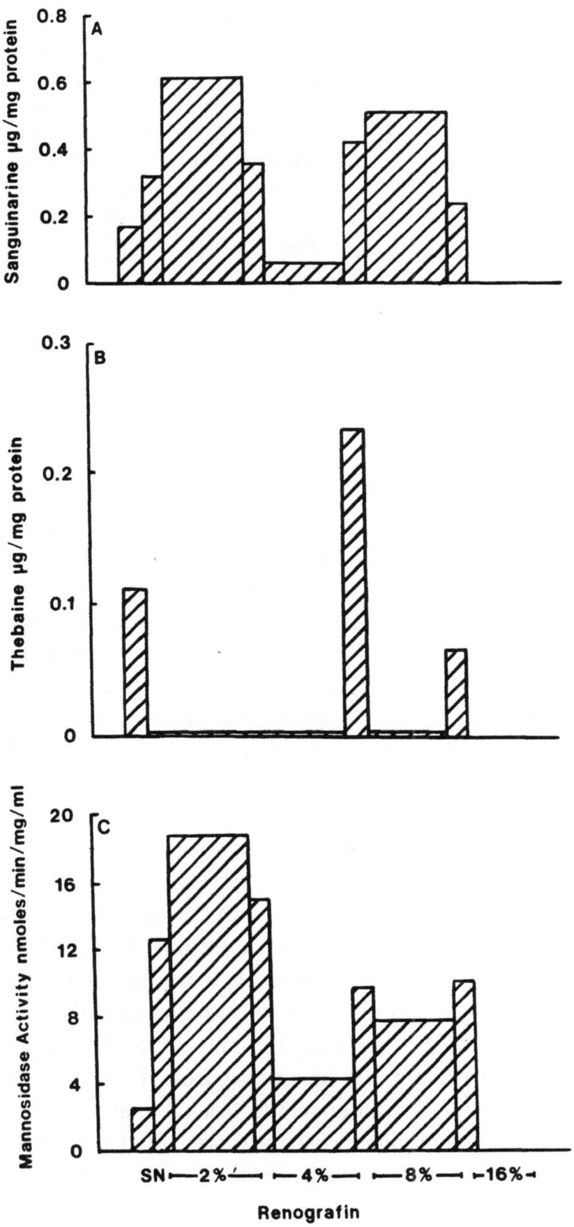

Fig. 4. Distribution of sanguinarine, thebaine and α-mannosidase on Renografin density gradients after cell wall hydrolysis. *P. bracteatum* static cultures maintained on MSRT supplemented with 0.1 mg l^{-1} 2,4-D and 0.5 mg l^{-1} BA in the light were transferred to hormone-free suspension culture (dark grown) nine days prior to enzymatic digestion. The cells were then collected and incubated with 10 mg ml^{-1} each of cellulase (*Aspergillus niger*), hemicellulase (*Aspergillus niger*) and pectinase (*Rhizopus sp.*). An enriched protoplast pellet (obtained by differential centrifugation at 100 *g* x 5 min and the pellet at 1,000 *g* x 20 min) was lysed and resolved on a renografin step gradient (3 ml each of 2%, 4%, 8% and 2 ml 16%) by centrifugation at 100,000 *g* for 2 h. The width of bars in graph is proportional to the fraction volume. **(A)** Sanguinarine and **(B)** thebaine content were assessed as described in Fig. 1. **(C)** α-Mannosidase activity in gradient fractions was determined by the absorbance at 405 nm of the p-nitrophenolate anion released upon hydrolysis of ρ-nitrophenyl-α-D-mannopyranoside (Boller and Kende 1979). Data shown are from the same experiment. In the original protoplasts, the concentrations of sanguinarine and thebaine were 4.4 and 0.082 µg mg^{-1} protein, respectively

We, therefore, adopted another approach (Sasse et al. 1979) entailing enzymatic digestion of cell walls, followed by density gradient centrifugation on Renografin (meglumine diatrizoate) (Figs. 4, 5, 6). To improve recoveries of metabolites in supernatant fractions, 20 mM ascorbate and 10 mM diethyldithiocarbamate (DIECA, an inhibitor of copper-containing enzymes) were included in all buffers. In the absence of these antioxidants, extravesicular dopamine was totally destroyed during the lengthy centrifugations used in these procedures. The following *Aspergillus niger*

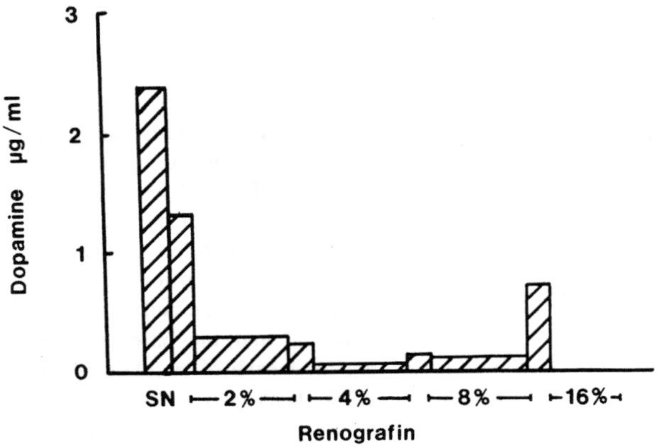

Fig. 5. Renografin density gradient profile of dopamine. Vacuole isolation was performed as described in Fig. 4 with pectinase from *Aspergillus niger* in the cell digestion medium. Gradient fractions were analyzed for dopamine content as described in Fig. 1

enzymes were used to hydrolyze cell walls: cellulase, hemicellulase and pectinase. In some experiments a *Rhizopus* pectinase was used (all enzymes were purchased from Sigma Chemical Co., St. Louis, MO). Resultant protoplasts were either placed in a dounce homogenizer to disrupt their cell membranes gently or were added directly to the top of a Renografin step gradient. Fractions were monitored by electron microscopy, and for thebaine, sanguinarine, dopamine as well as α-mannosidase, a soluble enzyme previously established to be a good marker for plant vacuoles (Boller and Kende 1979). As seen in Fig. 4C and Table 2, α-mannosidase was discovered at all of the interfaces of the gradient, but more of the enzyme was localized in the 2 and 8% bands. There was evidence for some disruption of vacuoles, as seen by the recovery of 18% of this enzyme in the supernatant fraction.

The distribution of sanguinarine closely paralleled the α-mannosidase activity both with respect to specific activities (Fig. 4A and C) and % recoveries (Table 2) for the various fractions of the Renografin gradient. The data suggest that sanguinarine is present in the same vacuoles as α-mannosidase and are not restricted only to the more dense laticifer vesicles. This is consistent with the distribution of sanguinarine (Table 1) in various tissues and electron microscopic surveys of the cell types present therein. Thus, sanguinarine may accumulate in sites that are both the same as, and different from, dopamine and thebaine (Fig. 4 and Table 2). By measuring protein levels of the fractions, it was possible to determine the alkaloid concentrations. Possible enrichment of various fractions with respect to the original protoplasts were then estimated. However, for sanguinarine the levels in the original protoplasts (4.4 $\mu g\ mg^{-1}$ protein) were higher than that of the individual fractions (Fig. 4A).

Thebaine was uniquely distributed on this gradient in three discrete bands, i.e., the 4/8 and 8/16% interfaces and the supernatant. As seen in Fig. 4 and Table 2, 40% of the total thebaine on the gradient appeared at the 4/8% interface which contains many dense small vesicles as shown by electron microscopy (Fig. 8A) and α-mannosidase. Calculation of concentrations based on protein, revealed that thebaine was enriched three fold in the vacuolar fraction at the 4/8% interface (Fig. 4). Electron microscopy revealed that the 8/16% interface band contained undisrupted protoplasts of laticifer cells (Fig. 7C) and other types of cells. Accordingly, the thebaine concentration of the band was not enriched over the original protoplasts. Although an equal

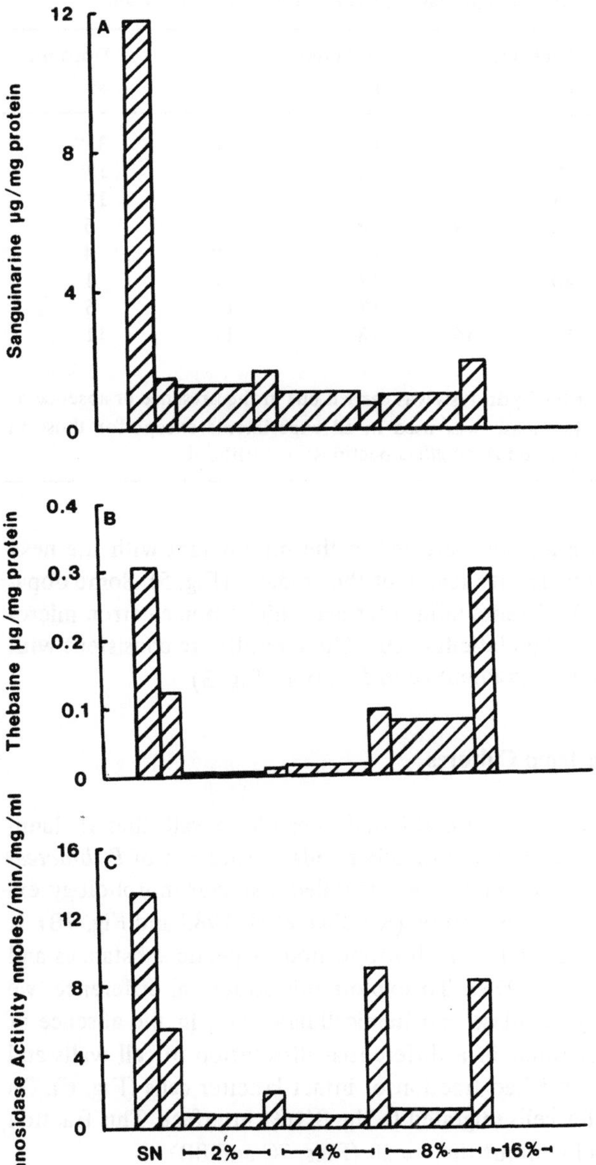

Fig. 6. Distribution of sanguinarine, thebaine and α-mannosidase on renografin density gradients after limited cell wall hydrolysis. Cell fractionation was carried out as described in Fig. 4 except pectinase was omitted from the digestion medium. The volume of fraction collected is reflected in the width of the bars in the graph. (**A**) Sanguinarine and (**B**) thebaine content and (**C**) α-mannosidase activity were determined as described in Fig. 4. Data shown are from the same experiment. In the original protoplasts, the concentrations of sanguinarine and thebaine were 11.6 and 0.013 $\mu g\ mg^{-1}$ protein, respectively

proportion of thebaine was recovered in the supernatant, the concentration of thebaine in this fraction was also no greater than that of the original protoplasts. Since laticifer alkaloidal vesicles are fragile (Roberts et al. 1983), the occurrence of thebaine in the supernatant could be explained by organelle disruption. The recovery of α-mannosidase activity in the supernatant (Fig. 4C and Table 2) and the absence of larger vesicles (Fig. 7A) is consistent with this line of reasoning.

Table 2. % recoveries of metabolites and α-mannosidase in Renografin gradient fractions

	Sanguinarine		Thebaine		α-Mannosidase		Dopamine
Fraction	+[a]	–	+	–	+	–	+
SN	8	25	29	10	18	17	33
SN/2%I	3	6	0	4	6	10	18
2%	13	13	8	14	6	9	17
2/4%I	5	12	6	9	4	5	6
4%	12	14	2	9	19	9	4
4/8%I	10	4	40	11	12	21	4
8%	34	14	0	9	17	12	6
8/16%I	15	13	16	34	18	17	11

[a]The + and – signs indicate whether the hydrolysis was carried out in the presence or absence of pectinase, respectively. *Rhizopus* pectinase was used in all experiments except for those in which dopamine was analyzed. In the latter *Aspergillus* pectinase was utilized.

The largest amount of dopamine was detected in the supernatant with the next highest concentrations in the low density region of the gradient (Fig. 5). Some dopamine was also present in the 8/16% renografin interface which upon electron microscopic inspection appeared to contain laticifer cells. These results are consistent with the data obtained by sucrose density gradient centrifugation (Fig. 3).

2.4 Separation of Laticifer Cells from Cultures

Cytological studies (Fig. 7A and 7B) reveal that laticifers can have walls that are laminated and conspicuously thicker than those of other cells. Inspection of *P. bracteatum* laticifers in culture by electron microscopy revealed a similar morphology except for transverse walls undergoing dissolution (Kutchan et al. 1983 and Fig. 7B). It has been reported that laticifers contain a high proportion of pectic substances and hemicellulose but no lignin (Fahn 1979). To exploit this structural difference, we prepared protoplasts using only cellulase and hemicellulase, i.e., in the absence of pectinase. As suspected, this resulted in a differential dissolution of cell walls and permitted the resolution of an enriched fraction of intact laticifer cells (Fig. 6). On the Renografin gradients laticifer cells migrated to the 8/16% interface. This fraction contained laticifer cells with and without their walls (Figs. 7C and 8B).

As seen in Fig. 6A and Table 2, more sanguinarine was found in the supernatant than in other fractions. The remainder appeared to be distributed throughout the gradient with no enrichment in any particular fraction. The larger amounts of sanguinarine in the supernatant may reflect the abundance of the older cells in dark brown tissue which appears to be necrotic (Fig. 6).

This limited digestion protocol also provided outstanding sequestration of thebaine. In the experiment shown in Fig. 6, concentrations of thebaine in the 8/16% interface were enriched 15-fold over the original levels in protoplasts. A high proportion of thebaine also appeared in the supernatant. As mentioned above, this could be due to vesicle disruption. However, some extravesicular thebaine arising due to the

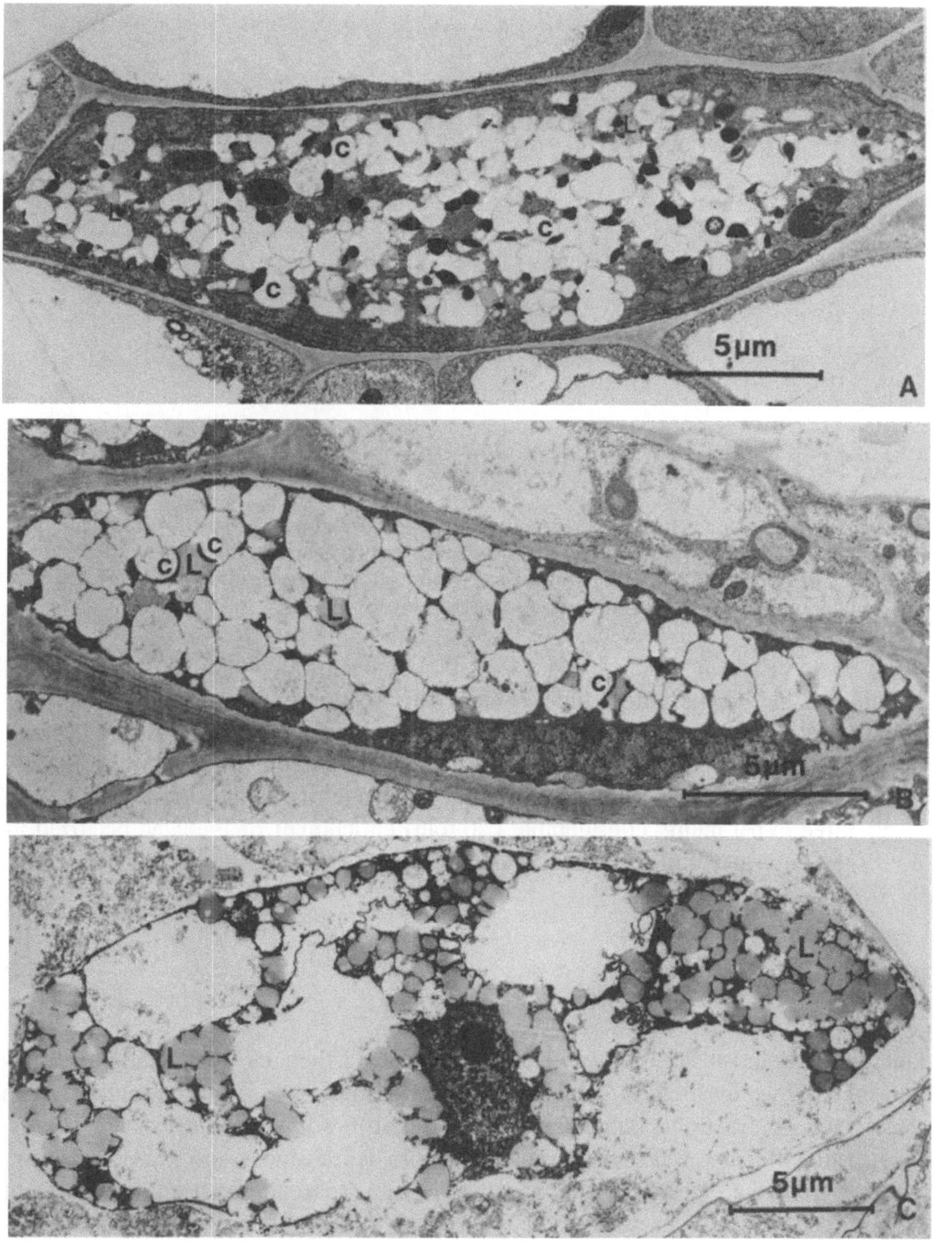

Fig. 7A–C. Transmission electron micrographs of laticifer cells from *P. bracteatum*. **(A)** Embryo, 4 days after seeds were sown. Many vacuoles contained electron dense deposits, caps (*c*), on the tonoplast. Some liposomes are also present. (x 6,000) **(B)** Shoot culture (MSRT, 5.0 mg l^{-1} BA, dark-grown) 10–12 days after transfer. Note caps (*c*) and liposomes (*L*). (x 7,000) **(C)** Protoplast sedimenting to the Renografin 8/16% interface following complete cell wall digestion of hormone-free suspension cultures. Protoplasts were layered directly onto the Renografin gradient. Cultured laticifer cells were frequently rich in lipid (*L*). (x 3,500)

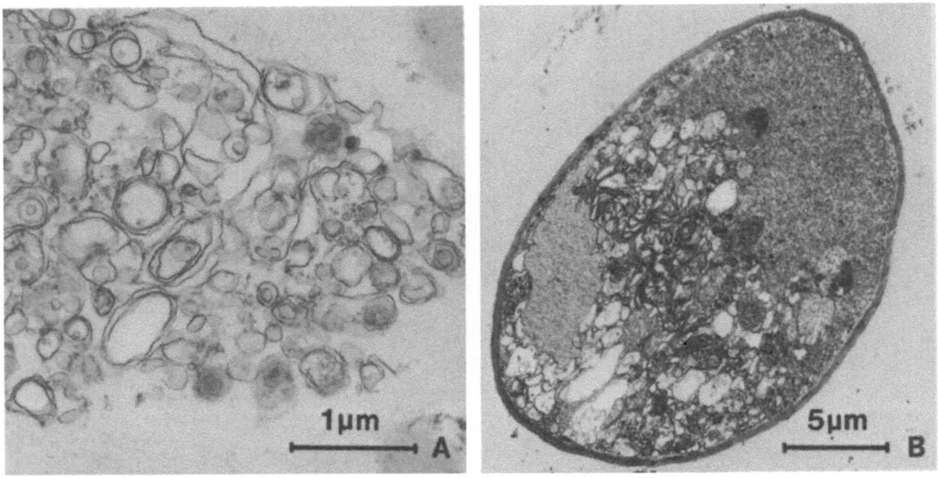

Fig. 8A,B. Transmission electron micrographs of Renografin gradient fractions. **(A)** 4/8% interace. Following complete cell wall digestion, protoplasts were layered directly onto the Renografin gradient. (x 24,900) **(B)** 8/16% interface. Partial cell wall digestion resulted in an enriched fraction of intact laticifer cells (viewed in transverse section) (x 4,000)

presence of older cells cannot be excluded. This experiment was repeated twice and Table 2 contains the average recoveries from the three experiments. Possibly because of the variability in the age of the cells used in each experiment, α-mannosidase and metabolite distributions changed somewhat from one run to another. Nevertheless, in all three experiments considerable (20-fold) enrichment of thebaine occurred in the 8/16% interface and this fraction always contained intact laticifer cells (Fig. 8B).

3 Conclusions

Our results indicate that dopamine, sanguinarine and thebaine accumulate at different sites. While the highest proportion of dopamine was localized in relatively less dense vacuolar fractions, sanguinarine was present in both light and heavy vacuoles. Most of the thebaine was sequestered in the more dense vesicles and if pectinase were omitted from the digestion medium, it was possible to isolate fractions rich in intact laticifer cells that proved to be enriched 20-fold in thebaine over the original protoplasts. In summary, laticifer cells appear to be a major site of accumulation of thebaine. While some dopamine and sanguinarine may also be present in these cells, most of these two metabolites are compartmentalized in different cellular and subcellular loci.

Acknowledgements. This work was supported by NSF Grant PCM 8011471 and NIH Training Grant 5 T32 HL-07050-09. We thank Elizabeth M. Coscia and Micheal D. Rush for their able assistance, Dr. Alan Kinnersley for *P. somniferum* cell cultures, and Schering, AG for Renografin.

References

Battersby AR, Foulkes DM, Binks R (1965) Alkaloid biosynthesis, part VIII. Use of optically active precursors for investigations on the biosynthesis of morphine alkaloids. J Chem Soc: 3323–3332

Battersby AR, Jones RGF, Kazlauskas R (1975) Experiments on the early steps of morphine biosynthesis. Tetrahedron Lett:1873–1876

Berlin J, Forche E, Wray V, Hammer J, Wolfgang H (1983) Formation of benzophenanthridine alkaloids by suspension cultures of *Eschscholtzia californica*. Z Naturforsch 38:346–352

Böhm H (1981) *Papaver bracteatum* Lindl. – Results and problems of the research on a potential medicinal plant. Pharmazie 36:660–667

Böhm H, Franke J (1982) Accumulation and excretion of alkaloids by *Macleaya microcarpa* cell cultures. I. Experiments on solid medium. Biochem Physiol Pflanz 177:345–356

Böhm H, Olesch B, Schultze CH (1972) Further investigations on the biosynthesis of alkaloids in isolated latex from opium poppy, *Papaver somniferum L.* Biochem Physiol Pflanz 163:126–136

Boller TH, Kende H (1979) Hydrolytic enzymes in the central vacuole of plant cells. Plant Physiol 63:1112–1132

Borkowski PR, Horn JS, Rapoport H (1978) Role of 1,2-dehydroreticulinium ion in the biosynthetic conversion of reticuline to thebaine. J Am Chem Soc 100:276–281

Brossi A (1982) Mammalian TIQ's: products of condensation with aldehydes or pyruvic acids? In: Bloom F, Barchas J, Sandler M, Usdin E (eds) Progress in clinical and biological research: Beta-carbolines and tetrahydroisoquinolines, vol 90. Liss, New York, p 125

Cramer CL, Ristow JL, Paulus TJ, David RH (1983) Methods for mycelial breakage and isolation of mitochondria and vacuoles of *Neurospora*. Anal Biochem 128:384–392

Fahn A (1979) Secretory tissues in plants. Academic, London New York, pp 223–243

Fairbairn JW, Kapoor LD (1960) Laticiferous vessels of *Papaver somniferum*. Planta Med 8: 49–61

Fairbairn JW, Steele MJ (1981) Biosynthetic and metabolic activities of some organelles in *Papaver somniferum* latex. Phytochemistry 20:1031–1036

Fairbairn JW, Djote M, Paterson A (1968) The alkaloids of *Papaver somniferum* L.–VII. Biosynthetic activity of the isolated latex. Phytochemistry 7:2111–2116

Felklova M, Babkova K (1958) Latex tubes in *Papaver somniferum* during the vegetative period. Pharmazie 13:220

Franke J, Böhm H (1982) Accumulation and excretion of alkaloids by *Macleaya microcarpa* cell cultures. II. Experiments in liquid medium. Biochem Physiol Pflanz 177:501–507

Ikuta A, Syono K, Furuya T (1974) Alkaloids of callus tissues and redifferentiated plantlets in the Papaveraceae. Phytochemistry 13:2175–2179

Kamimura S, Nishikawa M (1976) Growth and alkaloid production of the cultured cells of *Papaver bracteatum*. Agric Biol Chem 40:907–911

Kamimura S, Akutsu M, Nishikawa M (1976) Formation of thebaine in the suspension culture of *Papaver bracteatum*. Agric Biol Chem 40:913–919

Kettenes-Van Den Bosch JJ, Salemink CA, Khan I (1981) Biological activity of the alkaloids of *Papaver bracteatum* Lindl. J Ethnopharmacol 3:21–38

Kutchan TM, Ayabe S, Krueger RJ, Coscia EM, Coscia CJ (1983) Cytodifferentiation and alkaloid accumulation in cultured cells of *Papaver bracteatum*. Plant Cell Rep 2:281–284

Matile P, Wiemken A (1974) Vacuoles and spherosomes. In: Fleischer S, Packer L (eds) Methods in enzymology, vol 31. Academic, London New York, p 572

Nessler CL, Mahlberg PG (1978) Laticifer ultrastructure and differentiation in seedlings of *Papaver bracteatum* Lindl., Population Arya II (Papaveraceae). Am J Bot 65:978–983

Phillipson JD (1983) Infraspecific variation and alkaloids of *Papaver* species. J Med Plant Res (Planta Med) 48:187–192

Roberts MF (1974) Oxidation of tyrosine by *Papaver somniferum* latex. Phytochemistry 13: 119–123

Roberts MF, Antoun MD (1978) The relationship between L-DOPA decarboxylase in the latex of *Papaver somniferum* and alkaloid formation. Phytochemistry 17:1083–1087

Roberts MF, McCarthy D, Kutchan TM, Coscia CJ (1983) Localization of enzyme and alkaloidal metabolites in *Papaver* latex. Arch Biochem Biophys 222:599–609

Rueffer M, El-Shagi H, Nagakura N, Zenk MH (1981) (S)-Norlaudanosoline synthetase: the first enzyme in the benzylisoquinoline biosynthetic pathway. FEBS Lett 129:5–9

Sasse F, Backs-Hüsemann D, Barz W (1979) Isolation and characterization of vacuoles from cell suspension cultures of *Daucus carota.* Z Naturforsch 34:848–853

Schuchmann R, Wellmann E (1983) Somatic embryogenesis of tissue cultures of *Papaver somniferum* and *Papaver orientale* and its relationship to alkaloid and lipid metabolism. Plant Cell Rep 2:88–91

Shamma M, Moniot JL (1978) Isoquinoline alkaloids research: 1972–1977. Plenum, New York London

Spenser ID (1969) The biosynthesis of alkaloids and of other nitrogenous secondary metabolites. In: Florkin M, Stotz EH (eds) Comprehensive biochemistry, vol 20. Elsevier, Amsterdam New York, p 231 and references cited therein

Wilson ML, Coscia CJ (1975) Studies on the early stages of *Papaver* alkaloid biogenesis. J Am Chem Soc 97:431–433

Yoshikawa T, Furuya T (1982) Morphinan alkaloid production by opium poppy suspension culture. In: Fujiwara A (ed) Proc 5th Int Congr Plant Tissue Cell Culture. Abe Printing, Tokyo, p 307

Subject Index[1]

---

1 We thankfully acknowledge the help of Dr. N. G. Bisset in the preparation of this index

MIX
Papier aus verantwortungsvollen Quellen
Paper from responsible sources
FSC® C105338

If you have any concerns about our products,
you can contact us on
ProductSafety@springernature.com

In case Publisher is established outside the EU,
the EU authorized representative is:
Springer Nature Customer Service Center GmbH
Europaplatz 3, 69115 Heidelberg, Germany

Printed by Libri Plureos GmbH
in Hamburg, Germany